REMAKING AMERICAN SECURITY

SUPPLY CHAIN VULNERABILITIES & NATIONAL SECURITY RISKS ACROSS THE U.S. DEFENSE INDUSTRIAL BASE

BRIGADIER GENERAL JOHN ADAMS, U.S. ARMY (RETIRED)

ALLIANCE FOR
american
manufacturing

A Report Prepared for AAM by

GUARDIAN SIX
Research . Analysis . Strategy

Contributing Authors:

Paulette Kurzer, Ph.D. (Senior Vice President of Guardian Six Consulting LLC) • Amber Allen
Colonel Peter Aubrey, U.S. Army (Retired) • Eric Auner • Ryan G. Baird, Ph.D. • Chris Beecroft
Nathan Donohue • Keith A. Grant, Ph.D. • Ari Kattan • Janne E. Nolan, Ph.D.

First published May 2013 by the Alliance for American Manufacturing

ISBN 978-0-9892574-0-4 (paperback)

Printed in the United States of America.

April 15, 2013
Tucson, Arizona

As a 30-year veteran of the U.S. Army, I know that our national survival depends upon the readiness and skill of our armed forces. Our soldiers, sailors, airmen, and Marines stand guard around the world in defense of our nation and our Constitution. Yet all too often, we take for granted that our warriors will always have the equipment they need to win. All too often, we take for granted those who build the equipment that our warriors depend upon to perform their vital duties.

And all too often, we fail to recognize that weapons that can't be built can't be fired.

Our defense industrial base workers also stand guard—on our shop floors and in our factories, our chip foundries, and our shipyards. They ensure that our warriors have the world-class weapons and equipment they need to win on the battlefield. They, too, guard our nation.

When I began this report in mid-2011, I knew that some of the results of our investigation would be disturbing. The current level of risk to our defense supply chains and to our advanced technological capacity is not a good news story, to say the least. However, what I have learned has inspired rather than discouraged me. First, I confirmed that the men and women who keep our defense industrial base running—at all levels—are dedicated, competent, patriotic, and genuinely determined to ensure the United States' survival in a complex and often dangerous world. Second, I've come to understand that the United States has realistic options for preserving our defense industrial base—a vital national asset. Third, I have had the privilege of meeting many of those who work to sustain our defense industrial base and have heard how they too are concerned about its vitality. They have invested their professional and personal lives in their work for the sake of our great nation.

Of course, it should come as no surprise that we have such committed and capable people in our defense industrial base. They are, after all, Americans—the most ingenious, hard-working people on the planet. For this reason alone, I am confident in our efforts to preserve, strengthen, and—as necessary—revive our defense industrial base.

Working together, we will succeed. But we must not delay.

Respectfully,

John Adams
Brigadier General, U.S. Army (Retired)
President, Guardian Six Consulting LLC

ABOUT THE AUTHOR

Brigadier General John Adams, U.S. Army (Retired), is President of Guardian Six Consulting LLC. General Adams served his final military assignment as Deputy U.S. Military Representative to the NATO Military Committee in Brussels, Belgium. He retired from the U.S. Army in September 2007.

On September 11, 2001, General Adams was stationed at the Pentagon as Deputy Director for European Policy in the Office of the Secretary of Defense (OSD). He participated in immediate disaster recovery operations at ground zero and coordinated international support for the U.S. diplomatic and military response.

He is a veteran of Operation Desert Storm (1991) and Operation Guardian Assistance in Rwanda (1996). During more than 30 years of service in command and staff assignments, he spent nearly 18 years in Europe, Asia, the Middle East, and Africa, including assignments with U.S. Embassies in Belgium (1994-1997), Rwanda (1996), Croatia (1998-2001), and South Korea (2002-2003). As an Army Aviator, General Adams has more than 700 hours as pilot-in-command in fixed- and rotary-wing aircraft.

General Adams' military awards and decorations include the Defense Superior Service Medal (with Oak Leaf Cluster), the Bronze Star Medal, the Army Aviator Badge, Parachute Badge, and Ranger Tab. He is a recipient of the Military Intelligence Corps Association's Knowlton Award and the Director of Central Intelligence's Exceptional Collector National HUMINT Award for excellence in gathering human intelligence.

Born and raised in the Washington, D.C., area, General Adams is a Ph.D. candidate in Political Science at the University of Arizona and holds Masters Degrees in International Relations (Boston University), English (University of Massachusetts), and Strategic Studies (US Army War College).

ABOUT AAM

The Alliance for American Manufacturing (AAM) is a nonprofit, non-partisan partnership formed in 2007 by some of the United States' leading manufacturers and the United Steelworkers to explore common solutions to challenging public policy topics such as job creation, infrastructure investment, international trade, and global competitiveness.

We believe that an innovative and growing manufacturing base is vital to the United States' economic and national security, as well as to providing good jobs for future generations. AAM achieves its mission through research, public education, advocacy, strategic communications, and coalition-building around the issues that matter most to the U.S. manufacturing sector.

AAM's capacity includes a Washington D.C.-based lobby and research operation and a national field staff. Current platforms include promoting strong Buy American provisions at both the federal and state levels, advocating for fair trade, and promoting revitalization through a long-term national manufacturing strategy.

The blueprint for the future was built by AAM.

ABOUT GUARDIAN SIX

Guardian Six Consulting is a defense and national security consulting firm that advises governments, businesses, and nonprofits. Our experts have decades of practical, military, and academic experience in national security, defense policy, intelligence, international relations, and economic analysis.

We provide in-depth policy research, rigorous analysis, and innovative strategic advice on national security issues, bringing our expertise to bear on sensitive issues that demand the utmost pragmatism and insight. We distill complex issues and deliver usable products to our clients. Guardian Six is ideally suited to help companies with defense-related products and services compete in the growing global defense and national security marketplace. We address problems such as defense critical requirements, strategic threats, defense industrial policy, arms control, non-proliferation, terrorism, and new security threats.

TABLE OF CONTENTS

NATIONAL SECURITY
REQUIRES A STRONGER
U.S. MANUFACTURING SECTOR

The United States' national security is threatened by our military's growing and dangerous reliance on foreign nations for the raw materials, parts, and finished products needed to defend the American people. The health of our manufacturing sector is inextricably intertwined with our national security, and it is vital that we strengthen the sector.

This report—prepared by Guardian Six Consulting LLC for the Alliance for American Manufacturing—recommends 10 actions to make America less dependent on foreign nations for the vital products that enable America's soldiers, sailors, airmen, and Marines to be the most powerful and effective fighting force in the world.

The recommendations (detailed below) call for a joint strategy by government, industry, academic research institutions, and the military to increase U.S. domestic production of manufactured items and recovery of natural resources that the armed forces require. In addition, the recommendations emphasize the importance of investment today in the technological innovation, education, and training needed to keep America secure tomorrow.

This report also calls for properly enforcing existing and internationally accepted laws that give U.S. defense manufacturers certain preferences over foreign competitors. This enforcement will ensure a level playing field, high-quality materials and products, and a healthy U.S. defense industrial base. The report further recommends federal investment in America's high-technology manufacturing infrastructure, especially in advanced research and manufacturing capabilities.

Another recommendation calls for increasing U.S. production of certain key raw materials needed for the nation's defense to supplement our imports. The recommendation also proposes stockpiling these raw materials to ensure an adequate supply.

DANGERS OF MILITARY DEPENDENCY

With the closing of factories across the United States and the mass exodus of U.S. manufacturing jobs to China and other nations over the past 30 years, the United States' critically important defense industrial base has deteriorated dramatically. As a result, the United States now relies heavily on imports to keep our armed forces equipped and ready. Compounding this rising reliance on foreign suppliers, the United States also depends increasingly on foreign financing arrangements.

In addition, the United States is not mining enough of the critical metals and other raw materials needed to produce important weapons systems and military supplies. These products include the night-vision devices (made with a rare earth element) that enabled Navy SEALs to hunt down Osama bin Laden.

Consequently, the health of the United States' defense industrial base—and our national security—is in jeopardy. We are vulnerable to major disruptions in foreign supplies that could make it impossible for U.S. warriors, warships, tanks, aircraft, and missiles to operate effectively. Such supply disruptions could be caused by many factors, including:

- Poor manufacturing practices in offshore factories that produce problem-plagued products. Shoddy manufacturing could be inadvertent, could be part of a deliberate attempt to cut costs and boost profits, or could be intentionally designed to damage U.S. capabilities. Motivated by expected gains in cost, innovation, and efficiency, the Department of Defense (DoD) began a decided shift from parts made to military specifications (Mil-Spec) to commercial-off-the-shelf (COTS) parts and equipment two decades ago. However, COTS parts often lack the quality control and traceability necessary to ensure that parts used in the defense supply chain meet the rigorous standards we expect of equipment vital to our national security. Faulty and counterfeit COTS parts are already taking a toll on readiness in several defense sectors.

- Natural disasters, domestic unrest, or changes in government that could cut or halt production and exports at foreign factories and mines.

- Foreign producers that sharply raise prices or reduce or stop sales to the United States. These changes could be caused by political or military disputes with the United States, by the desire of foreign nations to sell to other countries, by the need to attract foreign investment and production, or by foreign nations wanting to keep more of the raw materials, parts, and finished goods they produce for their own use.

VITAL AND VULNERABLE
THE U.S. DEFENSE INDUSTRIAL BASE

VITAL TO OUR NATIONAL SECURITY

The U.S. workforce provides the tools for ready warfighters

 + +

U.S. WORKERS   **U.S. MANUFACTURING**   **READY WARFIGHTERS**

CREATING GREATER
**U.S. NATIONAL
SECURITY**

IMPORT DEPENDENCE

The United States currently relies heavily on the foreign supply of imports to manufacture many essential military systems

SUPPLY CHAIN VULNERABILITIES

The defense industrial base faces multiple supply chain vulnerabilities

FOREIGN EXPLOITATION   **LOSS OF INNOVATION**   **LACK OF RAW MATERIALS**   **LOSS OF DOMESTIC CAPACITY**

SUPPLY CHAIN COMPLEXITY

DEPARTMENT OF DEFENSE
LACKS VISIBILITY OF
DEFENSE SUPPLY CHAIN
LOWER TIER SUPPLIERS

DEFENSE PROCUREMENT

PRIME CONTRACTOR

SUBCONTRACTOR   SUBCONTRACTOR   SUBCONTRACTOR

SUBCONTRACTORS   SUBCONTRACTORS   SUBCONTRACTORS

?   **?**   **?**

RESEARCH TODAY
CAPABILITIES TOMORROW

Investment in research, coupled with advanced domestic manufacturing, will directly contribute to future warfighter success

 + =

RESEARCH AND DEVELOPMENT   **ADVANCED DOMESTIC MANUFACTURING**   **FUTURE WARFIGHTER SUCCESS**

SUPPORTING DEFENSE

American manufacturing supports critical U.S. defense needs

FACILITIES   **CAPITAL**   **EXPERTISE**

RECOMMENDATIONS

This report's 10 recommendations to make the United States less dependent on the importation of products essential to our national security are based on the premise that the U.S. defense industrial base is a vital national asset that is no less critical to our national security than our men and women in uniform. The recommendations call for:

 Increasing long-term federal investment in high-technology industries, particularly those involving advanced research and manufacturing capabilities. The distinguishing attribute of the U.S. defense industrial base is technological innovation. As foreign nations continue manufacturing an ever-larger share of America's defense supplies, the United States increases its risk of diminishing its capacity to design and commercialize emerging defense technologies. To help ensure that our armed forces dominate the future battlefield, Congress should provide funding for American manufacturers to develop and implement advanced process technologies.

 Properly applying and enforcing existing laws and regulations to support the U.S. defense industrial base. Domestic source preferences already enacted into law, such as those that apply to the steel and titanium industries under the Specialty Metals Clause, must be retained to ensure that important defense capabilities remain secure and available for the U.S. armed forces.

3 Developing domestic sources of key natural resources required by our armed forces. Right now the United States relies far too heavily on foreign nations for certain key metals and other raw materials needed to manufacture weapons systems and other military supplies. For example, most rare earth elements, which are essential components of many modern military technologies, currently must be purchased from China. The U.S. government and industry must stockpile these vital raw materials, strengthen efforts to resume mining and transformation of the materials in the United States, improve recycling to make more efficient use of current supplies, and identify alternate materials.

4 Developing plans to strengthen our defense industrial base in the U.S. National Military Strategy, National Security Strategy, and the Quadrennial Defense Review process. This would make creating and sustaining a healthier defense industrial base a higher national priority, with a focus on increasing support for the most important and vulnerable industrial sectors.

5 Building consensus among government, industry, the defense industrial base workforce, and the military on the best ways to strengthen the defense industrial base. These sectors must work collaboratively to successfully address the concerns of all defense industrial base stakeholders.

6 Increasing cooperation among federal agencies and between government and industry to build a healthier defense industrial base. The Departments of State, Treasury, Energy, Commerce, Homeland Security, and others in the Executive Branch should join the Department of Defense in working to bolster the defense industrial base.

7 Strengthening collaboration between government, industry, and academic research institutions to educate, train, and retain people with specialized skills to work in key defense industrial base sectors. The loss of U.S. manufacturing jobs has reduced the size of the workforce skilled in research, development, and advanced manufacturing processes.

8 Crafting legislation to support a broadly representative defense industrial base strategy. Congress and the Administration must collaborate on economic and fiscal policies that budget for enduring national security capabilities and sustain the industrial base necessary to support them.

9 Modernizing and securing defense supply chains through networked operations. These operations should be built on the excellent work that the DoD and industry are already doing to map and secure defense supply chains. The operations would provide ongoing communications between prime contractors and the supply chains they depend on. Closer communications, patterned on the networked operations of U.S. military forces around the world, would help managers identify and solve recurring problems with military supplies.

10 Identifying potential defense supply chain chokepoints and planning to prevent disruptions. This recommendation would require determining the scope of foreign control over critical military supply chains and finding ways of restoring U.S. control.

EXAMINING THE DEFENSE INDUSTRIAL BASE

This report examines defense industrial base nodes that are vital to U.S. security. Rather than focusing on final high-cost manufactured products (such as aircraft, ships, missiles, or tanks) the sectors we studied deal with 14 lower-tier commodities and raw materials, subcomponents, and end-items needed to build and operate the final systems. Some nodes are essential for foundational military capabilities, and others provide the tactical and logistical advantages necessary for our modern military. Still others provide niche capabilities that enable members of the military to operate in environments that would otherwise be inaccessible or exceedingly dangerous.

Studying the nodes allows us to conduct a bottom-up review of key sectors of the defense industrial base. This report devotes a chapter to each of the nodes, looking at the critical role each node plays in our national security, each node's contribution to U.S. military capabilities, and the consequences losing these capabilities would have on our defense. The nodes we analyze typically escape notice in Washington, D.C., but they are vital nonetheless. (For example, the absence or failure of a tiny fastener or semiconductor can hobble an aircraft that costs tens of millions of dollars.)

The commodities and raw material nodes examined in this report are steel armor plate, specialty metals, titanium, and high-tech magnets. The subcomponent nodes examined are fasteners, semiconductors, copper-nickel tubing, lithium-ion batteries, HELLFIRE missile propellant, advanced fabrics, and telecommunications. The end-item nodes examined are night-vision devices, machine tools, and biological weapons defense.

MILITARY EQUIPMENT CHART
THESE DEFENSE SYSTEMS FACE A RANGE OF SUPPLY CHAIN VULNERABILITIES

DEPARTMENT	WEAPON SYSTEMS	PLATFORMS	OTHER SYSTEMS
ARMY	■ Joint Direct Attack Munition (JDAM) Precision Guidance Kit (Semiconductors, Fasteners, High-Tech Magnets) ■ AGM-114 HELLFIRE Air-to-Surface Missile (HELLFIRE Missile Propellant, High-Tech Magnets, Machine Tools) ■ M4 Carbine (Fasteners, Machine Tools)	■ M1 Abrams Main Battle Tank (Steel Armor Plate, Semiconductors, Machine Tools, High-Tech Magnets, Fasteners, Specialty Metals) ■ UH-60 Blackhawk Helicopter (Fasteners, Semiconductors, Machine Tools, Titanium) ■ Mine-Resistant Ambush-Protected (MRAP) Vehicle (Steel Armor Plate)	■ Night-Vision Devices (Specialty Metals) ■ Laser Range-Finders (Specialty Metals) ■ Medical Counter-Measures (Biodefense) ■ Flame-Resistant Army Combat Uniform (FR-ACU) (Advanced Fabrics) ■ Communications Systems (Semiconductors and Telecommunications)
MARINE CORPS	■ Joint Direct Attack Munition (JDAM) Precision Guidance Kit (Semiconductors, Fasteners, High-Tech Magnets) ■ AIM-120 Advanced Medium-Range Air-to-Air Missile (AMRAAM) (Lithium-Ion Batteries, High-Tech Magnets) ■ M4 Carbine (Fasteners, Machine Tools)	■ F-35B Joint Strike Fighter (Titanium, Lithium-Ion Batteries, Machine Tools, High-Tech Magnets, Semiconductors) ■ V-22 Osprey Aircraft (Titanium, Semiconductors, Machine Tools)	■ Night-Vision Devices (Specialty Metals) ■ Interceptor Body Armor (Advanced Fabrics) ■ All Devices Powered by Lithium-Ion Batteries (Specialty Metals) ■ Communications Systems (Semiconductors and Telecommunications)
NAVY	■ AIM-120 Advanced Medium-Range Air-to-Air Missile (AMRAAM) (Lithium-Ion Batteries, High-Tech Magnets) ■ Submarine-Launched Ballistic Missiles (SLBMs) (Specialty Metals)	■ Guided Missile Destroyer (Steel Armor Plate, Copper-Nickel Tubing) ■ Nimitz-Class Nuclear-Powered Aircraft Carrier (Steel Armor Plate, Titanium, Copper-Nickel Tubing, Machine Tools, High-Tech Magnets) ■ Littoral Combat Ship (LCS) (Steel Armor Plate, Titanium, Copper-Nickel Tubing, High-Tech Magnets)	■ Night-Vision Devices (Specialty Metals) ■ Copper-Nickel Tubing for all Navy Vessels ■ Communications Systems (Semiconductors and Telecommunications)
AIR FORCE	■ AIM-9 Sidewinder Air-to-Air Missile (Lithium-Ion Batteries, High-Tech Magnets, Machine Tools, Titanium) ■ AGM-114 HELLFIRE Air-to-Surface Missile (HELLFIRE Missile Propellant, High-Tech Magnets, Machine Tools)	■ F-35A Joint Strike Fighter (Titanium, Lithium-Ion Batteries, Machine Tools, High-Tech Magnets, Semiconductors) ■ F-22 Raptor Fighter (Specialty Metals, Semiconductors, Machine Tools, Titanium, Fasteners, High-Tech Magnets) ■ MQ-1B Predator Drone (HELLFIRE Missile Propellant, High-Tech Magnets)	■ Night-Vision Devices (Specialty Metals) ■ Titanium for Aircraft Body and Armor ■ Communications Systems (Semiconductors and Telecommunications)

A CALL TO ACTION

This report identifies vulnerabilities created by the United States' growing reliance on foreign inputs to produce the military systems necessary to defend our nation and our people. It is a call to action to the United States' leaders in government and industry to reduce these vulnerabilities. Leaders must demand strategic thinking about the problems confronting the defense industrial base in the same way that they demand strategic thinking about the problems confronting our armed forces on the battlefield.

The United States needs a defense industrial base strategy that serves our most important security requirements. We need to review that strategy regularly to ensure that it keeps pace with rapidly shifting global trends and endures the test of time. As we shift our national security attention towards the Asia-Pacific region, we must ensure that our defense industrial base structure—especially our procurement policy—is consistent with our national security goals. For example, it makes little sense to depend on China for critical components of our defense industrial base. If we are to preserve the United States' status as the most powerful nation on the planet, we need to produce superior weaponry for today's warriors, as well as preserve our technological edge to ensure that those who will defend our nation in the next generation and beyond are equipped with the best weapons and systems available.

Without a healthy and technologically advanced defense industrial base, the United States will be unable to provide the weapons and advanced military systems our warriors require to defend the United States now and in the future. Nothing less than the survival of our nation is at stake. ■

SCOPE AND METHODOLOGY

"An industrial-base 'strategy' that seeks to preserve every sector deemed desirable by any of the war-fighting communities across the four military Services, the prime defense contractors, or their Congressional allies is not in fact a strategy and will not succeed. Indeed, even within the truly critical sectors, not every design or production capability will merit preservation. The sine qua non of the proposed guiding policy, then, is the imperative to make hard choices."[1]

The task of our report, "Remaking American Security: Supply Chain Vulnerabilities and National Security Risks across the U.S. Defense Industrial Base," is to survey the defense industrial base and identify vulnerabilities that could negatively impact the production of defense systems essential to U.S. national security. This report does not undertake the monumental task of mapping the defense industrial base in its entirety; although the authors believe such an endeavor is necessary for the long-term health of the defense industrial base. Instead, this report focuses on specific sectors (referred to in this report as nodes) of the defense industrial base that play critical roles in preserving and strengthening U.S. national security. The report emphasizes the challenges to and vulnerabilities of each of these nodes, with attention to each node's unique contribution to U.S. military capabilities and the consequences of losing those capabilities.

Some nodes are essential for foundational military capabilities, and without them the United States would be unable to field warships, tanks, and aircraft. Others enable the technological core of our advanced military, providing tactical and logistical advantages necessary for a modern military. Still others provide niche capabilities that enable the warfighter to operate in environments that would otherwise be inaccessible or exceedingly dangerous. Some sectors are particularly vulnerable in the short-term and require immediate attention, while others face future or long-term challenges. Certain vulnerabilities result from larger market trends, while other sectors experience competition from countries using opportunistic trade policies designed to manipulate global markets.

This study examines 14 defense industrial base nodes. Together they create a broad picture of the challenges facing the defense industrial base, while drawing attention to the likely consequences of ignoring those risks. Each chapter focuses on the lower tiers of the defense supply chains (the raw material or sub-component level), below the level of prime contractor or original equipment manufacturer.

These nodes have several attributes in common. They are all critical for meeting U.S. defense requirements, and all are at risk of greatly diminishing or disappearing entirely. Furthermore, because these nodes are located at intermediate and lower tiers of the defense industrial base, they are less visible and often overlooked as to their contributions to U.S. defense capabilities as well as the risks they face. Our selection of nodes is not meant to be representative of the entire U.S. economy or the entirety of the defense industrial base; only nodes where significant risks exist are included.

Each node report focuses on four topics. First, we briefly establish the context needed to understand the node. Although the context varies greatly among nodes, it tends to focus on the methods of production and trends in the related industries (output, profitability, domestic vs. foreign capacity, etc.). Next, we establish the node's contribution to U.S. national defense, and the potential consequences for U.S. defense capabilities of the unavailability of that input or end product. A third section highlights vulnerabilities in each node's supply chains, and the potential impacts of those vulnerabilities on military readiness. Finally, in light of the preceding analysis, we propose strategies to mitigate these risks.

We examine different kinds of vulnerability as well. The presence of foreign components in critical defense items constitutes not only a potential weakness in the supply chain due to the risk of substandard or unavailable parts, but also a vulnerability to foreign exploitation. Other important vulnerabilities examined in this report include the risks of lost knowledge and technological innovation, reliance on foreign-controlled designs, and dependence on imported inputs, without which advanced weapons and other defense systems will not work.

This report is derived from publically available sources including academic scholarship, industry and governmental reports, congressional policy statements and corresponding implementation guidelines from the Department of Defense (DoD) and other governmental agencies, articles from newspapers and other periodicals, and numerous interviews and correspondences with governmental, military, and industry experts (see the Annexes for experts consulted).

COMPARING NODES THROUGH DEFENSE CRITICALITY

We recognize that the challenges facing some defense industrial base sectors will be difficult to counteract. Efforts to address the risks to the defense industrial base must be prioritized according to the urgency and intensity of the risk, as well as the feasibility of a particular risk mitigation strategy.

To facilitate these judgments, we have constructed a matrix of defense criticality that compares the risk of and national security impact of supply chain disruptions, allowing for rough comparisons across the 14 nodes. We lack the information to make definitive comparisons of criticality across the 14 nodes; however, the criticality matrix is intended to provide rough comparisons at a glance.

We consider the following criteria when assessing risks to each node's supply chain:

- Speed with which a supply chain disruption would restrict U.S. military access to a significant commodity, technology, or end-product;

- Extent to which domestic production capacity is sufficient and/or could be developed, including whether commercial demand for the product could help facilitate this substitution;

- Adequacy and stability of long-term global supply and/or the relative geographic concentration of offshore production in regions where there is a potential for artificial supply manipulation; and

- Exposure of defense supply chains to unpredictable disruptions such as natural disasters, instability, and business failures.

DEFENSE CRITICALITY: RISKS

Nodes are placed into one of four risk categories. A *low* vulnerability node would be one in which adequate domestic sources are generally available, stable, and can be readily expanded to meet increasing demand. In *low* vulnerability situations, the relevant industry would be robust, competitive, and profitable. In a *low* risk setting, the risk of a supply disruption is unlikely,

and would likely require a series of coinciding, improbable events.

A *moderate* vulnerability exists either when domestic supply is insufficient to meet domestic demand or when the viability of the domestic industry faces significant challenges that may result in an inability to meet demand for a product or commodity, especially during an unanticipated surge in demand.

High vulnerability exists when a significant supply disruption could result from a single, improbable event, such as a natural disaster, artificial supply manipulation, or an inability to expand supply to meet increasing demand. *High* vulnerabilities involve significant dependence on foreign production in conjunction with one of the following conditions: lack of capacity for domestic production; rapidly increasing global demand resulting in scarcity; highly concentrated production or resource reserves; political or economic instabilities in a major producer nation; or actual market manipulation.

Extreme vulnerability is reserved for those nodes where there is strong evidence of an imminent shortage or the potential for a severe artificial shortage fabricated for either political or economic reasons. *Extreme* vulnerabilities may result from limited supply in conjunction with expanding global demand, which results in price instability and significant concerns over the short-term availability of a resource or product. *Extreme* vulnerabilities may also result from severe geographic concentration of a commodity lacking a close substitute, which creates the potential for artificial supply restrictions for either political or material gain. In these situations, it is often more a question of *when* a supply disruption will occur than *if* one will.

DEFENSE CRITICALITY: IMPACTS

Based on the specific defense purpose or purposes of each commodity, resource, or technology, each node's risk of supply chain disruption is paired with an assessment of its threat to U.S. defense capabilities. The first question of the uniqueness of each node's function is whether there is a suitable and available substitute, capable of being seamlessly introduced to overcome the shortage. If there is a readily available substitute, a shortage will have little or no real impact on defense readiness, suggesting only a *marginal* impact on U.S. capabilities. If a node lacks a suitable and cost-effective substitute, the impact of supply shortages must be evaluated according to the military capabilities no longer available. The impact of a non-substitutable product or commodity is determined by the scope or breadth of its usage in conjunction with the specific function it performs.

An *isolated* impact is one in which the non-availability of a commodity, resource, or technology will have a minor impact on military operations. Examples include the substitution of an item with an adequate but not ideal alternative, or when non-availability affects a very narrow range of operations. This category is reserved primarily for capabilities that are force multipliers rather than enablers: without these products, operations could be conducted but with reduced efficiency due to the lack of a tactical advantage or at higher risk to the warfighter.

When the non-availability of a product or commodity begins to take options off the table, it can be said to have a *significant* impact on national defense capabilities. For example, product non-availability restricts the use of mission-critical capabilities. In contrast to an isolated impact, which reduces the effectiveness of military platforms or operations, a *significant* impact would render a given capability unavailable. In these situations, substitutes may exist, but would be restrictive in cost or result in significant performance loss.

An *incapacitating* impact is one in which a broad segment of U.S. military capabilities effectively would be crippled. This classification is reserved for commodities, components, or end-items needed to produce and sustain military capabilities for which the U.S. military has widespread use. Examples include products critical to the construction of important aircraft and naval vessels. In the event of a shortage or loss of supply, substitutes would be altogether unavailable and national security severely compromised.

ENDNOTES

1 Watts, Barry D. and Todd Harrison, Center for Strategic and Budgetary Assessments, *Sustaining Critical Sectors of the U.S. Defense Industrial Base* (2009), xiii. http://www.csbaonline.org/publications/2011/09/sustaining-critical-sectors-of-the-u-s-defense-industrial-base/

CHAPTER 1 • INTRODUCTION

"Without our industry partners, we can't field an army."[1]

—Deputy Assistant Secretary of Defense for Manufacturing and Industrial Base Policy Brett Lambert

The United States' armed forces confront our enemies throughout the globe, protecting our nation. Our soldiers, sailors, airmen, and Marines constitute the best equipped, best trained, and most effective fighting force the world has ever seen. They are supported by a defense industrial base that is every bit as world-class, thanks to the power of innovative new technologies, superior application, and sheer U.S. ingenuity. The U.S. defense industrial base provides our warriors with the weapons they need to win on the battlefield.

The structure of the defense industrial base has changed dramatically over the past 30 years, along with the structure of U.S. manufacturing more broadly. Market forces and globalization are essential to the health of the U.S. economy and defense industrial base. However, globalization and two of its major attributes, outsourcing and offshoring, bring risks as well as benefits.

As a result of these risks, the health of our defense industrial base is now in jeopardy. The transformation of our manufacturing base has profoundly impacted the defense sector and the United States' ability to defend itself adequately. U.S. armed forces must be capable of deploying rapidly into crisis or conflict, capable of initiating operations without pause. Ground, naval, and air forces must operate without impediments caused by supply chain difficulties, poor quality control, or inferior parts. Many U.S. industries that have moved substantially or entirely offshore are critical to our national security, providing necessary components and items.

The supply chains that link our defense industrial base to our armed forces are vulnerable to disruption. Many of the tools we need to mitigate that vulnerability are at hand, but must be strengthened in light of the global forces that affect our entire economy, including our vital defense industrial base. The most urgent task is to galvanize our government and industry to act to address the problem.

In *The Rise and Fall of Great Powers*, historian Paul Kennedy argues that the rise and continued success of great powers hinges upon the strength of their economic base, of which the defense industrial base is a key, if not the most critical, component.

"[The] historical record suggests that there is a very clear connection in the long run between an individual Great Power's economic rise and fall and its growth and decline as an important military power… Technological and organizational breakthroughs…bring greater advantage to one society than another."[2]

Chief among the risks are those posed to defense industrial base supply chains. The preservation of supply chains must balance globalization's cost and efficiency advantages with the necessity for reliability, quality, and timeliness of production for national defense. Driven largely by globalization, many defense items' supply chains have shifted under the control of foreign strategic competitors and are now dependent at either the level of raw material or at the lower tiers of the supply chain for component parts of critical defense items. Should a strategic competitor choose to deny the resource or disrupt the component supply chain, the defense item may become unavailable to the U.S. military in a crisis.

Of equal importance are the risks to U.S. leadership in high-technology industries. The failure to address the health of the U.S. defense industrial base poses wider risks that the U.S. will lose competitive advantage, both in defense and industrial technology at large. Such a failure also risks the U.S. ability to mobilize and surge industrial production in the event of future conflict. Gerald Abbott and

Stuart Johnson, keen analysts of the U.S. defense industrial base, argue that:

"…the essential link between the productive base and national power was the ability to increase production runs of weapons through the course of a conflict. World War II and the Korean War are prime cases of needing time to close the gap between productive output and military requirements and paying for that time in blood and territory."[3]

The economic effects of globalization correspond with an increased dependence on foreign entities for products that may be essential to U.S. security. Reliance on external production, while in some cases cheaper (at least in the short term) than domestic production, risks entrusting U.S. security to foreign producers, regulated by foreign governments, who often do not share U.S. strategic interests. This risk introduces an element of supply-side uncertainty, as the domestic policies of those countries influence the availability and pricing of necessary inputs to critical U.S. defense systems. In the extreme, foreign entities can gain the ability to weaken the United States purposely and strategically by withholding critical and non-substitutable components, reducing or halting production of certain systems, developing counterforce capabilities by knowing the capabilities or weaknesses of our system, or even rendering critical defense systems inoperable in times of crisis.

U.S. warriors and workers are inseparable and equally essential elements of our national defense. This report is a call to action to ensure that successful linkage lasts far into the future. Nothing less than our national survival is at stake.

Key themes discussed in this report are:

- The U.S. defense industrial base is the key enabler of the world's most powerful military and is vital to national security. A weapon that can't be built can't be fired.

- U.S. manufacturing maintains the United States' comparative economic and technological advantage, inextricably linking strong national defense and innovation.

- The defense industrial base is a pillar of U.S. prosperity and security. It employs many American workers in firms of all shapes and sizes.

- Lower-tier defense industrial base firms play a critical and underappreciated role in producing U.S. defense capabilities.

- Our strong defense industrial base cannot be taken for granted. Especially in a tough fiscal environment, we must strategically apply government policies and legislative frameworks to preserve the critical elements of the defense industrial base.

- Globalization accelerates offshoring and outsourcing of critical defense technologies. This change in turn risks depriving U.S. industry of the capacity to design and commercialize emerging defense technologies.

- Foreign exploitation, natural disasters, and unexpected global events can disrupt defense industrial base supply chains, causing shortages of parts and products necessary for critical defense systems.

- Foreign control over defense industrial base supply chains increases the risk that those countries will restrict U.S. access to critical defense resources. It also places U.S. defense capabilities at

risk in time of crisis and enables foreign suppliers to leverage concessions in bargaining situations.

- Inattention to preserving U.S. access to natural resources places national defense capabilities at risk, because the United States has withdrawn from mining and extracting them. Lack of policy coordination has hampered the formulation of a coherent materials strategy to combat risks to mineral and material supply chains.

THE CHALLENGES OF GLOBALIZATION

Since the 1980s, and increasingly during the 1990s, U.S. manufacturing aggressively has moved abroad in order to take advantage of lower labor costs in emerging markets. The trend started with furniture, textiles, shoes, and electronic consumer goods, and presently encompasses virtually every type of consumer, capital, and defense good.[4]

U.S. official thinking, supported by an established policy network of government officials and private experts, advocates the idea that the world economy operates best when it is based on a "natural" division of labor: some countries produce goods while others supply services. According to the theories of comparative advantage and economic specialization, the relocation of manufacturing jobs offshore inexorably leads to a more efficient allocation of resources, taking advantage of cheaper labor markets and maximizing returns on resource endowments for the benefit of all. In turn, increased competition enhances productivity levels at home and creates a win-win situation, where consumers have access to a large selection of affordable

> The ability of our warriors to fight is directly linked to the ability of American industry to provide them with the weapons and equipment they need to win.

goods because producers are continuously competing and innovating to capture a greater market share.

These trends have had varied effects in the broader U.S. economy, but it is difficult to argue that they have been unambiguously beneficial for the U.S. defense industrial base.

The pace of decline in U.S. manufacturing abruptly accelerated since 2000. Between 2000 and 2009, the United States lost 31.2 percent of its manufacturing jobs, and in 2009 the manufacturing sector fell from 13.1 percent of total employment to 9.1 percent. During the same period, the manufacturing share of U.S. Gross Domestic Product (GDP) fell sharply, declining at nearly twice the rate of the previous 15 years.[5] The nation's manufacturing output grew by only 11 percent during this period, while GDP grew by 15.7 percent, leaving U.S. manufacturing's share of GDP to fall from 14.2 percent to 11 percent.[6]

Even as the scale of U.S. manufacturing declined, this sector continues to represent the bulk of U.S. exports as well as two-thirds of spending on research and development (R&D). The sector remains the leader in innovation, employing 36.4 percent of the nation's engineers while accounting for 70 percent of industry-funded R&D.

In his 2013 State of the Union Address, President Barack Obama pledged that "Our first priority is making America a magnet for new jobs and manufacturing." The Obama administration has presided over a small though notable improvement; more than half a million jobs in manufacturing have been created between January 2010 and January 2013.[7] However, this increase does not compensate for the shedding of millions of jobs since 2000, when manufacturing employment stood at 17.2 million jobs. Moreover, as the United States largely has abandoned job creation in manufacturing, vocational training also has eroded. As a result, there are many vacancies in U.S. industry for machinists and other highly trained workers. Unfortunately, there simply are not enough Americans with the training and skills to fill those jobs—a shortfall that further increases the likelihood of U.S. businesses migrating overseas. Even as Americans lose access to the kind of jobs that can support a middle-class lifestyle, the United States risks losing its knowledge base and capacity to manufacture high-tech products. Once lost, these manufacturing capabilities, and the jobs that come with them, will be very difficult to get back. This fact may explain why the United States Bureau of Labor Statistics predicts that employment in manufacturing in 2020 will be more or less the same as in 2010 (11.5 million jobs), thus an even smaller fraction of the overall labor market.[8]

The forces of globalization are often irreversible, but all too often, businesses put short-term profit maximization ahead of long-term competitiveness. Lower production costs based on outsourcing and offshoring may lead to higher profits, but they can undermine our national security interests by diminishing productive capacity, transferring technology, and risking access to materials and supplies.

Realism about the changing international system is not a reason for fatalism about the health and status of the U.S. defense industrial base. Other advanced industrialized countries, often with similar labor costs and world-class environmental, social, and health standards, have not experienced an equivalent decline of manufacturing output and employment relative to the United States. During the first decade of the 21st century, for example, Germany lost six percent of its manufacturing jobs compared to the 28 percent decline the United States experienced. Italy coped with serial fiscal and financial challenges and had to compete in global markets with an expensive currency, but lost only 14 percent of its manufacturing employment. Japan is the only country that came close to the United States in relative employment decline, losing 20 percent of its manufacturing jobs during this same period.[9] That said, in 2011, manufacturing employment still accounted for 16.8 percent of total employment in Japan, in contrast to 10.2 percent in the U.S.[10]

European countries have fared better than the United States despite their relatively higher labor costs (adjusted for purchase power parity), particularly Germany, Norway, Switzerland, France, and the Netherlands.[11] Unit labor costs do not account for why manufacturing jobs have declined more in the United States than other Organization for Economic Cooperation and Development (OECD) countries.

The conventional wisdom that U.S. manufacturing job loss is simply the result of productivity-driven restructuring–an old economy making way for a new vibrant and innovation-driven economy–also is incorrect.[12] Former President of the Federal Reserve Alan Greenspan often referred to his vision of the United States

naturally evolving from the production of manufactured goods to the provision of advanced knowledge and services. The success of high-technology innovation hubs like Silicon Valley is typically offered as the prime example of how the United States has entered the era of "dotcom" enterprises, with U.S. firms leading and dominating international markets in the exciting field of information and computer technology, while less advanced partners inherit simpler, more traditional forms of enterprise.

This benign view of the United States' economic evolution is unfortunately not borne out by empirical reality. Aggregate losses in manufacturing span across the board and job loss has occurred not only in traditional manufacturing sectors, such as the automotive industry, but in advanced technological industries such as information and communications technologies (ICT). The Census Bureau, which regularly collects data on 22,000 different product items, labels 500 products as "advanced technology." The Census Bureau considers a product advanced if it is derived from a recognized high-technology field (biotechnology, life sciences, nuclear technology, or advanced materials) or, alternatively, if that product constitutes a leading edge technology in a particular field (for example, electronic components that result in improved performance and capacity, miniaturization, or ICT products that are able to process increased volumes of information in shorter periods of time). According to the Census Bureau's statistics, the balance of trade in advanced technological products remained positive until 2002 (until the bust of the dotcom bubble), after which the balance has been consistently negative.

Without a clear understanding of the challenges to the health of the defense industrial base, we incur the ultimate risk: cutting too sharply in the areas that provide the core of both our defense industrial base and the economy at large. It is imperative to bolster the health of those sectors responsible for the innovative technologies that characterize American competitive advantage within the world economy.

More surprisingly, trade deficits in advanced technological products widened even as the value of the dollar declined during this period.[13] Competing interpretations are offered to explain this trend, several of which correspond with the arguments presented in this study. These include:

First, technological convergence has taken place as more countries have caught up in areas where the U.S. previously enjoyed clear advantages. These countries have built from the ground up to achieve technologically advanced export sectors that are eclipsing the United States.[14]

Second, U.S. companies have steadily moved large chunks of their manufacturing supply chains offshore, a trend which started in the 1980s as a way to take advantage of cheaper labor costs in newly industrializing countries. In the 2000s, offshoring culminated in the relocation of high-tech manufacturing to emerging markets in order to take advantage of the special credits, grants, and subsidies these states offered, as well as to gain proximity and greater access to other manufacturing facilities or large and growing consumer markets.

Third, and in no small way, U.S. policy responses to global economic challenges and associated growing pains have been awkward, insufficient, and often counter-productive. A simplistic view of neo-classical economics (rife with assumptions from earlier times, including the idea that the division of labor always yields a net social gain) has resulted in muted responses to the gradual hollowing out of the United States' high-tech manufacturing base. Although there have been voices arguing for a more active and forward-looking approach to nurture and protect high-tech manufacturing, the U.S. federal government response has tended to be ambivalent, to say the least.

The ideological framework within which much of U.S. policy discourse takes place continues to assume that the United States automatically reaps benefits when its manufacturing sector seeks lower costs and moves offshore. According to this view, the global redistribution of manufacturing follows a "law" that reallocates resources where they will be most efficiently used, inevitably raising economic welfare for all. According to this scheme, specialization is inevitable, as low-cost manufacturing moves to low-income countries, leaving high-income countries to concentrate on areas of production that yield higher wages and push up the overall productivity of the economy.[15]

As we survey the first decade of the 21st century, we see the fallacy of this theory. Job losses have remained constant; unproductive firms have gone under, and there have been no replacements for traditional manufacturers that closed because of competition from lower labor cost countries.[16] Large corporations have responded opportunistically to the competitive pressures of globalization by moving activities to countries where the policy and corporate environments appear more receptive to competition, innovation, and manufacturing.

Smaller firms that are less nimble or able to adapt to these emerging challenges have been exposed to the full impact of global competition. The lack of a coherent government response to the adverse effects of increased—and often unfair—competition on smaller firms has been especially painful for this sector. High-technology manufacturing and innovation raise high entry barriers for small firms, and they clearly would benefit from a more consistent response and policy intervention by the U.S. federal government. There are many ideas for ways policy reforms can protect these firms from disproportionate penalties; for example, federal or local incentives that support innovation clusters can help small firms attract the skills, capital, and market outlets necessary to become not just viable but genuinely competitive.

THE DEFENSE INDUSTRIAL BASE – AN ESSENTIAL PART OF THE FORCE STRUCTURE

We cannot ignore globalization's impact on the United States' ability to sustain and meet its national security needs. The defense industrial base is a segment of the U.S. economy that is inextricably connected to the rest of the industrial base and indeed to an increasingly globalized economy. As such, the defense industrial base is shaped by globalization. Indeed, because of the scale of the defense sectors in certain industries, the effects of globalization on some defense industries may be much more than others. This disparity exists because, depending on the industry, the defense-unique sector may be quite small in comparison to the larger industrial base. For example, the defense sector of the five-axis machine tools industry is a relatively small portion of the larger machine tools industry. On the other hand, some industrial products (for example, steel armor plate, which is a defense-unique product) belong almost entirely to the defense sector, even as the capacity to manufacture those products is embedded in the larger industrial base. The defense component of the U.S. aerospace industry is relatively large as well, and contributes a significant proportion of the overall U.S. aerospace industry exports and employment.

DoD demand provides an important cushion for industries exposed to global competition and volatile price developments. Defense demand softens the impact of global downturns in sectors like aerospace and shipbuilding. An example is titanium, which requires large capital investments and long lead times to produce. In spite of the complexity of extracting, processing, and fabricating titanium alloys, the global market regularly undergoes periods of famine and feast. While defense applications account for a relatively small share of the total output of U.S. titanium (used in military aviation and armored vehicles), this demand supports the survival of a defense-critical sector at critical times in the business cycle.

> "Essentially, the industrial base is part of our force structure and we have to treat it like it is."[a]
>
> —Under Secretary of Defense (AT&L) Frank Kendall

On the other hand, though defense needs are not subject to cyclical downturns, they are subject to the vagaries of the acquisition process, budget cuts, and profound uncertainties stemming from the present dysfunction in the federal budget process.

The U.S. federal government and the Department of Defense (DoD) play a pioneering role in supporting technological innovation. In the 1950s and 1960s, DoD spurred innovation in semiconductors through procurement and targeted research programs. In the 1960s through the 1980s, DoD- and NASA-sponsored research heavily contributed to building American science and engineering capabilities in chip design, aeronautics, and satellite communications.

Especially in a time when the Pentagon budget faces the largest cuts in more than a decade,[17] it is important to understand the risks to the defense industrial base and the risk to securing the new frontier of technological innovation. It is equally vital to know which risks are unacceptable from a national security perspective.

Unfortunately, the unique challenges to and vulnerabilities in the defense industrial base are not well understood, even among those charged with preserving its health. Using national security as a focus, this report aims to identify key vulnerabilities and recommend mitigation strategies for key sectors of the defense industrial base. Erosion of the defense industrial base, due in large part to the disappearance of manufacturing output and production, and sharp reductions in domestic investments in advanced manufacturing technologies, undermines the capacity of the military services to deploy and protect troops abroad, undertake offensive and defensive operations, invest in innovative weapons systems, and retain a technological skill base. In short, the U.S. defense industrial base is an essential component of U.S. national security, not merely as a source of weapons systems and industrial support for the warfighter, but because the defense sector is of vital importance for innovation and the development of emerging technologies.

Moreover, synergy exists between defense manufacturing technology and innovation. For decades the U.S. defense industry reliably has produced the best weapons systems in the world, and a significant amount of U.S. industrial innovation over the past few decades has either originated or has been strongly propelled by the defense sector. Notably, innovations in emerging technologies and aerospace, as well as communications and advanced materials continue to be pioneered by the U.S. defense sector. Particularly important in the development of emerging technologies is the role of small- and medium-sized firms. But small firms need to be embedded in a cluster or geographic network, which supports the diffusion of knowledge, manufacturing technologies, and skill formation, and bolsters relationships with assemblers, suppliers, and customers.[18]

Outlined below are just two of the many prominent examples of how defense-related investment and development led to major breakthroughs that occurred after the manufacturing technology and know-how had moved offshore. These illustrate that once the knowledge base moves offshore, further innovation and technological applications are also at risk of moving offshore.

Example 1: DoD and the U.S. Department of Energy (DoE) were large initial investors in high-density rechargeable batteries. Lithium-ion (Li-ion) batteries are built on complex chemistries that offer superior weight savings per unit of energy density. They last a long period of time during disuse and are low-maintenance. Although the original invention of the Li-ion battery took place in U.S. laboratories housed in U.S. universities funded by the U.S. federal government, the commercialization of rechargeable batteries moved offshore to Japan and South Korea, both of which are now leaders in the advanced battery industry.

As a result of the shift overseas of the advanced battery industry, most innovation in the field has likewise taken place overseas. A prime example of this phenomenon is the development of electric car batteries, in which considerable investments have been made in the U.S. since 2009. Nonetheless, U.S. companies struggle to compete against Korean and Japanese ones because the latter continue to enjoy a comparative advantage due to their earlier start. Ironically, the first generation of Li-ion battery design was developed in the United States and has led to many other high-tech battery applications—but the innovation is not U.S.-driven and takes place offshore.

Example 2: High-tech permanent magnets pack enormous power in a very small size. Often referred to as NdFeB magnets because they are composed mainly of neodymium (Nd), iron (Fe) and boron (B), high-tech permanent magnets are widely used in the production of electronics, machinery, communication equipment, weapons, and military aviation systems. Although U.S. scientists were among the first to recognize the unusual properties of the rare earth element (REE) neodymium, the fabrication, design, and production of these magnets now takes place outside the United States. The decline of the magnetic material industry also has resulted in the closing of select university laboratories devoted to studying REEs and the technology for designing high-tech permanent magnets.

There are numerous other negative ramifications when manufacturing moves offshore. When a major player in an industry moves abroad, it often cuts funding for long-term research. The move allows the company to enjoy quick cost advantages due to lower labor costs, subsidized start-up expenses, lower regulatory standards or lax enforcement, and other benefits of operating offshore. Rivals have to follow suit to keep up and domestic employment opportunities in the sector steadily shrink. The reservoir of skilled workers and scientists are forced to move to other fields or abroad, and eventually the previous knowledge base on which the U.S. industrial sector thrived will be reduced to a few remaining survivors who are cut off from the most exciting and innovative new frontiers of research and manufacturing.[19]

Increasingly, manufacturing and innovation take place in geographic clusters, bringing together producers, suppliers, customers, scientists, workers, and funding.[20] The virtue of the cluster dynamic is that groups of suppliers, clients, and producers work closely together and interact frequently, thereby strengthening innovation and improving quality. Widespread offshoring means that regional innovation clusters emerge outside the United States, depriving U.S. corporations, scientists, investors, and workers access to competitive knowledge networks.[21]

As production goes overseas, the United States not only loses immediate access to products necessary for defense, but also risks losing institutional memory and know-how. Patents for emerging technologies move offshore as well. The U.S. Geological Survey warns that "[l]arge reductions in American high-skilled production and science and engineering workforces leads to loss of technological know-how critical to U.S. leadership in critical technologies."[22]

Such trends endanger the United States' capacity to make the products necessary for the country to mobilize its defense industrial base in a future conflict. Leveraging superiority in the application of advanced materials and sophisticated electronics, communications, and satellite technologies will win future conflicts.

ABANDONING MINING: THE OVER-RELIANCE ON IMPORT OF CRITICAL RESOURCES

When considering potential supply chain disruptions, another major concern is the vulnerability created by the limited domestic supply of rare earth (RE) minerals, combined with the increased reliance on them. Many advanced products rely on obscure chemical elements that are found in either high concentration in a few countries or in limited deposits in many countries. The United States used to have relatively easy access to many mineral ores, but this situation has changed dramatically as the U.S. has neglected to preserve its mining base, and global demand for minor and unusual chemical elements has surged.

Demand for REs has surged largely for two reasons. First, as mentioned above, many advanced electronics, communications, and green technologies require RE minerals. Second, the rapid pace of development in China and India, which together account for a third of the world's population, has led to an explosion in demand for the conveniences of modern life.

For REEs, the result of reduced U.S. production and increased global demand is that the United States now relies on imports for at least 60 different elements, with a total lack of domestic production for 19 of them.[23]

The importance of China, a major producer of REEs and RE minerals, cannot be overstated. China has been intensely concerned about the trends in pricing and supplies of less common chemical elements and has pursued an explicit policy of gaining control over raw materials and the processing of minerals into finished products. A decision by China or another strategic rival to restrict access to the supply of minerals necessary for advanced weapon systems, communication networks, electronics, nuclear energy, and green technology could compromise U.S. national security.

To encourage local Chinese mining operators to move up the value-added chain, the Chinese authorities sought to induce foreign and domestic fabricators to refine and process the raw minerals in China itself. Accordingly, many Western and Japanese companies have felt great pressure to relocate to China to gain access to these critical materials and compete with Chinese producers. As U.S. companies move to China to gain access to these critical minerals, they also knowingly and unknowingly transfer technology to Chinese competitors, who then compete with established Western companies.

The U.S. government, the defense establishment, and analysts have raised alarm about the RE situation and encouraged the re-opening of RE mining in the United States. However, in the past year, the softening of economies of Europe, Japan, and the United States has led to a fall in demand for REEs. At the same time, many non-Chinese mining companies have rushed to open new mines outside China. At this point, there is sufficient supply of REEs on the market, but the fabrication and manufacturing of defense items and gadgets continues to take place outside the United States, in China.

PROTECTING OUR DEFENSE INDUSTRIAL BASE – THE ROLE OF POLICIES, LEGISLATIVE FRAMEWORKS, AND FEDERAL AGENCIES

Former U.S. Ambassador to the United Nations John Bolton speaks of the necessity to draw a line between "sweeping too broadly" and "too narrowly" in the context of how we should approach restrictions on arms exports, and specifically exports of critical technologies. Indeed, experts in the U.S. government and industry have wrestled with that analytical task since the early 20th century. The U.S. government, in cooperation with U.S. industry, has created various Executive Branch agencies and legislative frameworks to protect advanced defense technologies, U.S. production, and U.S. innovation writ large. Unfortunately, many of these legislative frameworks have eroded over time, sometimes because the technological context has changed, sometimes because the dynamics of globalization have changed the contours of the industry itself. Fortunately, Congress recently appears intensely interested in learning more about contemporary supply chain risks, and updating the array of Executive and Legislative Branch programs and agencies accordingly. For example, in the

"Sweeping too broadly justifiably raises concerns about an under-the-table industrial policy that acts as a hidden tariff barrier against the disfavored. Sweeping too narrowly, however, risks the unintended transfer abroad of key texhnologies or placing at risk our supplies of critical national security assets at decisive moments. Unless one is prepared to argue that everything our military and intelligence services require can be outsourced abroad, there is no way to avoid drawing this line."[b]

– Former U.S. Ambassador to the United Nations John Bolton

FY2012 National Defense Authorization Act (NDAA), Congress instructed DoD to provide a full assessment of the supply chains for key defense items in order to understand the extent to which the United States depends on foreign suppliers.[24]

This report refers repeatedly to the array of Executive Branch agencies and legislative frameworks constituted to protect the U.S. defense industry, equipment, and technologies. The U.S. government has long been aware of the need to 1) prevent the unauthorized transfer of critical defense technologies; 2) ensure a reliable supply of critical and strategic materials for defense applications; and 3) safeguard against the risk of deficient supplies due to shortages and foreign dependence, especially in times of national emergency. (Export controls are aimed at preventing undesirable technology transfer.) The two dominant approaches to addressing the risks to defense industrial base supply chains have been through domestic preference legislation such as the 1933 Buy American

Act, the 1941 Berry Amendment, and the 1973 Specialty Metals Clause (SMC), and the practice of stockpiling strategic and critical materials. The following section briefly reviews these specially constituted agencies and frameworks.

DOMESTIC SOURCE PREFERENCES

In 1933, Congress passed the Buy American Act (41 U.S.C 10b), which mandated that acquisitions made using federal funds follow a preference for acquiring domestically produced goods and products, except in certain instances. The law states that a product is American-made if 51 percent of the cost of producing the product was incurred in the United States.

In order to protect the U.S. industrial base so that it could meet DoD requirements during periods of crisis, in 1941 Congress passed the Berry Amendment to the Buy America Act (later codified as 10 U.S.C. 2533a). The Berry Amendment restricted DoD food and textile (notably uniform) acquisitions: "to ensure that United States troops wore military uniforms wholly produced in the United States and to ensure that U.S. troops were fed food products wholly produced in the United States." The amendment removed many of the exceptions present in the Buy American Act and modified the majority domestic provision of the Buy American Act to require 100 percent domestic origin for food and textile purchases.[25] With the passage of the FY2002 NDAA in December 2001, the Berry Amendment was made a permanent part of the U.S. Code. The Berry Amendment allows the Secretary of Defense to waive the amendment's domestic procurement requirements under

certain conditions, such as domestic non-availability.[26]

In 1973, the SMC (later codified as 10 U.S.C. 2533b) was added to the Berry Amendment and applied to the Defense Federal Acquisition Regulation Supplement (DFARS) in order to apply domestic source preferences to certain specialty metals. The SMC prohibits DoD from acquiring end-units or components for aircraft, missile and space systems, ships, tank and automotive items, weapon systems, or ammunition unless these items have been manufactured with specialty metals that have been melted in the United States or by "qualifying countries (primarily NATO Allies)."[27] The SMC's objective is to mandate domestic procurement of key metals such as military-grade steel and titanium, and to offset painful contractions in global demand by guaranteeing a flow of defense contracts for these critical domestic specialty metals producers. As with the Berry Amendment, the SMC provides the Secretary of Defense with the authority to waive the requirement to buy domestically if the proper metals "cannot be procured as and when needed."[28]

Domestic source preferences for military-grade steel armor plate and titanium have sustained a stable legislative framework that helps safeguard a domestic production capability for these critical defense materials. In turn, this framework creates a predictable business and investment climate and provides incentives for U.S. production and R&D. The SMC, which currently exists as a standalone section in the U.S. code, is presently the main domestic sourcing requirement governing steel armor plate and titanium procurement.

STOCKPILING AND PROTECTING THE PRODUCTION CAPABILITY FOR STRATEGIC MATERIALS

Just prior to U.S. involvement in World War II, Congress passed the Strategic Materials Act authorizing the Departments of War and the Interior in conjunction with Army and Navy Munitions Board to create a stockpile of strategic raw materials. Following the war, Congress created the National Defense Stockpile (NDS) (50 U.S.C. 98 §2b) to maintain a stockpile of critical and strategic materials and create a sort of insurance policy for defense operations against global supply shortages, including "cartel-like" behavior of foreign exporters.

NDS inventory peaked in 1989, with holdings of 62 different material types worth almost $10 billion. Since the end of the Cold War in 1991, policy-makers have decided to eliminate the U.S. defense stockpile and sell off most its inventory. Accordingly, DoD determined that the methodology used to make stockpiling decisions was outdated, and nearly the entire stockpile inventory was judged to be in excess of forecasted strategic requirements. In the FY1993 NDAA, Congress authorized the disposal of large portions of stockpile holdings, which began the over $6 billion decrease in stockpile inventory. No new additions have been made to the stockpile since 1997.[29]

Recognizing that stockpiling strategies needed revision in light of the growing importance of specialty metals to contemporary defense technologies, a 2008 National Resource Council study concluded that the "design, structure, and operation of the NDS render it ineffective in responding to modern needs and threats", and that DoD "appears not to fully understand its need for specific materials or to have adequate information on their supply."[30]

In the FY2007 NDAA, coincident with the revision of and separation of the SMC from the Berry Amendment, Congress created the Strategic Materials Protection Board (SMPB) as the successor to the NDS. The SMPB was charged with determining the need for protection of supply chains of materials critical for national defense, assessing risk associated with the non-availability of those materials, and advising policy-makers on how to ensure that supply. The SMPB was initially required to meet a minimum once every two years, to publish recommendations regarding materials critical to national security, and notably to vet the list of the most salient contemporary issue, specialty metals.

The SMPB met in 2008 and issued its report and recommendations in December 2008 and February 2009. The SMPB established "critical" materials as those essential for important defense systems lacking viable alternatives, provided that DoD acquisitions dominate the market to the extent that DoD business shapes the direction of that market, and that external markets face significant risk of disruption. Central to the SMPB's report was the statement: "reliable access does not always necessitate a domestic source," and most problematically for domestic producers of specialty metals, that although specialty metals are "essential," they are not "materials critical to national security." Rather, according to the SMPB the report, "strategic materials" warrant monitoring but not domestic source restrictions, and that specialty metals restrictions should be loosened to reduce costs. However, the SMPB did point out that:

"Foreign sources may pose an unacceptable risk when there is a high 'market concentration' combined with political or geopolitical vulnerability. A sole source supplier existing in one physical location and vulnerable to serious political instability may not be available when needed."[31]

In other words, the SMPB's restrictive definition of "materials critical to national security" constrained the DLA Strategic Materials from acting until there were no longer any U.S. domestic suppliers of a strategic material and therefore already in a crisis situation. Meanwhile, absent any action to address the inherent vulnerability of foreign supply for specialty metals, some of the U.S. military's most important and advanced systems came to rely almost entirely on foreign nations, notably China. [32]

However, in the FY2013 NDAA, Congress provided DLA Strategic Materials with the ability to mitigate supply chain vulnerabilities for materials critical for national defense. DLA Strategic Materials can now act when the supply chain depends upon a "single point of failure," as opposed to waiting until no domestic sources remain. In short, DLA Strategic Materials now can anticipate and manage material shortages.

Although the FY2013 NDAA represents an improvement in DoD's mandate to address risks to the strategic materials supply chain, it remains to be seen whether the materials' supply chain vulnerabilities will be effectively mitigated. (The supply chains for specialty metals have already been severely disrupted.) Moreover, engineering skills and manufacturing technology already have moved to locations where

the raw materials were extracted in the first place. Thus, while the United States now mines some REEs, for example, the processing of the elements into fabricated products takes place overseas. Furthermore, U.S. government agencies have to agree on a formula for determining whether a metal or material is at risk. To date, different federal agencies use their own criteria to specify whether a particular mineral is at high risk for supply disruptions or is critical to national security.[33]

THE DEFENSE INDUSTRIAL BASE AS A PUBLIC GOOD

In the modern U.S. defense industrial base, payment that goes to a prime contractor is distributed to a whole network of subcontractors, each of which may in turn have its own network of subcontractors. This diffusion means that the supply chain for the majority of defense end-items, considering both raw materials and subcomponents, is literally global. For example, the material and subcomponents for the F-35 Joint Strike Fighter come from 1,300 different suppliers in nine countries. To further complicate visibility of the supply chain, the lower tiers of the supply chain below prime contractors largely are opaque. The supply chain is made even more opaque by mergers among prime contractors during the past 30 years, due in part to U.S. government policy. As the contours of the defense industrial base change, the relationships of component suppliers to the primes change as well, decreasing visibility of the supply chain even more.

Figure 1: The F-35 Strike Fighter Global Supply Chain

ROLLS-ROYCE HOLDINGS
Lift fan
$11.5 Million
Bristol, U.K., and Indianapolis

PRATT & WHITNEY
Engine
$11 Million
East Hartford, Conn.

KONGSBERG GRUPPEN
Rudders and vertical leading edges
$800,000
Kongsberg, Norway

KULITE SEMICONDUCTOR PRODUCTS
Pressure Sensors
$100,000
Leonia, New Jersey

MARTIN-BAKER AIRCRAFT
Ejection seat
$200,000
Higher Denham, U.K.

LAI INTERNATIONAL
Exterior titanium and aluminum panels
$100,000
Phoenix

ALENIA AERONAUTICA
Wing box
$900,000
Campania, Italy

Source: Bloomberg Business Week, "The F-35s Global Supply Chain." (September 1, 2011). http://www. businessweek.com/magazine/the-f35s-global-supply-chain-09012011-gfx.html

Recognizing that the defense industrial base is a public good, an essential force multiplier, and a key element of national security, DoD is devoting significant efforts to understand the defense industrial base vulnerabilities, as well as taking measures to mitigate the risks. Because DoD's knowledge across industrial base sectors and down into the lower tiers of the supply chain is limited, simply defining the scope of the vulnerabilities to defense industrial base supply chains is a challenge. Efforts to gain a greater understanding of supply chains are underway, notably with DoD's "Sector-by-Sector, Tier-by-Tier" (S2T2) program, which aims to build a database of the prime contractors and sub-contractors, mapping supplier relationships at all tiers.[34] Previous government-industry collaboration to investigate weapons systems supply chains have sometimes foundered on the clashing rocks of classification and legal reviews, making a determination of supply chain risks to defense systems a frustrating exercise at best.[35] When complete, DoD's S2T2 is intended to highlight over-reliance on foreign suppliers and areas of limited competition, and identify "single points of failure" within DoD's supply chains.[36] S2T2 focuses mostly on large combat platforms and weapons systems. In the words of Brett Lambert, Deputy Assistant Secretary of Defense for Manufacturing and Industrial Base Policy, the data will assist DoD in "getting out of the role of firefighter,

> "... as formerly cutting-edge technologies become commoditized it is easy to imagine that a second-tier supplier for a maintenance contract for a U.S. military system would find the cheapest source of a component of its offerings, and that this source might not be friendly to the United States."[c]

waiting for a building to be on fire before we respond."[37] However, S2T2 is a database rather than a management system, designed to map the supply chains as a management tool. The very fact that S2T2 is a new effort shows that DoD does not yet possess the sufficient detail about the kinds of information that it needs to understand the risks to defense industrial base supply chains.[38] Nevertheless, efforts such as S2T2 promise to shed important light on the supply chain risks, enabling the analysis necessary to devise mitigation strategies. Such efforts ultimately will lead to effective collaboration between government and industry.

A major goal of this report is to identify trade-offs between retaining the efficiencies of globalization and the imperative of preserving a strong defense industrial base that underpins U.S. national security. Only

> "... every dollar the United States spends on old and unnecessary programs is a dollar we lose from new, necessary strategic investments."[d]
>
> –Deputy Secretary of Defense Ashton Carter

by identifying these trade-offs will it be possible to chart a course that effectively and efficiently mitigates the risks. A major benefit of realizing supply chain efficiency and reliability is that the United States can save billions in defense costs, which will reduce the pressure for defense cuts in other areas—for example R&D aimed at addressing future U.S. defense needs.[39]

A FEW WORDS ABOUT SEQUESTRATION

Finally, as the current debate features great concern over the impact of sequestration, we must realize that sequestration not only will impact hundreds of thousands of jobs in the defense industry, but also threatens to damage the sinews of our defense industrial base. Under the Budget Control Act, because the U.S. government failed to reach a deal to reduce the U.S. Federal deficit by $1.2 trillion, the federal budget now faces approximately $109 billion in automatic cuts per year over the next decade, divided evenly between defense and non-defense discretionary spending. Unless the law is changed, these cuts will take the form of percentage reductions to every single program, project, and activity (PPA). As it now stands, DoD will lack the authority to prioritize among PPAs. Nothing is more foolish than to allow these across-the-board cuts to defense spending to remain in place—but defending every defense program is just as foolish. The prospect of declining budgets heightens the need for strategy, prioritization, and wise decision-making.

Will sequestration gut U.S. military and defense industrial capabilities overnight? No. However, if sequestration remains in place, many defense industrial base firms (especially at the lower tiers) may go out of business or move out of the defense field. Significant capacity and essential defense industrial capabilities may be lost. This loss may not be apparent until the next time the United States needs to rapidly surge production of a particular system, by which point it will be too late, as neglected defense industrial capacity can be lost.

The example of the United Kingdom is instructive. In the 1990s, as the United Kingdom planned a new nuclear submarine, it faced a six-year gap between the end of the previous submarine production line and the beginning of a new one. Unfortunately, in those six years, the technology and (more importantly) the technical skills required to build these advanced defense systems eroded to the point that the new class of submarines suffered from design problems, budget overruns, and delays. The United Kingdom lost the capacity to design, much less build a submarine. Ultimately, the submarine contractor (BAE) engaged the services of a U.S. company, General Dynamics Electric Boat, to assist them. Nevertheless, the submarine acquisition program was significantly delayed and ran $2 billion over budget, in large part because this important part of the UK defense industrial base was allowed to atrophy.[40]

We cannot afford, nor is it necessary, to protect every capability and sector in our defense industrial base; rather, we must wisely choose which sectors must be nurtured in order to safeguard U.S. national security interests. A clear understanding of the risks to critical defense sectors that our national security relies on is crucial to preserving and restoring those defense industrial base sectors' health, and preventing permanent damage to U.S. national security.

"A technologically advanced, vibrant, and financially successful defense industry is in the national interest … We'll be looking as we make changes for … any skillsets that are now in the defense industry that if we allow them to go away would be very difficult … or time-consuming or expansive to recreate … (and) can't be found in commercial industry. Those … we have an obligation to sustain."[e]

–Deputy Secretary of Defense Ashton Carter

This report will examine several of the most troubling vulnerabilities within our defense industrial base, and recommend strategies for government and industry to address the most urgent risks. Neglecting the health of our defense industrial base places U.S. national security at risk. If the dynamic and innovative U.S. defense industrial base is allowed to continue to wither, we risk our warriors' lives and their ability to carry out their missions. Moreover, the United States will lose the institutional knowledge, skill, and innovation that underpin the most important component of U.S. national power–our economy. We need strategy-driven, concerted, aggressive action on the part of industry and government to address the vulnerabilities of our defense industrial base.

ENDNOTES

a. Frank Kendall, quoted by Mark A. Gordon, "Hearing on 'American Manufacturing Competitiveness Act of 2012,'" Testimony before the House Energy and Commerce Committee (June 1, 2012), 4. http://energycommerce. house.gov/hearing/subcommittee-commerce-manufacturing-and-trade-legislative-hearing.

b. John R. Bolton, "The Hidden Security Risk," The Washington Times, June 17, 2008. http://www.aei. org/article/foreign-and-defense-policy/terrorism/ the-hidden-security-risk/.

c. Synthesis Parners, "Tracing the Supply Chains for the F-22 Raptor Fighter Aircraft, UH-60 Blackhawk Helicopter and the DDG 1000 Zumwalt-Class Destroyer" (November 9, 2007), 5. http://www.uscc.gov/Research/ tracing-supply-chains-f-22-raptor-fighter-aircraft-uh-60-blackhawk-helicopter-and-ddg-1000.

d. Deputy Secretary of Defense Ashton Carter, quoted in Emilie Rutherford, "Carter: DoD to Protect Vital Industry Skillsets in Next Year's Budget," Defense Daily, May 31, 2012. http://www.defensedaily.com/free/17916.html.

e. Emilie Rutherford, "Carter: DoD to Protect Vital Industry Skillsets in Next Year's Budget," Defense Daily, May 31, 2012.

1 Brett Lambert, "The Defense Industrial Base: The Role of the Department of Defense." Testimony before the House Armed Services Committee (November 1, 2011). http://armedservices.house.gov/index.cfm/2011/11/the-defense-industrial-base-the-role-of-the-department-of-defense.

2 Paul Kennedy. *The Rise and Fall of Great Powers: Economic Change and Military Conflict from 1500 to 2000* (New York: Vintage Books, 1989).

3 Gerald Abbott and Stuart Johnson, "The Changing Defense Industrial Base," *Strategic Forum,* 96 (November 1996), 1-5.

4 Clair Brown, Barry Eichengreen, and Michael Reich, eds. *Labor in the Era of Globalization* (New York: Cambridge University Press, 2010).

5 Joel S. Yudken, *Manufacturing Insecurity: America's Manufacturing Crisis and the Erosion of the U.S. Defense Industrial Base* (September 2010), 2-3. http:www.ndia.org/Divisions/Divisions/Manufacturing/ Documents/119A/1%20Manufacturing%20 Insecurity%20ES%20V2.pdf.

6 Susan Helper and Howard Wial, "Accelerating Advanced Manufacturing with New Research Centers," *Brookings-Rockefeller: Project on State and Metropolitan Innovation* (February 2011), 1. http://www.brookings.edu/~/ media/research/files/papers/2011/2/08%20states%20 manufacturing%20wial/0208_states_manufacturing_wial. pdf.

7 Bureau of Labor Statistics, *Employment, Hours, and Earnings from the Current Employment Statistics Survey* (March 2013). http://data.bls.gov/pdq/ SurveyOutputServlet.

8 Bureau of Labor Statistics, "Employment Outlook: 2010–2020," *Monthly Labor Review* (January 2012). http://www.bls.gov/opub/mlr/2012/01/art4full.pdf.

9 Bureau of Labor Statistics, "International Comparisons of Annual Labor Force Statistics, Adjusted to U.S. Concepts, 10 Countries, 1970-2009," Table 2-4. http://www.bls.gov/fls/flscomparelf/employment. htm#table2_4.

10 Ibid., Table 7.

11 "Unit Labour Costs - Annual Indicators (OECD STAT extracts)," Organization for Economic Cooperation and Development. http://stats.oecd.org/Index. aspx?queryname=430&querytype=view#.

12 Kevin Hassett, "Obama's Obsession Drives Progress in Reverse: Kevin Hassett," *Bloomberg.com*, August 15, 2010. http://www.bloomberg.com/news /2010-08-16/obama-s-obsession-drives-progress-in-reverse-commentary-by-kevin-hassett.html.

Jon Gertner, "Does America Need Manufacturing?" *The New York Times Magazine*, August 24, 2011. http://www.nytimes.com/2011/08/28/magazine/does-america-need-manufacturing.html?pagewanted=all.

13 U.S. Census Bureau, *2012: U.S. Trade in Goods with Advanced Technology Products.* http://www.census.gov/ foreign-trade/balance/c0007.html.

14 Dan Breznitz, *Innovation and the State: Political Choice And Strategies For Growth In Israel, Taiwan, And Ireland* (New Haven: Yale University Press, 2007).

15 Gregory Tassey, "Rationales And Mechanisms For Revitalizing, U.S. Manufacturing R&D Strategies," *Journal of Technology Transfers* 35 (2010), 283–333.

16 Robert Atkinson, Luke Stewart, Scott Andes, and Stephen Ezell, "Worse Than the Great Depression: What Experts Are Missing About American Manufacturing Decline," The Information Technology and Innovation Foundation, (March 2012).

17 C. Whitlock, "Pentagon Budget Set to Shrink Next Year," *Washington Post*, January 26, 2012. http://www. washingtonpost.com/world/national-security/pentagon-budget-set-to-shrink-next year/2012/01/26/gIQALpfNTQ story.html.

18 Susan Helper and Howard Wial, *Accelerating Advanced Manufacturing with New Research Centers* (Washington, DC: Brookings Institution, February 2011). http://www. brookings.edu/~/media/Files/rc/papers/2011/0208_ states_manufacturing_wial/0208_states_manufacturing_ wial.pdf.

19 Gary Pisano and Willy Shih, "Restoring American Competitiveness," *Harvard Business Review* 87: 7/8 (July/August 2009), 114-125.

20 "DOE to Support Competition to Strengthen Manufacturing Clusters Across the Nation," *EERE News*, May 29, 2012. http://apps1.eere.energy.gov/news/daily. cfm/hp_news_id=355.

21 Committee on Competing in the 21st Century, *Growing Innovation Clusters for American Prosperity: Summary of a Symposium*: National Research Council. Washington DC: National Academic Press (2011).

Peter Marsh, *The New Industrial Revolution: Consumers, Globalization, and The End of Mass Production* (New Haven: Yale University Press, 2012).

22 Joel S. Yudken, *Manufacturing Insecurity: America's Manufacturing Crisis and the Erosion of the U.S. Defense Industrial Base* (September 2010).

23 U.S. Geological Survey, Mineral Commodity Summaries (Reston, VA: USGS, 2012), 6. http://minerals.usgs.gov/ minerals/pubs/mcs/2012/mcs2012.pdf.

24 "Congress Orders Pentagon To Assess Health Of Defense Industrial Base, Consider A Rare-Earths Stockpile, and Find Out Why U.S. Is Dependent on Foreigners for Night Vision Systems," *Manufacturing and Technology News* 18, no. 20 (December 30, 2011). http://www.manufacturingnews.com /cgi-bin/backissues/backissues.cgi?flag=show toc&id_issue=327&id_title=1.

25 Valerie Bailey Grasso, "The Berry Amendment: Requiring Defense Procurement to Come from Domestic Sources" (Washington, D.C.: Congressional Research Service, January 13, 2012). http://www.hsdl.org/?view&did=697958.

26 Ibid., 6.

27 Specialty metals covered by this provision include certain types of cobalt, nickel, steel, titanium and titanium alloys, and zirconium and zirconium base alloys.

28 Valerie Bailey Grasso, "The Specialty Metals Provision and the Berry Amendment: Issues for Congress" (Washington, D.C.: Congressional Research Service, October 5, 2010), 7. www.fas.org/sgp/crs/misc/RL33751.pdf.

29 Under Secretary of Defense for Acquisition, Technology, and Logistics, "Strategic and Critical Materials Operations Report to Congress" (January 2011). https://www.dnsc.dla.mil/Uploads/Materials/dladnsc2_9-13-2011_15-9-40_FY10%20Ops%20Report%20-%2005-06-2011.pdf.

30 National Research Council, Committee on Assessing the Need for a Defense Stockpile, *Managing Materials for a Twenty-first Century Military* (Washington, DC: The National Academies Press, 2008), 2-3. www.nap.edu/catalog/12028.html.

31 Department of Defense, "Report of Meeting, Department of Defense, Strategic Materials Protection Board, December 12, 2008." http://www.acq.osd.mil/mibp/docs/report_from_2nd_mtg_of_smpb_12-2008.pdf.

"Analysis of National Security Issues Associated With Specialty Metals," *Federal Register* 74, no. 34 (February 23, 2009), 8061-64. http://edocket.access.gpo.gov/2009/pdf/E9-3708.pdf.

32 Jeffery A. Green, "Congress Finally Tackles Strategic Materials Reform," *National Defense Magazine*, March 2013. http://www.nationaldefensemagazine.org/archive/2013/march/Pages/CongressFinallyTacklesStrategicMaterialsReform.aspx.

33 Daniel McGroarty and Sandra Wirtz, "Reviewing Risk: Critical Metals & National Security," American Resources Policy Network (June 6, 2012). http://americanresources.org/wp-content/uploads/2012/09/ARPN_Fall_Quarterly_Report_WEB.pdf.

34 Department of Defense Office of Manufacturing and Industrial Base Policy, "Sector-by-Sector, Tier-by-Tier (S2T2) Industrial Base Review" (2011). http://www.acq.osd.mil/mibp/s2t2.shtml

35 Synthesis Partners, "Tracing the Supply Chains for the F-22 Raptor Fighter Aircraft, UH-60 Blackhawk Helicopter and the DDG 1000 Zumwalt-Class Destroyer" (November 9, 2007), 7. http://www.uscc.gov/Research/tracing-supply-chains-f-22-raptor-fighter-aircraft-uh-60-blackhawk-helicopter-and-ddg-1000.

36 Brett Lambert, "Presentation to National Defense Industrial Association," (August 2011), 4. http://www.ndia.org/Advocacy/Resources/Documents /NDIA_S2T2_Briefing_AUG11.pdf.

37 Brett Lambert, "Testimony before the Panel on Business Challenges within the Defense Industry of the House Committee on Armed Services: he Defense Industrial Base: The Role of the Department of Defense" (November 1, 2011), 9. www.gpo.gov/fdsys/pkg/CHRG.../pdf/CHRG-112hhrg71456.pdf.

38 Synthesis Partners, "Tracing the Supply Chains for the F-22 Raptor Fighter Aircraft, UH-60 Blackhawk Helicopter and the DDG 1000 Zumwalt-Class Destroyer" (November 9, 2007), 11. http://www.uscc.gov/Research/tracing-supply-chains-f-22-raptor-fighter-aircraft-uh-60-blackhawk-helicopter-and-ddg-1000.

39 Greg Parlier, *Transforming U.S. Army Supply Chains: Strategies of Management Innovation* (Williston, VT: Business Expert Press, 2011).

40 National Defense Industrial Association, "Shipbuilders Forecast Exodus of Submarine Designers," by Grace Jean, *National Defense Magazine*, November 2007. http://www.nationaldefensemagazine.org/archive/2007/November/Pages/Shipbuilders2446.aspx.

CHAPTER 2 • STEEL ARMOR PLATE

EXECUTIVE SUMMARY

Steel armor plate is a critical structural component of nearly all advanced armored ground vehicles and the hulls of most U.S. naval vessels. The U.S. steel industry manufactures steel armor plate to precise military chemical and physical specifications. The continued ability of the U.S. defense industrial base to produce steel armor plate for U.S. combat platforms is important for the country's national security.

Its importance was seen recently in the response to the improvised explosive device (IED) threat. Beginning in 2006, this threat prompted the rapid development and deployment of the Mine-Resistant Ambush-Protected (MRAP) vehicle, which required the swift production of large quantities of steel armor plate. The U.S. defense industrial base was able to respond quickly and flexibly, assisting in the deployment of a platform that then-Secretary of Defense Robert Gates said saved thousands of lives.

Today, the main risk to steel armor plate production capacity, aside from the broader defense drawdown, comes from attempts to weaken the Specialty Metals Clause (SMC). The SMC mandates that all steel armor plate used by the U.S. military must come from domestic sources—although there are numerous exceptions to the statute. Until 2008 the SMC had been understood to require that the melting phase—the most capital-intensive phase of steel armor plate production—must be carried out within the United States. However, the Department of Defense (DoD), driven by concerns about a lack of capacity in the U.S. defense industrial base, has explored whether a redefinition of the SMC is warranted to allow steel armor plate melted abroad but heat-treated in the United States to count as having been "produced" domestically.

Given that current U.S. capacity is sufficient to meet demand from DoD, and that DoD has preexisting authority to temporarily waive SMC restrictions if domestic capacity is at some point insufficient, a permanent redefinition of the SMC is unnecessary. The permanent redefinition of the SMC could undercut the U.S. defense industrial base's ability to carry out all phases of steel armor plate production and provide protection to the U.S. warfighter.

STEEL ARMOR PLATE
PROTECTING U.S. FORCES

MANUFACTURING SECURITY

Steel armor plate is essential for U.S. combat platforms

 +

U.S. GROUND FORCES   **U.S. NAVAL FORCES**

PROTECTING WARFIGHTERS

Steel armor plate protects Mine-Resistant Ambush-Protected (MRAP) vehicles

MRAPS HAVE HELPED SAVE **"THOUSANDS OF LIVES"**

Armored platforms have shielded U.S. forces in Iraq and Afghanistan

 +

IRAQ   **AFGHANISTAN**

SPECIALTY METALS CLAUSE

The SMC should not be weakened to allow foreign-melted steel armor plate

 + =

FOREIGN PRODUCTION   **UNSECURE SUPPLY**   **DIMINISHED U.S. CAPACITY**

PRODUCTION

 = 1%

U.S. steel production manufactured for homeland security and national security applications

3%
OF U.S. STEEL PRODUCTION

MANUFACTURING PROCESS

The U.S. should be able to perform all phases of steel armor production necessary for U.S. defense products

  MELTING

  ROLLING

  HEAT TREATING

  STEEL ARMOR PLATE

MITIGATING RISKS

Avoiding uncertainty in U.S. armor plate supply

STRONG LEGAL FRAMEWORK   **UNDERSTANDING SUPPLY CHAINS**   **PARTNERSHIP BETWEEN DOD & INDUSTRY**

MILITARY EQUIPMENT CHART
SELECTED DEFENSE USES OF STEEL ARMOR PLATE

DEPARTMENT	PLATFORMS
ARMY	A wide range of Army platforms including: ■ Mine-Resistant Ambush-Protected (MRAP) vehicle ■ M1A2 Abrams main battle tank ■ Stryker fighting vehicle
MARINE CORPS	A wide range of Marine Corps platforms including: ■ Mine-Resistant Ambush-Protected (MRAP) vehicle ■ M1A2 Abrams main battle tank ■ Stryker fighting vehicle
NAVY	A wide range of Navy platforms including: ■ Freedom-class Littoral Combat Ship (LCS) ■ SSN-774 Virginia-class nuclear-powered attack submarine ■ Nimitz-class nuclear-powered aircraft carrier

INTRODUCTION

American military dominance requires global force protection and the ability to sustain military operations in hostile and volatile environments. The U.S. military has excellent long-range and precision strike capabilities. However, certain kinds of missions, such as the ongoing conflict in Afghanistan, also require American forces to engage with adversaries at close range.

Steel armor plate, a product of the U.S steel industry, has many force protection applications and is used in many U.S. ground combat platforms. In Iraq and Afghanistan, American ground troops have been equipped with Mine-Resistant Ambush-Protected (MRAP) vehicles that use steel armor plate to increase resistance against enemy mines and improvised explosive devices (IEDs).[1]

Steel armor plate also protects American naval vessels, from large naval platforms such as Nimitz-class aircraft carriers, to smaller, more nimble platforms such as the Littoral Combat Ship (LCS). While the U.S. Navy does not face an immediate or near-term threat from peer competitors, it must nevertheless be prepared for asymmetric

and future threats and challenges by having its ships fitted with appropriate armor.[2]

Sturdy, armored naval platforms are especially vital in the context of the current U.S. rebalancing to Asia, which places an emphasis on naval deployments to bolster allies and partners and assure U.S. access and influence.[3]

Naval vessels protected with steel armor plate are also essential in U.S. plans for ballistic missile defense (BMD) under the phased adaptive approaches currently being implemented in collaboration with North Atlantic Treaty Organization (NATO) and other allies. The Aegis air and missile defense system with the Standard Missile

PROTECTING U.S. TROOPS (a notional though realistic situation)

A Mine-Resistant Ambush-Protected (MRAP) vehicle with seven U.S. troops inside was on a routine patrol outside of Baghdad, Iraq in February 2008. MRAPs are protected by steel armor plate. The MRAP had been deployed to Iraq only two months earlier. As the vehicle turned a corner, a member of a local insurgent group remotely detonated a roadside improvised explosive device. The low explosion violently shook the vehicle and its passengers; however, the vehicle's v-shaped hull and steel armor protected against the blast, and all inside survived. Without the protection provided by the MRAP, that attack would almost certainly have been fatal.

3 (SM-3) interceptor is deployed on U.S. Navy cruisers and destroyers, which are protected by steel armor plate. Under the strategy envisioned in the U.S. Ballistic Missile Defense Review (BMDR) these assets will be deployed in various theaters as part of a flexible response to evolving missile threats.[4]

As the United States struggles to deal with deficits and debt, the U.S. military faces increasingly constrained and uncertain resources. In such a fiscal environment the U.S. military will have to make tough choices about its acquisition and modernization programs, including the ones that use steel armor plate. This chapter argues that the imperative to cut budgets should not drive the U.S. to weaken an important part of the defense industrial base, which, once lost, will be difficult and expensive to reconstitute.

Key themes discussed in this chapter are:

- Steel armor plate is a vital force protection tool for U.S. ground- and sea-based combat platforms.

- The U.S. steel industry has the proven capacity and flexibility to rapidly respond to complex military requirements. This capacity and ability cannot be taken for granted.

- It is unnecessary and counterproductive to permanently weaken U.S. domestic sourcing requirements and allow steel melted abroad to be used for U.S. combat applications.

A NOTE ON CRITICALITY

Steel armor plate is an important component for armored ground vehicles, including personnel carriers and tanks and the armored hulls of nearly all U.S. Navy

vessels. The inability to utilize domestically produced steel plate would *incapacitate* U.S. military capabilities, rendering the United States unable to construct and repair many military platforms used by the U.S. Army, U.S. Marine Corps, and U.S. Navy.

While a shortage of steel armor plate would be damaging to U.S. military capabilities, challenges facing the sector of the defense industrial base that produces steel armor plate constitute a *moderate* risk. Despite an increased military demand for steel armor plate throughout the latter part of the last decade, in light of the recent economic downturn and foreign competition, the U.S. steel industry has struggled with reduced commercial demand.

BACKGROUND

Steel armor plate differs from other plate steels that are used for applications such as bridge-building. Its special chemical and physical properties allow it to withstand explosions and gunshots, and it is manufactured using specialized equipment and precise manufacturing processes.

Steel armor plate is a critical input to the supply chains that produce and maintain certain wheeled and tracked ground combat vehicles. It is also an input into the shipyards that produce U.S. Navy surface ships and submarines. Without steel armor plate, U.S. vehicle manufacturers and shipyards could not produce platforms such as the MRAP in compliance with U.S. military requirements. As then-Secretary of Defense Robert Gates told *USA Today* in 2011, MRAPs have saved "thousands and thousands of lives."[5]

Steel armor plate represents a small portion of total U.S. steel industry output; the

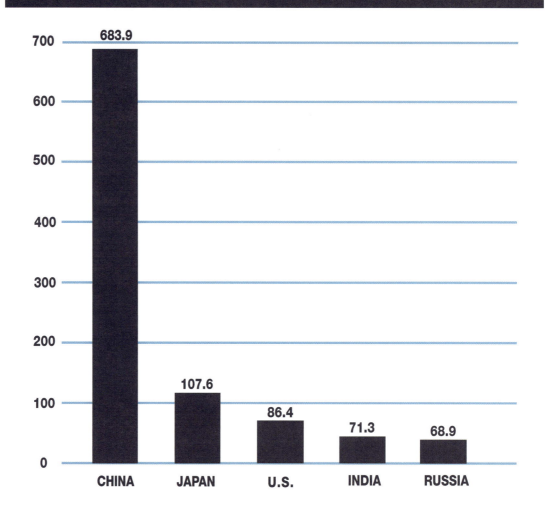

Figure 1: Top Five Steel Producers
(in million metric tons)

CHINA	683.9
JAPAN	107.6
U.S.	86.4
INDIA	71.3
RUSSIA	68.9

Source: World Steel Association, World Steel in Figures 2011.
http://www.worldsteel.org /dms/internetDocumentList/bookshop/ WSIF_2011/document/
World%20Steel%20in%20Figures%202011.pdf

majority of steel produced in the United States is for commercial applications.[6] The United States is the third largest producer of steel in the world, behind Japan and, the largest, China (see Figure 1).[7] Despite its high production, China exports a relatively a small percentage of its total steel output. In 2009 China exported 4.2 percent of its total production, compared to 15.9 percent by the United States and 38.1 percent by Japan.[8] In part, China's exports to the United States have been constrained by U.S. trade laws—vital measures that have limited imports into the U.S. of unfairly traded steel.

China has increased its production capacity for steel at a rate that far exceeds its domestic consumption patterns, thereby putting pressure on non-Chinese international steel producers. Many of these producers, such as those in the United States, do not benefit from government subsidies.

Given that steel armor plate is a relatively small portion of the total output for any particular manufacturer in the United States, commercial sales make up a majority of orders. Therefore, a high level of commercial demand is necessary to keep the specialized facilities used to manufacture steel armor plate economically viable. Out of total U.S. steel shipments in 2010, only three percent were for defense and homeland security applications. The two main uses for U.S. steel were construction (42 percent) and the automotive industry (24 percent).[9]

STEEL ARMOR PLATE AND U.S. DEFENSE CAPABILITIES

Steel armor plate is a vital defense product with a proven record of saving lives. Steel armor plate, when used as a component of U.S. ships and ground-based platforms, enhances the durability of these platforms and increases the likelihood of survival for the U.S. service members they protect. The U.S. military requires certain grades of steel plate for particular U.S. platforms, and their testing ensures that the armor plate meets protection requirements. Each piece of steel armor plate must be precisely the correct height, width, gauge, and flatness in order to be properly integrated into the final product.

The ability to quickly and flexibly produce steel armor plate was critical to the success of the MRAP program, which required large quantities of steel armor plate in a short time span. The U.S. defense industrial base was able to meet this need because of the specialized equipment, capacity, and knowledge possessed by U.S. steel plate producers. The ability of the U.S. steel industry to respond rapidly to the demand generated by the MRAP program does not mean that it automatically will be able to respond to future crises or surges in demand. It also cannot be taken for granted.

The United States does not currently maintain a significant inventory of steel armor plate, due in part to the sheer variety of steel plates needed for U.S. platforms.

Steel armor plate, when used as a component on U.S. ships and ground-based platforms, enhances the durability of these platforms and increases the likelihood of survival for the U.S. service members they protect.

STEEL ARMOR PLATE PRODUCTION

Various facilities in the United States and Canada complete the multiple, complicated, and capital-intensive steps required to produce steel armor plate for the U.S. military. (Canada is treated like the 51st state by U.S. laws that govern armor plate production.) Currently ArcelorMittal USA, a division of Luxembourg-based ArcelorMittal, is the largest supplier of steel armor plate to the U.S. military.[10] ArcelorMittal USA carries out all phases of steel armor plate production, including melting, rolling, and heat treating. Other companies also produce steel armor plate in the United States, including Nucor, which entered the armor plate production business to help increase the production of MRAPs needed for the Iraq War, and Allegheny Technologies Incorporated (ATI). Gaining the capability to carry out each phase of production requires specialized equipment and significant capital investment, especially for the melting phase.

The melting stage is the first phase of steel armor plate production and the part in the process when the most significant percentage of the capital is expended. This steel scrap comes from a variety of sources including demolished automobiles and buildings. Steel scrap prices fluctuate according to various factors including automotive sector trends and foreign demand.[11] Almost all scrap used for U.S. steel armor plate production is acquired domestically.

The molten scrap metal is refined and purified in a furnace, and nickel, chromium, and molybdenum are added in precise amounts to create an alloy with the desired chemical properties. Specialized equipment removes impurities from the molten steel, and the chemistry of the metal is adjusted if necessary. At the end of this phase, the molten metal is either cast as slabs or poured into ingot molds for thicker plates. Although described simply and briefly here, the melting phase of the steel armor plate production process is highly technical, complex, and costly.

In the next phase of the steel armor plate production process, the slabs or ingots are heated to a specific temperature for rolling. The rolling process, aided by sophisticated computer programs, achieves the precise plate thickness and flatness.

Once the steel plate has been rolled, it is ready to be heat treated. Heat treatment is necessary for higher-grade steels, because it alters the physical properties to achieve the physical characteristics necessary to protect U.S. troops. The steel is heated and held at a high temperature, adding strength, and is then quenched (cooled rapidly) to make the steel even harder. The next step is tempering, a process that reheats the steel slightly to reduce brittleness.

The manufacturer tests the plates in-house to ensure that they meet military chemical and physical specifications. U.S. government facilities conduct ballistic testing of each lot before accepting the final product.

ALTERNATIVES TO STEEL ARMOR PLATE

Due to its low cost, durability, ability to withstand multiple hits, and effectiveness against a broad spectrum of threats, steel armor plate has been, and will likely remain, the default material for most land- and sea-based platform armor needs.[12] The main drawback of steel armor plate is its high weight relative to other materials, which can limit mobility. Weight is especially restrictive in the transport and deployment of heavy ground vehicles such as the M1 Abrams tank, which weighs approximately 70 tons and often must be air-transported one at a time. Weight is also becoming more relevant for naval vessels due to the growing need to operate in littoral zones (sea-based areas close to the shore).

Ceramic or composite armors are lighter weight alternatives to steel armor plate. Ceramic materials are non-metal, inorganic materials often formed through advanced heating and cooling processes. Advanced ceramics are engineered through a multi-phased process, culminating in their exposure to extreme heat that causes molecular changes to the ceramic, including the elimination of pores that result in a denser and more resilient product. Most advanced ceramics are produced through a technique called hot pressing, which involves heating ceramic powders at temperatures exceeding 2,000 °C (3,673 °F) while squeezing the materials together at high levels of pressure.[13]

Ceramic armor has advantages and disadvantages compared to steel armor plate. With a backing of advanced synthetic fabrics such as Kevlar and Spectra,[14] which absorb the force of a projectile,[15] ceramic armor possesses stopping power

> Ceramic armor is also more fragile than steel and may fracture if dropped or mishandled. Unlike steel armor plate, which can withstand multiple attacks, ceramic armor tends to weaken with each progressive attack, especially if hit in rapid succession.

comparable to that of steel plate. In contrast to steel armor, which has a general density of 7 to 8g/cm, ceramic armor has a general density of only about 4g/cm. Replacing metal armor with ceramic plate can in some cases significantly reduce vehicle weight, which is important when considering aerial transportation, fuel efficiency, and payload capacity concerns.[16]

However, ceramic armor has certain drawbacks, and it is generally less robust than steel plate. Unlike steel armor plate, ceramics are not suitable to bear large weights, and they cannot be incorporated directly into the structure of a given platform.[17] Ceramic armor is also more fragile than steel and may fracture if dropped or mishandled. Unlike steel armor plate, which can withstand multiple attacks, ceramic armor tends to weaken with each progressive attack, especially if hit in rapid succession.[18]

Concerns over durability and cost mean that ceramic armor is unlikely to replace steel plate in many military applications. However, ceramic armor can be used in conjunction with steel plate armor to augment resilience and survivability and decrease weight.

RECENT DEVELOPMENTS

The most recent surge in armor plate production coincided with the decision to rapidly field the MRAP.[19] MRAPs were deployed in large numbers to counter enemy IEDs, which killed and wounded significant numbers of U.S. and coalition troops in Iraq.[20]

In response to the landmine and IED threat, the U.S. Marine Corps began acquiring the Cougar, an MRAP-type vehicle, between 2004 and 2006.[21] As the IED threat increased, the Marine Corps established the Office of the Program Manager, MRAP, in 2006. That year the Marine Corps solicited and received proposals from industry for ways to meet MRAP requirements. Source selection took place on an accelerated basis. In May 2007, then-Secretary of Defense Robert Gates deemed the MRAP program the highest priority DoD program.[22] From June 2007 to December 2007, monthly MRAP production increased from 82 vehicles per month to 1,300 per month.[23] The MRAP production line was closed in October 2012.[24]

In 2007, DoD conducted an assessment of U.S. industrial capacity to produce steel armor plate, and supply concerns motivated the department to reevaluate domestic sourcing requirements.[25] DoD proposed a new rule modifying the definition of specialty steel "produced" in the United States in 2008.[26] The proposed rule was part of DoD's larger effort to implement the FY2007 National Defense Authorization Act (NDAA). That law had separated specialty metals such as armor-grade steel from the purview of the Berry Amendment, which requires that DoD acquire goods such as food and textiles from completely domestic sources.

Domestic sourcing requirements for certain key metals were recodified under the Specialty Metals Clause (SMC), part of the U.S. code.[27] The statute states that specialty metals procured by DoD must be "melted or produced" in the United States. The word "produce" is not defined in the statute, opening the door for DoD's 2008 proposed rule.

Under the new definition, steel armor plate would be considered as having been "produced" in the United States as long as "certain significant production processes" such as heat treating, quenching, and tempering occurred domestically. This definition allows the U.S. military to use steel melted and rolled anywhere in the world, as long as it undergoes finishing processes in the United States. The U.S. steel industry took issue with DoD's assessment of domestic armor plate production capacity, and they argued that DoD had the option to temporarily waive domestic sourcing requirements in the case of domestic non-availability of sufficient quantities, rather than permanently altering the rules.[28] Furthermore, certain "qualifying countries" with whom the U.S. maintains defense cooperative agreements may supply specialty metals, notwithstanding the domestic sourcing requirement. The U.S. steel industry continues to argue that DoD should retain the original meaning of "produced." Indeed, expanding the terms for eligibility may very well undermine domestic production capabilities by making potential demand more uncertain.

Many Members of Congress and the key jurisdictional committees with responsibility for the law have taken an interest in DoD's definition of the term "produced" as it applies to steel armor plate. In February 2012, Sen. Sherrod Brown (D-OH), Sen. Amy Klobuchar (D-MN), Sen. Al Franken (D-MN), Sen. Kirsten Gillibrand (D-NY),

Sen. Chuck Schumer (D-NY), Sen. Robert Casey (D-PA), and Sen. Kay Hagan (D-NC) introduced the "United States Steel and Security Act of 2012," which would require military steel to be "100 percent made in America."[29] The bill was referred to committee and was never voted on.

The issue of the definition of "produced" has been raised during hearings of the Congressional Steel Caucus, chaired by Rep. Tim Murphy (R-PA), including during the 2012 "State of Steel" hearing. At the hearing, Murphy said that he hopes to "ensure the Pentagon follows the law—and uses steel armor plate that is truly made and melted in America."[30]

The FY2011 NDAA mandated a review and, if necessary, revision of the regulation to ensure the definition of the term "produced" was consistent with Congressional intent. Subsequently, in July 2012, DoD proposed amending the definition of "produce" to encompass all stages of armor plate production, including melting.[31] The final rule was published March 28, 2013 restoring the original definition of "produce" and bringing DoD practice in line with the original intention of domestic sourcing restrictions for steel armor plate.[32]

In the FY2013 defense budget submitted in February 2012, the Pentagon proposed a cut in procurement spending of approximately 5.5 percent compared to FY2012 (10 percent when Overseas Contingency Operation spending is considered.) The U.S. Army, the most significant user of armored ground combat vehicles, had already received the most significant cut as a part of FY2012 spending, and received over 50 percent of total proposed cuts in 2013.[33]

Across the board the Army and the other services face significant further cuts

under sequestration, which took effect on March 1, 2013. Sequestration imposes mandatory cuts to defense and domestic discretionary spending under the 2011 Budget Control Act. These cuts will continue unless Congress finds an alternative method to reduce the deficit by $1.2 trillion over ten years, or change the law, which has not happened as of this writing. The politics of sequestration create an environment of substantial uncertainty for DoD and the defense industrial base that complicates long-term military planning.

Defense cuts and persistent budgetary uncertainty mean that the Army and the other services will be unlikely to procure large numbers of armored platforms in the near term, as illustrated in a recent debate about whether to idle production at the armored vehicle plants in Lima, Ohio, and York, Pennsylvania.[34]

ISSUES AFFECTING STEEL ARMOR PLATE AVAILABILITY

U.S. government policies have a significant effect on U.S. armor plate production capacity. Armor plate and other defense applications represent approximately three percent of U.S. steel shipments.

The military's demand for steel armor plate is too small, in relative terms, to make a significant difference to the overall health of the U.S. steel industry. However, U.S. government policies that influence the industry such as taxation, support for investment in infrastructure, and trade policies can have an important effect on armor plate production capacity.

The U.S. government maintains policies that specifically govern steel armor plate acquisition, especially domestic sourcing requirements. Federal restrictions on acquisition of steel armor to protect domestic sources have been in place since 1973, initially to ensure the availability of domestic materials during the Vietnam War.[35]

The Specialty Metals Clause (SMC):

The SMC mandates domestic procurement of military-grade steel as well as other key metals such as titanium (see Chapter 4 for this report's discussion of titanium).[36] As noted above, the domestic sourcing restriction for specialty metals was originally contained in the Berry Amendment.

The SMC, under Title 10, section 2533b of the U.S. Code, prohibits DoD from acquiring aircraft, missile and space systems, ships, tanks and automotive items, weapons systems, or ammunition "containing a specialty metal not melted or produced in the United States." DoD can obtain an exemption to this restriction if the proper metals "cannot be procured as and when needed."[37]

DoD has explored weakening the domestic sourcing requirement under the SMC through a redefinition of what it means for steel to be "produced" in the United States. The new proposed definition would allow steel melted outside the United States to be purchased by DOD, as long as late stage processes such as heat treating and testing were carried out in the United States.

The U.S. steel industry has generally opposed this redefinition, arguing that it violates the SMC's original intent. The United Steelworkers, the largest North American industrial labor union, stated in a September 2011 letter that DoD's definition of the term "produced" is "improper, flouts over 35 years of legal interpretation and administrative practice, and is contrary to Congressional intent." The letter goes on to argue that the definition "puts in jeopardy the health of the domestic armor plate industry and its workers" and "is likely to increase our reliance on imported metals and, as a result, threatens this nation's defense industrial base."[38] In July 2012, DoD proposed amending the "produced" definition to restore the original intent of the SMC and cover all stages of steel armor plate production. In a letter to former Secretary of Defense Panetta, Sen. Brown and other advocates for domestically produced steel armor plate applauded the move. "The revised definition will help ensure that steel armor plate is produced right here in the United States, to the benefit of the domestic armor plate industry, its workers, and this nation's national security," the senators wrote.[39]

Exports of U.S. defense platforms:

The U.S. steel industry does not export significant amounts of armor plate, although some exports have been made to allied countries such as Israel.[40] However, armor plate is an input for platform manufacturers. If these platforms are exported, it will generate additional business for U.S. armor plate manufacturers. Iraq, for example, has announced that it will purchase U.S. armored platforms.[41] Such exports could help compensate for shortfalls in DoD demand.

VULNERABILITIES IN STEEL ARMOR PLATE SUPPLY CHAINS

DoD does not purchase steel armor plate directly. Armor plate is a lower tier input to U.S. shipyards and vehicle manufacturers, and DoD does not consistently and actively monitor products that are lower tier inputs into the equipment that it eventually purchases. This is due in part to DoD's general preference for relying on the free market to supply inputs for defense products, and in part because of the sheer difficulty of monitoring a vast network of complex supply chains.

Weakening of the SMC: Domestic sourcing requirements for military grade steel armor plate have helped to sustain a stable legislative framework to guide steel producers. This framework in turn creates a predictable business and investment climate and incentivizes production and research and development (R&D) in the United States. The SMC is currently the main domestic sourcing requirement governing U.S. steel armor plate procurement.

As discussed above, there is a risk that the SMC will be weakened through a redefinition of what constitutes steel "produced" in the United States. Recent statements from DoD indicate that this harmful redefinition will be reversed, but sustained attention is necessary to ensure that strong domestic sourcing rules are sustained and enforced.

While changing the definition of "produced" could create some business for those U.S. firms that only perform the later stages of armor plate production, it would reduce the incentive for U.S. firms to invest in all phases of steel armor plate production, especially the rolling and

melting phases. Currently, U.S. melting capacity is more than sufficient to meet U.S. military needs, but U.S. firms still need to attract investment to maintain, upgrade, and expand existing facilities. Permanently weaker domestic sourcing requirements for steel armor plate would make this maintenance and improvement more difficult. Diluting the law will also depress R&D in this area, as investment returns will become increasingly uncertain. Furthermore, this redefinition would diminish the ability of the United States to monitor and regulate all stages of armor plate production.

If the United States loses its capacity to melt steel for armor plate, the capital expenditures associated with rebuilding that capacity down the road will almost certainly be prohibitive. Doing so would also likely take several years—far too long to respond to any future surge in demand.

This delay would have implications for the United States' ability to quickly expand production in a time of crisis. DoD currently has the ability to issue "rated" orders under the Defense Production Act, which compel U.S. companies to prioritize DoD orders over orders from their other clients. Such orders were placed during the MRAP production surge, and U.S. steel firms fulfilled them at the expense of their commercial clients. Foreign firms will have little or no incentive to prioritize DoD armor plate orders, and DoD will not be able to compel them to do so.

In the event of a future need to rapidly surge production to protect U.S. troops, DoD should not have to wait in line.

> It is in the national security interest of the United States to retain the capability to produce sophisticated and durable armored platforms to meet future security challenges around the world.

Unpredictable DoD demand:

Planning investments and articulating requirements is difficult in a time of evolving threats and budgetary uncertainty. As a result, DoD demand for the defense platforms that require armor plate, especially in the future, is unpredictable. This uncertainty affects the companies that produce steel armor plate.

Questions remain about several key acquisition programs that require steel armor. The U.S. Army, for example, plans to acquire the armored Ground Combat Vehicle (GCV), which is intended to replace the armored Bradley Fighting Vehicle. However, GCV production has faced delays and questions over its affordability, and demand for armor associated with the program may materialize later or at a lower level than is currently anticipated.[42] Indeed, the technology development phase for the program was recently extended by six months.[43] Under sequestration, the future of this program would be even more uncertain.

Negative trends in the U.S. steel industry:

Given that a only about three percent of U.S. steel production as of 2010 was for "National Defense and Homeland Security" applications, the market for U.S. steel is therefore primarily affected by trends in the larger economy. Defense-related orders are insufficient to sustain the sector.[44] Furthermore, policies that affect demand for U.S. steel, such as decisions about infrastructure spending, can significantly affect the steel industry.[45] There is currently no mechanism for coordinating these decisions across DoD, much less across the government as a whole.

The recent economic downturn significantly hurt the U.S. steel industry. U.S. consumption of steel mill products went from approximately 9 megatonnes (million metric tons) in the first half of 2008 to a low of approximately 4 megatonnes in the middle of 2009. Consumption has recovered to approximately 8 megatonnes.[46] Overall U.S. steel production followed a similar pattern, with a sharp decline until mid-2009 followed by a gradual recovery (see Figure 2).

At the end of 2012, the steel industry experienced a decrease in capacity utilization, a reversal of some of the post-recession gains. Capacity utilization was 71.7 percent in December 2012. In late 2008 and early 2009 capacity utilization hovered around 40 percent. By historical standards, capacity utilization in the U.S. steel industry remains low according to the U.S. Department of Commerce.[47]

By way of comparison, China's steel output is significantly higher than that of the United States. As of December 2012, China produced 47 percent of total global output; in contrast, the United States produced 6 percent.[48]

Figure 2: U.S. Crude Steel Production (2006-2011)
(in million metric tons)

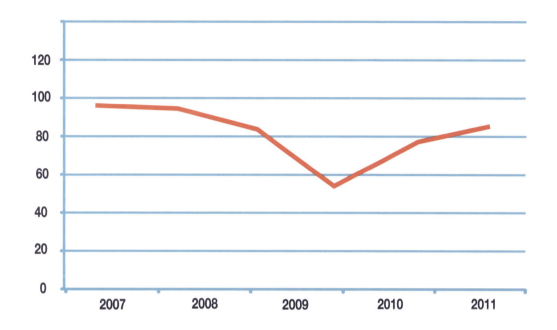

World Steel Association, World Steel Production – Summary (January 23, 2012).
http://www.worldsteel.org/dms/internetDocumentList/press-release-
downloads/2012/2011-statistics-tables/document/2011%20statistics%20tables.pdf

Even as the U.S. steel industry recovers from the recession-induced collapse in demand, it is clear that the industry, including its capacity to produce steel armor plate, is vulnerable to macro-economic shocks, especially in a time of increased global competition. The recent economic downturn happened to coincide with the MRAP-related surge in steel armor plate demand, which helped to sustain armor plate capacity. This may not be the case during a future downturn.

MITIGATING THE RISKS

The United States should make sure that the proper policies and frameworks are in place to ensure a robust and flexible capacity to domestically carry out all phases of steel armor plate production and ensure ongoing investments in R&D and surge capacity. It is in the national security interest of the United States to retain the capability to produce sophisticated and durable armored platforms to meet future security challenges around the world. The section below describes steps that can be taken to mitigate vulnerabilities to future U.S. armor plate manufacturing capacity.

Support and expand efforts to gain insight into the supply chains that support armored platform production.
In keeping with its market-based approach to the defense industrial base, DoD does not currently gather comprehensive information on the supply chains that support most defense goods and weapons systems on an active basis. This list includes the supply chains that produce U.S. armored ships and ground vehicles.

Efforts are underway to address this lack of awareness. One of these is the Sector-by-Sector, Tier-by-Tier (S2T2) defense industrial base review. DoD undertook S2T2 with the intention of using survey data to map out the supply chains in several key defense industrial base sectors. Information collected under S2T2 is intended to enable "fact-based" analysis of globalization's role in the defense industry and other key issues. Among the sectors that S2T2 is investigating are shipbuilding and ground vehicles, two sectors for which steel armor plate is an input.

But S2T2 is insufficient on its own. It will not provide up-to-date information, and the collected information may not be useful for guiding policy without the proper context, which may or may not be available to DoD. S2T2 should be part of a broader DoD and U.S. government effort to gain greater awareness of supply chain issues and potential vulnerabilities as they appear.

Retain the original meaning of armor plate "produced" in the United States under the Specialty Metals Clause.
The U.S. steel industry has sufficient capacity to supply DoD demand for the foreseeable future. Capacity has increased since the dramatic surge in demand for MRAPs. In the event of a sudden spike in demand, as was the case with the rapid acquisition of MRAPs during the Iraq War, current rules include mechanisms that allow for the temporary use of non-domestic steel armor plate.

The proposed redefinition of the SMC, which DoD has indicated will not be adopted, would encourage the offshoring of the melting phase of steel production and the deterioration of U.S. capacity. Retaining the original definition of "produced" will help maintain U.S. capabilities to carry out all phases of steel armor plate production. If the U.S. steel industry is unable to attract investment in this critical phase of production, the United States risks a major degradation in its ability to respond flexibly to new challenges as they develop and in new technologies to address future threats. Foreign firms may lack the ability or inclination to work closely with DoD during a future crisis, especially if it coincides with a peak in global demand.

Build effective partnerships with U.S. armor plate producers.

The ability of the U.S. defense industrial base to effectively produce steel armor plate to meet U.S. defense needs depends in large part on effective working relationships between DoD and industry. The U.S. steel industry possesses a wealth of knowledge about armor plate, based not only on a technical understanding of the plate itself, but also on years of experience working with DoD during past production surges. DoD will benefit by working more closely with industry to take advantage of this knowledge and experience. This collaboration will serve three main purposes:

- Give DoD ongoing feedback on the state of the industry and the challenges that it faces. Newer entrants into the field especially may face difficulties in communicating what they can offer, as well as their concerns, to DoD.

- Provide DoD with information and context with which to make decisions about the defense industrial base.

- Better acquaint steel firms with DoD requirements, priorities, and practices.

Strengthened partnerships should be accompanied by efforts to simplify the process of doing business with DoD. A series of House Armed Services Committee hearings last year addressed this very issue, culminating in a report that described a wide array of challenges faced by firms in the defense industrial base.[49] This issue is complicated and much discussed, and there is no single fix that will make the acquisition process accessible and transparent for all entrants.

Learn the lessons of MRAP in collaboration with industry.

The multifaceted MRAP program responded to a complex strategic and operational challenge. The Government Accountability Office ultimately concluded that the "use of a tailored acquisition approach to rapidly acquire and field MRAP vehicles was successful."[50] A part of this approach was the rapid production, testing, and acquisition of many sizes and grades of steel armor plate. Even though this effort succeeded overall, it was preceded by concerns that the U.S defense industrial base lacked sufficient capacity to respond to DoD demand. Both DoD and industry should learn the lessons of this experience so that necessary relationships, practices, and understandings are already in place before the next unexpected production surge.

Take measures to reduce uncertainty in demand.

The United States is in an uncertain and constrained fiscal environment, especially given the failure to prevent the mandatory cuts under sequestration. These changes will create new realities to which all participants in the defense industrial base will have to adjust. The nation also faces an uncertain international security environment. Nevertheless, DoD should still take steps to reduce uncertainty in demand for armor plate and ameliorate what Deputy Assistant Secretary of Defense for Manufacturing and Industrial Base Policy Brett Lambert has called the "peaks and troughs" in demand "that really impact the second and third tiers" of the defense industrial base.[51]

CONCLUSION

The United States cannot predict the future of combat. U.S. military planners in 2000 could not have anticipated the need to defend against IED threats during a protracted occupation of Iraq. However, it *is* clear that U.S. forces will need to be protected regardless of how threats evolve in coming years. Steel armor plate, manufactured by U.S. steel companies, will have an important part to play in providing that protection.

This chapter has outlined some of the potential vulnerabilities faced by the sector of the U.S. defense industrial base that supplies U.S. armored platform manufacturers with the range of grades, shapes, and sizes of steel plates required to meet U.S. military requirements. It has also provided recommendations for mitigating these vulnerabilities and preserving a vital capability to strengthen U.S. national security.

The short-term pressure to reduce U.S. defense spending in a time of fiscal austerity, especially in an era of declining U.S. military commitments in the Middle East and Central Asia, should not cause the United States to neglect the long-term maintenance of this important defense industrial base capability.

ENDNOTES

1 For a discussion of the debate surrounding the MRAP, see: Christopher J. Lamb, Matthew J. Schmidt, and Berit G. Fitzsimmons, "MRAPs, Irregular Warfare, and Pentagon Reform," *Joint Forces Quarterly* 55 (Winter 2009): 76-85.

2 Two versions of the LCS currently are in production. The Freedom Class LCS (LCS-1) has a steel hull, while the Independence Class LCS (LCS-2) has an aluminum hull. The two versions are intended to fulfill similar requirements. Christopher P. Davis, "LCS 1 Vs. 2: Both Meet the Requirements, But Similarities End There," *Defense News* (May 3, 2010). http://www.defensenews.com/article/20100503/DEFSECT03/5030307/LCS-1-Vs-2-Both-Meet-Requirements-Similarities-End-There.

3 Jonathan W. Greenert, "Keynote Address – Cooperation from Strength: The U.S., China, and the South" (Speech, Center for New American Security, Washington, D.C., January 11, 2012). Video of the speech available here: http://www.cnas.org/node/7668.

4 U.S. Department of Defense, *Ballistic Missile Defense Review Report* (February 2010). http://www.defense.gov/bmdr/docs/BMDR%20as%20of%2026JAN10%200630_for%20web.pdf.

5 Tom Vanden Brook, "Gates: MRAPs Save 'Thousands' of Troop Lives," *USA Today* (June 27, 2011), http://www.usatoday.com/news/military/2011-06-27-gates-mraps-troops_n.htm.

6 *U.S. Bureau of Labor Statistics*, "May 2010 National Industry-Specific Occupational Employment and Wage Estimates: NAICS 331100 - Iron and Steel Mills and Ferroalloy Manufacturing," http://www.bls.gov/oes/current/naics4_331100.htm

7 World Steel Association, "World Steel in Figures 2011," 9. http://www.worldsteel.org/dms/internetDocumentList/bookshop/WSIF_2011/document/World%20Steel%20in%20Figures%202011.pdf

8 Rachel Tang, "China's Steel Industry and its Impact on the United States: Issues for Congress," (Washington, D.C.: Congressional Research Service, R41421, September 21, 2010), 19. http://www.fas.org/sgp/crs/row/R41421.pdf.

9 American Iron and Steel Institute, "2010 Steel Shipments by Market Classification." http://www.steel.org/About%20AISI/Statistics/Market%20Applications%20in%20Steel.aspx.

10 ArcelorMittal USA, *Armor: Steels for National Defense* (Chicago, IL: May 2012), 1. http://www.arcelormittalna.com/plateinformation/documents/en/Inlandflats/ProductBrochure/ARCELORMITTAL%20ARMOR.pdf.

11 To see the countries to which the U.S. exports iron and steel scrap, see Mike Sebany, "U.S. Iron and Steel Scrap Exports by Country for October," *Bloomberg*, February 24, 2012, http://www.bloomberg.com/news/2012-02-24/u-s-iron-and-steel-scrap-exports-by-country-for-october-table-.html.

12 Industrial College of the Armed Forces, *Industry Study: Strategic Materials* (Washington, D.C.: National Defense University, Spring 2007). http://www.ndu.edu/icaf/programs/academic/industry/reports/2007/pdf/icaf-is-report-strategic-materials-2007.pdf.

13 Mitch Jacoby, "Science Transforms the Battlefield," *Chemical and Engineering News* (August 11, 2003). http://pubs.acs.org/cen/coverstory/8132/8132science.html.

14 Spectra is a polymer similar to Kevlar, though lighter, displays strengths of at least 10 times stronger than steel per weight. Spectra is a trademark of Honeywell International Inc. Kevlar is a trademark of E.I. du Pont de Nemours and Company.

15 ICAF, *Industry Study: Strategic Materials*.

16 Jacoby, "Science Transforms the Battlefield."

17 Sandra Erwin, "Search Continues for Lighter Alternatives to Steel Armor," *National Defense Magazine*, February 2008, http://www.nationaldefensemagazine.org/archive/2008/February/Pages/SearchContinuesForLighter.aspx.

18 Mingwei Chen, James McCauley, and Kevin Hemker, "Shock-Induced Localized Amorphization in Boron Carbide." *Science* 299, no. 5612 (2003): 1563—66.

19 Thomas H. Miller, "Does MRAP Provide a Model for Acquisition Reform?" *Defense AT&L* (July-August 2010). http://www.dau.mil/pubscats/Lists/ATL%20Database/Attachments/715/Miller_jul-aug10.pdf.

20 Thomas L. Day, "Rare IED Success: MRAPs Cut U.S. Death Rate in Afghanistan," *McClatchy Newspapers*, January 19, 2010, http://www.mcclatchydc.com/2010/01/19/v-print/82618/new-vehicles-help-reduce-us-deaths.html.

21 Thomas H. Miller, "Does MRAP Provide a Model for Acquisition Reform?" 17.

22 Andrew Feickert, *Mine-Resistant Ambush-Protected (MRAP) Vehicles: Background and Issues for Congress*, (Washington, D.C.: Congressional Research Service RS22707, January 18, 2011). http://www.fas.org/sgp/crs/weapons/RS22707.pdf.

23 Miller, "Does MRAP Provide a Model for Acquisition Reform?" 18.

24 Richard Sisk, "Pentagon Shuts MRAP Production Line," *DOD Buzz*, October 1, 2012. http://www.dodbuzz.com/2012/10/01/pentagon-shuts-mrap-production-line/.

25 Austin Wright, "Has Steel Become A Product of Confusion?" *Politico*, November 9, 2011, http://www.politico.com/news/stories/1111/67978.html.

26 U.S. National Archives and Records Administration, Defense Acquisition Regulations System, *Federal Register* 73, no. 40 (July 2008): 42300.

27 Valerie Bailey Grasso, *The Specialty Metal Provision and the Berry Amendment: Issues for Congress* (Washington, D.C.: Congressional Research Service, October 28, 2008).

28 Austin Wright, "Has Steel Become A Product of Confusion?" *Politico*, November 9, 2011, http://www.politico.com/news/stories/1111/67978.html.

29 "Sen. Brown Leads Group of Senators on Legislation to Ensure That the U.S. Military Buys Steel Made in America--Not in China--For Armor Plates," Senator Sherrod Brown press release, February 9, 2012, on his official website, http://brown.senate.gov/newsroom/press_releases/release/?id=ad96009a-28f5-46de-b1c0-a2314e3ca25b.

30 "State of Steel," Congressman Tim Murphy press release, March 21, 2012, on his official website, http://murphy.house.gov/index.fm?sectionid=29&parentid=1&sectiontree=29&itemid=1972.

31 Federal Registry 77(142), pp. 43474-43477 (July 24, 2012).

32 78 FR 18877, Federal Register, 18877 -18879 (March 28, 2013). https://federalregister.gov/a/2013-07107.

33 "Defense Industrial Base Implications of the FY13 Budget" (Video presentation, the Center for Strategic and International Studies, Washington, D.C., February 15, 2012). http://csis.org/event/defense-industrial-base-implications-fy13-budget.

34 Sydney J. Freedberg Jr., "Why Senate, House Authorizers Both Added Dough for Armor," *AOL Defense*, May 25, 2012, http://defense.aol.com/2012/05/25/why-senate-house-authorizers-both-added-dough-for-armor.

35 Valerie Bailey Grasso, *The Specialty Metal Provision and the Berry Amendment: Issues for Congress* (Washington, D.C.: Congressional Research Service, October 28, 2008), 7.

36 Under the SMC, steel "with a maximum alloy content exceeding one or more of the following limits: manganese, 1.65 percent; silicon, 0.60 percent; or copper, 0.60 percent" or containing more than .25 percent of aluminum, chromium, cobalt, columbium, molybdenum, nickel, titanium, tungsten, or vanadium is considered a specialty metal.

37 *U.S. Code* 10 § 2533b.

38 Holly Hart, "Comments of the United Steel, Paper and Forestry, Rubber, Manufacturing, Energy, Allies Industrial and Service Workers International Union (USW) on the Definition of 'Produced' in DFARS 225.7003, *Restriction on the Acquisition of Specialty Metals* (76. Fed. Reg. 44,308)," September 8, 2011.

39 Sherrod Brown, Robert P. Casey, Jr., Charles E. Schumer et al., letter to Leon Panetta, September 26, 2012. http://www.brown.senate.gov/download/comment-period-armor-plate-letter-to-panetta.

40 ArcelorMittal USA, *Armor: Steels for National Defense* (Chicago, IL: May 2012), 1. http://www.arcelormittalna.com/plateinformation/documents/en/Inlandflats/ProductBrochure/ARCELORMITTAL%20ARMOR.pdf.

41 Jim Michaels, "Iraq Buys U.S. Drones to Protect Oil," *USA Today*, May 20, 2012, http://www.usatoday.com/news/world/story/2012-05-20/iraq-oil-drones/55099590/1.

42 Andrew Feickert, *The Army's Ground Combat Vehicle (GCV) Program, Background and Issues for Congress* (Washington D.C.: Congressional Research Service R41597, January 2, 2013).

43 Paul McLeary, "Ground Combat Vehicle Development Delayed," *Army Times*, January 17, 2013, http://www.armytimes.com/news/2013/01/dn-ground-combat-vehicle-development-delayed-011713/.

44 American Iron and Steel Institute, "Profile of the American Iron and Steel Institute 2010-2011,"
http://www.steel.org/~/media/Files/AISI/About%20AISI/Profile%20Brochure%20F-singles_CX.ashx.

45 Louis Uchitelle, "Steel Industry, in Slump, Looks to Federal Stimulus," *The New York Times*, January 1, 2009, http://www.nytimes.com/2009/01/02/business/02steel.html?pagewanted=all.

46 U.S. Department of Commerce, International Trade Administration, *Steel Industry Executive Summary*, February 2013. http://hq-web03.ita.doc.gov/License/Surge.nsf/webfiles/SteelMillDevelopments/$file/exec%20summ.pdf?openelement.

47 "Steel Industry Executive Summary," 9.

48 "Steel Industry Executive Summary."

49 House Armed Services Panel on Business Challenges in the Defense Industry, "Challenges to Doing Business with the Department of Defense (March 19, 2012). http://armedservices.house.gov/index.cfm/files/serve?File_id=f60b62cb-ce5d-44b7-a2aa-8b693487cd44.

50 Government Accountability Office, "Defense Acquisitions: Rapid Acquisition of MRAP Vehicles," by Michael J. Sullivan, (October 8, 2009).

51 Graham Warwick, "Pentagon Seeking Out Supply-Chain Weaknesses," *Aviation Week*, August 4, 2011.

CHAPTER 3 • SPECIALTY METALS

EXECUTIVE SUMMARY

Specialty metals are used in countless ways, including high-strength alloys, semiconductors, consumer electronics, batteries, and armor plate, to name a few. The United States possesses significant reserves of many specialty metals, with an estimated value of $6.2 trillion. However, it currently imports over $5 billion worth of minerals annually, and is almost completely dependent on foreign sources for 19 key specialty metals.

Industrial metals are a group of specialty metals that are most often added to base metals to form alloys. These metals play critical roles in many steel alloys, adding hardness, heat resistance, and strength. They are often highly reactive transition metals and require complex and expensive extraction processes. In a few cases they can only be extracted as byproducts of other metals. As such, production is dictated by production of their carrier metals, resulting in limited supply and mounting demand.

Rare earth elements (REEs), a second important group, have unique properties that make them essential for many defense products, especially high-technology ones. Currently, China dominates REE production, controlling 90 percent of global supply. This market share was achieved in part by undercutting competitors through overproduction, which drove U.S. and other mines out of business. Upon obtaining a near monopoly, Chinese producers have scaled back production to inflate prices through restricted supply. Quotas limiting the amount of raw REEs that may be exported have been used to force foreign investment in Chinese manufacturing, while exports to Japan were halted temporarily in 2010 after a diplomatic incident. Western companies scrambled to invest in REE mining to secure supplies just as the speculative bubble burst in fall 2011, sending prices downward and leaving the industry outside China in disarray. China still controls the global supply chain of REE oxides.

Production of the platinum group metals (PGMs) is dominated by South Africa. The country possesses more than 90 percent of known PGM reserves, and accounts for almost 40 percent of global palladium production and 75 percent of world platinum production. PGMs are commonly used in automotive engines and advanced electronics, and do not have viable substitutes. South Africa's dominance over PGM production threatens the integrity of defense industrial base supply chains, as political and economic instabilities within South Africa could restrict U.S. access to these metals. Recent

SPECIALTY METALS
HARNESSING THE PERIODIC TABLE

MANUFACTURING SECURITY

Specialty metals are crucial to U.S. national security and are used in a wide range of military end-items

COMPUTER CHIPS AND DEVICES

MISSILE GUIDANCE

AIRCRAFT COMPONENTS

U.S. CONSUMPTION

 = 1,000

Total U.S. annual consumption of new non-fuel minerals per capita

25
THOUSAND POUNDS

RARE EARTH VULNERABILITY

China is the leading supplier of rare earth elements essential to national security

CHINA PRODUCES 90% OF THE WORLD'S **SUPPLY OF REES**

U.S. DEPENDENCY

The U.S. is wholly reliant on foreign suppliers for 19 key minerals used in specialty metals

100%
IMPORT-RELIANT FOR
19 KEY MINERALS

VALUABLE METALS

Many specialty metals are vastly more valuable than other commonly used metals

RHENIUM

VS.

ALUMINUM

$3600
PER KILOGRAM

$1.60
PER KILOGRAM

MITIGATING RISKS

Protecting U.S. access to specialty metals

STRENGTHEN NATIONAL STOCKPILE

DEVELOP DOMESTIC CAPACITY

INTERAGENCY & INTERNATIONAL COORDINATION

MILITARY EQUIPMENT CHART
SELECTED DEFENSE USES OF SPECIALTY METALS

DEPARTMENT	WEAPON SYSTEMS	PLATFORMS	OTHER SYSTEMS
ARMY	■ Missile guidance systems (gallium, neodymium, and rhenium) ■ BGM-71 TOW Anti-Tank missile (tantalum)	■ Platforms that use Steel Armor Plate (molybdenum) ■ M1 Abrams main battle tank (tantalum)	■ Lithium-ion batteries (lithium) ■ Night Vision devices (lanthanum and gallium) ■ Laser rangefinders (neodymium)
MARINE CORPS	■ Missile guidance systems (gallium, neodymium, and rhenium) ■ Submarine-launched ballistic missiles (tungsten)	■ Platforms that use Steel Armor Plate (molybdenum)	■ Lithium-ion batteries (lithium) ■ Night vision devices (lanthanum and gallium) ■ Laser rangefinders (neodymium)
NAVY	■ Missile guidance systems (gallium, neodymium, and rhenium)	■ Platforms that use Steel Armor Plate (molybdenum)	■ Lithium-ion batteries (lithium) ■ Night vision devices (lanthanum and gallium) ■ Laser Rangefinders (neodymium)
AIR FORCE	■ Missile guidance systems (gallium, neodymium, and rhenium) ■ GBU-28 laser-guided bomb	■ Jet engines (rhenium and tungsten) ■ MQ-1B Predator drones (indium) ■ F-22 Raptor fighter (yttrium) ■ C-17 Military transport aircraft (yttrium)	■ Lithium-ion batteries (lithium) ■ Night vision devices (lanthanum and gallium) ■ Laser Rangefinders (neodymium)

reforms have increased taxes on PGM mines and introduced Chinese investment into those mines, increasing scarcity and forcing prices to rise while creating uncertainty over the future availability of the commodities.

Mitigating these risks is complex, and strategies will vary among commodities. The United States should maintain strategic reserves of those defense-critical elements that face likely shortages (REEs and PGMs) while seeking alternative sources. Congress is beginning to give this issue the necessary attention, and is shifting towards a more bottom-up approach to securing the supply chains of key materials—but more must be done. The federal government has not formulated a comprehensive and coherent policy approach to address the national security risks of inadequate access to many key minerals and metals. Strengthening efforts to identify substitutes and improve recycling will help mitigate these risks.

INTRODUCTION

This chapter will investigate "specialty metals," categories of metals that are also known as industrial, rare, or precious metals. Other common names for these types of metals include military, green, clean, critical, minor, technology, and strategic metals. It should be noted that specialty metals are *not* base metals (e.g. iron, copper, nickel, lead and zinc), or metals that oxidize, tarnish, or corrode easily. In addition, specialty metals are *not* energy metals (e.g. uranium and thorium). This chapter will examine specialty metals, comparing their properties and assessing their vulnerabilities with respect to U.S. military capabilities and U.S. economic

competitiveness associated with the extraction and production of these metals.

It is currently estimated that an average U.S. consumer's lifestyle requires roughly 25,000 pounds of non-fuel minerals per year, requiring massive efforts to either extract or import these materials.[1] Each year, the U.S. Department of Defense (DoD) acquires nearly 750,000 tons of minerals for an array of defense and military functions.[2] For example, tungsten, which is almost as hard as diamond, has the highest melting point of all non-alloyed metals, and is commonly used in turbine blades, missile nose cones, and other applications requiring exceptional heat resistance. Other minerals acquired are Rare earth elements (REEs) (some of which are used to fabricate permanent magnets), which maintain their magnetic fields even at high temperatures and are used in missile guidance and nearly every other small motor. Yet another example is palladium, which is part of the platinum metals group (PGMs), and is used in catalytic converters.

Despite possessing an estimated $6.2 trillion worth of key minerals reserves, the United States recently recorded a small surplus on the trade balance of raw mineral materials: it exported $9 billion and imported $8 billion of unprocessed minerals in 2012. However, the United States runs a deficit of $27 billion on the balance of processed mineral materials because it exported $120 billion and imported $147 billion in 2012.[3] In short, although the U.S. is self-sufficient in many minerals and has the chemical engineering know-how to process them, to some extent, it has chosen to rely on imports.

Increasingly, it is recognized that minerals are central to modern life and modern defense preparedness. Yet the federal government has not formulated a comprehensive and coherent response to the mineral/materials supply vulnerabilities, and there is no standard definition of which minerals or materials are critical and strategic and how the government should improve access to key minerals.[4]

The Defense Logistics Agency (DLA) Strategic Materials stores 28 commodities at 15 locations. In FY2012, DLA Strategic Materials sold $1.5 million of minerals and materials from its stockpile. At the end of the fiscal year, mineral materials valued at $1.4 billion remained. The stockpile is meant to help remedy the fact that the U.S. is completely import-dependent for 19 key minerals (including arsenic, asbestos, bauxite, graphite, fluorspar, indium, manganese, mica, niobium, tantalum, yttrium, and all REEs).[5] (The DLA Strategic Materials stockpile does not adequately compensate for the import dependence on a host of minerals because it emphasizes zinc, cobalt, chromium, and mercury, which are mined or recycled in the United States.) The stockpile is meant to protect against domestic and foreign supply constraints, spiking prices, and excessive speculation. However, because the U.S. government lacks a working understanding of which minerals are absolutely critical and which are strategic, the selection of metals for inclusion in the future stockpile managed by the DLA seems somewhat arbitrary.

In the past, a global abundance of minerals has been more than able to meet U.S. demand. However, as mineral-producing countries begin to consume more of their domestic production to fuel their own growing economies, the quantities available in the global marketplace have decreased. The increased demand for minerals has encouraged resource nationalism, where countries seek to exert greater control over the extraction and processing of key elements. Furthermore, many minerals are mined in only a few countries (some of which are politically unstable), exposing the United States and other importing countries to potential supply disruptions and other risks.

This situation is widely recognized as critical. In the words of one observer, "the whole periodic table is under siege… the growing demand for complex materials is leading to exploding demand for elements that are now used in only small quantities."[6]

The metals in this chapter fall into three different groups. The first group is industrial metals (e.g. antimony, manganese, tungsten, molybdenum, vanadium, and magnesium), which are usually mixed with base metals to create alloys to manufacture different kinds of steel products. Demand has risen for these alloyed metals because of their special properties that make them essential in aviation, engine turbines, green technology, and nuclear energy. Many of these metals are scarce because they are the byproduct of the other processes and because they are expensive to produce. Moreover, processing these metals involves advanced industrial chemistry and metallurgy that is more complex than extracting copper, zinc, and iron ore.

The second group consists of REEs, which are found across a surprisingly wide variety of applications and devices that enhance modern life in advanced industrialized countries. REEs are almost exclusively mined in China, which has by far the largest concentration of these elements. Mining REEs requires a more complex process than that used to mine gold or zinc, for example. From initial extraction to production, the process takes approximately 10 days. REEs are separated based on atomic weight, with actual processing duration based on the specific element. The most abundant REE is cerium. Terbium, a heavy REE, is more difficult to extract, and its extraction can take an additional 30 days.[7] Neodymium is also found with cerium, but the mine must first separate cerium and then extract the neodymium. This explains the length of production time and the costs. Importantly, companies cannot know beforehand whether valuable REEs are mixed in with the more common kinds, as each individual mine is different. Geologists and mining engineers must study each mine to find out which elements are available. The many engineering and processing challenges make REE mining among the most difficult types of mining operations.[8]

Mine operators need to know in advance how the REEs are going to be used so that they can determine the appropriate extraction and refining process. (Different processes must be used depending on the intended end-use of the REE.[9]) In fact, REEs are not inherently rare, but they are costly to mine and process because they are found in minute quantities mixed in with other ores. As Table 4 shows, REEs are used in a strikingly diverse range of products, including high-tech permanent magnets (see this report's chapter on magnets) and night vision devices (see this report's chapter on night vision devices).

The third group of specialty metals is very small and consists of the platinum group metals (PGMs), which are used in a range of applications such as vehicle production, future power sources, and many key military technologies. Palladium and platinum are used in catalytic converters. The largest concentrations of these deposits and reserves are found in South Africa.

Key themes discussed in this chapter are:

- Within the past decade, many countries rich in natural resources have taken a stance of "resource nationalism" and are attempting to control and manipulate extractive mining by threatening to impose extra taxes, reduce exports, nationalize mining operators, and restrict licensing.

- Western countries and mining operators face competition from less developed countries for access to specialty metals as well as from China, which has moved aggressively offshore to guarantee access to natural resources.

- Advanced industrialized countries, including the United States, have abandoned mining and mining exploration, even though global demand for economically and militarily significant ores and chemical elements has risen and will continue to rise.

- Many specialty metals are found in only a handful of countries, and often in regions that are politically and economically unstable.

- The risk of disruptions to the supply chains that use specialty metals is high, jeopardizing U.S. national security.

- Various U.S. agencies recognize the risks, but they provide different and divergent answers and solutions. The lack of a mechanism to coordinate policies among agencies hampers the development of a comprehensive and coherent strategy.

A NOTE ON CRITICALITY

Access to many natural resources is largely a function of geography. Although different types of specialty metals face different levels of risk (as described below), PGMs are consistently classified as facing the highest risks. Global reserves are situated almost exclusively in South Africa, which is the only country possessing significant long-term production capability. Limited global production capacity is coupled with high and increasing demand for PGMs, leading to high, unstable prices. Any number of events could create temporary or protracted shortages of PGMs, the most likely of which being internal political and economic instabilities associated with the South African government. The geographic concentration of PGM reserves, the high potential for disruption to the primary global provider, and the scarcity imposed by heightened demand indicate an *extreme* risk of these metals becoming unavailable.[10]

An insufficient supply of PGMs would have a *significant* impact on national defense capabilities. Although PGMs are most commonly known for their role in catalytic converters that reduce emissions from internal combustion engines, they also play an important role in advanced electronics used by the military (such as guided missile systems) due to their exceptional performance and ability to withstand high temperatures.

BACKGROUND

The issue for most advanced industrialized countries is that demand for rare elements has risen, while proven reserves and mining operations are increasingly

THE COST OF FAILURE TO ADDRESS POTENTIAL SPECIALTY METALS SUPPLY CHAIN DISRUPTIONS (a notional though realistic scenario)

The inauguration of the new South African president has led to a strengthening of ties between the Republic of South Africa and the People's Republic of China. In return for financial assistance in achieving its internal developmental policies and goals, South Africa has agreed to export manganese exclusively to China. Department of Defense supply chain specialists have begun to seek other sources of the metal; however, the effect on the market of this exclusive deal is expected to be pronounced. South Africa possesses one of the largest deposits of this mineral, and the removal of this source is expected to significantly increase prices for remaining sources. Reduced manganese supply means increased defense costs, as the U.S. military is a major consumer of manganese as a component of a variety of weapons systems and capabilities, including in the manufacture of steel armor plate and munitions.

concentrated in a handful of countries that have sought to exploit their geological advantages and their desire to meet their own growing domestic needs. In 2011, the British Geological Survey published a "risk list" that employed four variables (detailed below) to assess the risk factors of 52 elements or element groups with economic value.[11] The variables they used were scarcity (or the abundance of elements in the earth's crust); production concentration (the location of current production); reserve base distribution (the location of reserves); and governance (the political stability of those locations). Using these categories, experts determined that the chemical elements or element groups with the highest supply vulnerabilities were antinomy, which is produced in China and is used in micro-capacitors; PGMs, which

As the United States frees itself from fossil fuel dependence, it may replace it with dependence on energy sources from power-generating equipment that relies on specialty metals.

are produced in South Africa and used in automobile catalytic converters, fuel cells, seawater desalination equipment; mercury, which is produced in China; tungsten, which is produced in China and is a hard metal used in all cutting tools; REEs, which are produced in China; and niobium, which is produced in Brazil and used in MRI scanners, touch screens, micro capacitors, and ferroalloys.[12]

The German government also has expressed concern, as the country's large manufacturing base requires substantial amounts of REEs. As demand from emerging economies has risen, the German government has been aggressive in securing access to REEs in regions or countries other than China. The German government entered into multiple agreements with Kazakhstan to give German companies better access to REEs.[13]

Last but not least, the European Union has pushed the governments of its member states to agree to a "critical metals" list, and to approve new policies to ensure continuous access to gallium, indium, tantalum, and tungsten, in addition to REEs. One of the measures on the agenda is to establish a critical metals stockpile, which would include gallium, indium, tantalum, and tungsten.[14]

It is not surprising that the U.S. Geological Survey (USGS), DoD, the Department of Energy (DoE), and the Congressional Research Service have joined the chorus of concerned voices by publishing numerous reports and presenting long lists of critical minerals. Critical minerals are indispensable to modern life and security, yet they may be at risk because of their geographic availability, the costs of extraction and processing, the dearth of (manmade) substitutes, and limited potential for recycling. USGS puts REEs highest on their list,[15] followed by cobalt, indium, and tellurium, which are needed for many important applications including magnets for motors and super alloys common in turbine blades and other aeronautical functions. In light of the rapid growth in demand for advanced batteries, most of which require minute amounts of lithium, USGS also has raised concerns about the possibility of depleting all known reserves of the element (see this report's chapter on lithium-ion batteries).

Tables 1 and 2 demonstrate the trend of the last 10 years during this relatively short period of time, U.S. import dependence has radically increased across the board.

A wide variety of metals are plagued by the same issues that account for this current state of affairs. For one, political leaders of advanced industrialized countries have abandoned mining in light of the substantial negative externalities and pollution of waterways, soil, and air. Take the example of REEs. In reality, they are abundant in the earth's crust, but they tend to be found in small concentrations and deposits. They rarely exist in pure form and must be extracted from other oxides, which increases the costs of processing. More importantly than the expense of extraction, RE mining also creates radioactive environmental pollutants.[16] In every mining operation, the extraction process results in tailings (ground rock, processing agents, and chemicals), which cannot be fully reclaimed or reused or recycled. Frequently, the unrecoverable and uneconomic metals, minerals, chemicals, and process water are discharged, normally as slurry, to a final storage area. RE mining, however, produces tailings that contain radioactive uranium and thorium, which pose additional environmental threats beyond the risks associated with normal mining waste. In Western countries, governments and the public essentially have decided that it is easier to offshore this process to localities with less vocal and organized citizens or less democratic and transparent regimes. China, for example, has witnessed extreme degradation of its soil, water, and air quality to a degree that would not be tolerated in advanced industrialized countries.[17]

Another issue is that global demand is being driven higher by new discoveries of these metals' special properties, and by new technological innovations in how to design, fabricate, and incorporate them into consumer and military products. For example, neodymium (an REE) combined with iron and boron was discovered to possess strong magnetic properties, and it became the foundation of the high-tech permanent magnet sector (discussed in this report's chapter on high-tech magnets). Other examples include: gallium and tellurium, which are used in completing types of solar panels; rhenium, used in the super alloys employed in jet turbines; indium, which is used in flat panel displays; and graphite, used in lithium-ion (Li-ion) batteries. Green technology (such as hybrid cars, wind turbines, electric motors, and lightweight metals) relies heavily on specialty metals and REEs.

Many technological devices consume tiny amounts of specialty metals, without which the product would not operate or would need to be much larger and heavier. For example, every guided missile requires modest amounts of oxides, the form in which REEs occur in the mineral ore. While the amount of REEs used in a guided missile is genuinely small in quantity, without them the missiles would be heavier, less precise, and less advanced. In a similar vein, some metals must be able to withstand high temperatures, which are primarily achieved by adding minor elements to steel.

Additionally, more than two billion people (notably, the populations of China and India) are moving towards higher standards of living more closely resembling those in advanced industrial nations such as the United States and those in Europe. This development means that demand for electronic devices, green technology, and other advanced applications will continue to rise and in spite of economic crises in Europe, the United States, and Japan.

Table 1: U.S. Net Import Reliance for Selected Nonfuel Mineral Materials in 2000

Material	Percent
ARSENIC (TRIOXIDE)	100%
China, Chile, Mexico	
ASBESTOS	100%
Canada	
COLUMBIUM (NIOBIUM)	100%
Australia, Guinea, Jamaica, Brazil	
BAUXITE & ALUMINA	100%
Brazil, Canada, Germany, Russia	
FLUORSPAR	100%
China, South Africa, Mexico	
GRAPHITE (NATURAL)	100%
China, Mexico, Canada	
MANGANESE	100%
South Africa, Gabon, Australia, France	
MICA, SHEET (NATURAL)	100%
India, Belgium, Germany, China	
QUARTZ CRYSTAL	100%
Brazil, Germany, Madagascar	
STRONTIUM	100%
Mexico, Germany	
THALLIUM	100%
Belgium, Canada, Germany, United Kingdom	
THORIUM	100%
France	
YTTRIUM	100%
China, Hong Kong, France, United Kingdom	
GEMSTONES	100%
Israel, India, Belgium	
BISMUTH	95%
Belgium, Mexio, United Kingdom, China	
ANTIMONY	94%
China, Mexico, South Africa, Bolivia	
TIN	86%
China, Brazil, Peru, Bolivia	
PLATINUM	83%
South Africa, United Kingdom, Russia, Germany	
STONE	80%
Italy, Croatia, Spain, India	
TANTALUM	80%
Australia, China, Thailand, Japan	
CHROMIUM	78%
South Africa, Kazakhstan, Russia, Zimbabwe	
TITANIUM CONCENTRATES	76%
South Africa, Australia, Canada, India	
COBALT	74%
Norway, Finland, Zambia, Canada	
RARE EARTHS	72%
China, France, Japan, United Kingdom	
BARITE	71%
China, India, Mexico, Morocco	

Source: U.S. Geological Survey, Mineral Commodity Summaries
2000 (Washington DC: U.S. Geological Survey).

Table 2: U.S. Net Import Reliance for Selected Non-fuel Mineral Materials in 2011

Material	Percent
ARSENIC (TRIOXIDE) Morocco, China, Belgium	100%
ASBESTOS Canada, Zimbabwe	100%
BAUXITE & ALUMINA Jamaica, Brazil, Guinea, Australia	100%
CESIUM Canada	100%
FLUORSPAR Mexico, China, South Africa, Mongolia	100%
GRAPHITE (NATURAL) China, Mexico, Canada, Brazil	100%
INDIUM China, Canada, Japan, Belgium	100%
MANGANESE South Africa, Gabon, China, Australia	100%
MICA, SHEET (NATURAL) China, Brazil, Belgium, India	100%
NIOBIUM (COLUMBIUM) Brazil, Canada, Germany, Russia	100%
QUARTZ CRYSTAL (INDUSTRIAL) China, Japan, Russia	100%
RARE EARTHS China, France, Estonia, Japan	100%
RUBIDIUM Canada	100%
SCANDIUM China	100%
STRONTIUM Mexico, Germany	100%
TANTALUM China, Germany, Kazakhstan, Australia	100%
THALLIUM Russia, Germany, Kazakhstan	100%
THORIUM France, India, Canada, United Kingdom	100%
YTTRIUM China, Japan, France, United Kingdom	100%
GALLIUM Germany, Canada, United Kingdom, China	99%
IODINE Chile, Japan	99%
GEMSTONES Israel, India, Belgium, South Africa	98%
GERMANIUM China, Belgium, Russia, Germany	90%
BISMUTH China, Belgium, United Kingdom	89%
DIAMOND (DUST, GRIT, & POWDER) China, Ireland, Republic of Korea, Russia	89%

Source: U.S. Geological Survey, Mineral Commodity Summaries 2012 (Washington D.C.: U.S. Geological Survey).

Finally, metal and mineral suppliers have witnessed booming mining sectors due to rising prices. Thanks to the rising value of natural resources, producing countries have pursued a policy of resource nationalism. Many of the most sought-after elements are found in developing countries that face multiple economic and political challenges. To finance development projects or to extract rents, governments of these countries might be tempted to push for a greater share of the profits made by mining companies. Examples of this trend are ubiquitous. Ghana has been reviewing mining contracts, and may renegotiate existing arrangements to increase governmental revenue. Zambia doubled its copper royalty to six percent. Guinea, which controls the largest known reserves of both bauxite and iron ore, has taken a 15 percent stake in mining operations. In Namibia, a state-owned company controls all new mining and exploration. Foreign mining operations in Zimbabwe must cede a 51 percent stake to local owners.[18]

To ensure the country benefits from its mineral wealth, South Africa may impose a 50 percent windfall tax on mining profits and a 50 percent capital gains tax on prospecting rights. The ruling African National Congress wants to collect a larger share of the resource boom. Even Australia, an advanced industrialized country, plans to impose a new, $8 billion tax on mining.[19]

This state of affairs has not gone unnoticed. Since the mid to late 2000s, increased scrutiny and heightened alarm surround the fact that the U.S. economy and national security depend on specialty metals—many of which are vulnerable to supply threats resulting from sovereign risk and resource nationalism, geological scarcity, lack of viable substitutes, byproduct sourcing, and inadequate post-consumer recycling and recovery programs.[20]

In 2008, the National Research Council Committee on Critical Mineral Impacts on the U.S. Economy (Committee on Earth Resources) compiled a statistical approximation to assess supply restrictions impact on the entire U.S. economy and defense capabilities. The report also took into consideration the technical substitution potential of a mineral.[21]

The National Research Council report presented a criticality matrix that juxtaposed the probability of a supply disruption with the overall economic impact of that supply disruption. Supply disruptions can be caused by the physical unavailability of a commodity or by increasingly restrictive prices as a result of scarcity or of artificial means. The study considered five factors that contribute to availability: geological; technical; social and environmental; economic; and political. Economic impact was assessed by the availability of a close substitute, the costs associated with that substitution, and the consequences of the supply restriction. The committee examined 11 metals or metal groups: copper, gallium, indium, lithium, manganese, niobium, PGMs (including iridium, osmium, palladium, platinum, rhodium, ruthenium), REEs, tantalum, titanium, and vanadium to determine their criticality. The study's conclusions are presented in Figure 1.

Indium, manganese, niobium, PGMs, and REEs fall in the "critical" zone of the matrix.[22] They are considered critical because of the importance of their applications in catalytic converters, industrial chemical production, electronics, batteries, liquid crystal displays, and hardeners or strengtheners in steel and iron alloys. In addition, if a physical disruption or sudden price surge jeopardizes supplies, there are no readily available mineral substitutes for these applications.

Figure 1: Specialty Metals Criticality Matrix

	SUPPLY RISK low → high			
IMPACT OF SUPPLY RESTRICTION (low → high)	**1**	**2**	**3**	**4**
4 (high)			• Manganese	• Rhodium
3	• Copper	• Tantalum	• Indium • Niobium	• Palladium • Platinum • Rare Earth Elements
2		• Vanadium • Lithium • Titanium	• Gallium	
1 (low)				

Source: National Research Council, Minerals, Critical Minerals, and the U.S. Economy (Washington, D.C.: National Academies Press, 2008), 165.

However, the study concludes that essentially any mineral could be considered critical, because both economic importance as well as factors influencing availability could change. Additionally, the report stresses that import dependence alone is not means for alarm; however, the concentration of supplies in a small number of countries plagued by political instability could be disastrous. Alternatively, rapid growth in the internal demand of exporting countries could limit the quantities available on the global market, resulting in rising prices and restricted supply.

INDUSTRIAL METALS

Industrial metals (also called minor metals) are in vogue because new uses for these metals are discovered frequently. They are classified as minor metals because until recently they were largely ignored by industry. They are not readily available or mined in the United States. Often, the elements are in fact rare and are not abundant in the earth's crust, with only a few parts per million of recoverable ore, even in the geologically significant deposits. As many of these elements are only found in a few dense concentrations globally, extraction may be dominated by a handful of countries. Subsequently, the price and supply of the element may be subject to export controls, price manipulation, and sudden disruptions. In some cases, elements are in fact a byproduct of a primary ore and are uneconomical to extract independent of the refining process for those other ores. These metals are therefore relatively costly and challenging to produce. Finally, the time required to adapt to new production and utilization processes is long, making planning and investment difficult.

The United States (along with almost all Organization for Economic Co-operation and Development [OECD] countries) relies heavily on imports for these materials, while the main producers are often countries with rapidly expanding economies (such as China, Russia, Chile, and South Africa) with sizeable and increasing domestic demand for these metals. Because certain metals are only commercially produced in a few countries, they can claim near monopolies over global reserves and influence pricing and availability.

The evolution of computing circuitry over the past three decades clearly illustrates the critical importance of industrial metals. The number of elements used in computer circuitry has expanded from 12 in the 1980s, to 16 during the 1990s, to over 60 today.[23] These circuits are found in nearly every piece of modern technology, and especially in highly specialized, high-tech defense applications.

The summary of the industrial metals sector below includes an overview of the different metal groups, selected elements, their most significant uses, and some of the concerns surrounding these commodities. The next section presents a more general discussion of the dominant risks facing this sector. The critical importance of these metals should be readily apparent. At the most basic level, many of them are used in heat-resistant, hard metal alloys that are used in aircraft, ships, submarines, and countless other defense-related applications. Other metals are at the core of solar energy, which is necessary for defense satellites and has a growing importance for civilian energy. Others still are used in electronic components such as rechargeable batteries, which are essential to consumer electronics, communication, and hybrid engines.

THE UNIVERSE OF INDUSTRIAL METALS

Most of the elements in industrial metals are used in alloys in order to improve heat resistance, reduce the weight of a metal item, or harden steel. (Table 3 provides an overview of the different metals and their defense applications and describes the particular risks or vulnerabilities associated with each industrial metal.) Many industrial metals are in demand in consumer electronics, high-energy rechargeable batteries, and the computer industry. They also are indispensable in numerous and wide-ranging military defense applications. Radar systems, airframes and engines, optical equipment, armor plating, coatings, electronic display screens, solar cells, and military batteries rely on small but vital quantities of industrial metals.

The universe of industrial metals can be divided into different chemical classifications. Each chemical group possesses different properties and advantages, which are further discussed below.

ALKALI AND ALKALI EARTH METALS

Alkali and alkali earth metals are located in the first two columns of the periodic table (excluding hydrogen). They are highly reactive elements, and as such, are not found in their elemental form, but instead as compounds in the earth's crust. Alkali metals (such as lithium) are relatively soft with low melting points, and form weak bonds with other elements because they have only one electron available for bonding. Alkali earth metals (such as beryllium) are harder and denser than the alkali metals, though not to the same extent as the transition metals.

LITHIUM

Lithium (Li) is a light and highly reactive metal, and is a key component of the rechargeable, high-energy lithium-ion (Li-ion) batteries that are widely used in the military and have a bright future as the main power source for electric or hybrid vehicles. Chile, Australia, Argentina, and China are the leading producers of lithium; almost the entirety of the U.S. import market comes from Argentina and Chile. Chile possesses over half of the world's known lithium reserves and is the main producer, extracting lithium from the Atacama Desert.[24] U.S. production of lithium is insignificant.[25] Because lithium is highly

reactive and reacts with water, producing the pure form of lithium is very complex and requires a dry environment.[26]

Increase in demand for lithium, especially from China, have caused a recent expansion of production in many countries. Production of lithium was reported to have increased 20 percent in both Australia and Chile in 2011, while Chinese production was reported to have increased 30 percent.[27] This expansion corresponds to the growing demand for high-purity lithium for use in Li-ion batteries.

Analysts in the advanced battery sector and green technology community express considerable concern about the world's reliance on lithium, because most of the reserves are concentrated in two countries (Chile and Argentina) and may outstrip global demand as soon as 2017. Currently, there is no substitute for lithium, which is the ideal material to create rechargeable batteries and energy network stations to store surplus power from solar and wind power (see this report's chapter on lithium-ion batteries).[28] Unlike with other specialty metals, the main concern about lithium is not price or the potentially monopolistic behavior by foreign governments but rather that the world may face supply restrictions as reliance on technologies that require lithium increases and the world's known reserves of lithium are depleted.[29]

BERYLLIUM

Beryllium (Be) currently is considered a material critical to U.S. national defense, and is retained in the DLA Strategic Materials stockpile. Beryllium is critical to many military systems, including the airborne Forward-Looking-Infrared (FLIR) system, missile guidance systems, and surveillance satellites. There are no

Table 3: Industrial Metals
Properties, Uses, and Defense Applications

Element	Atomic Symbol	Atomic Number	Uses and Applications	Significant Producers
Lithium	(Li)	3	Batteries	Chile, Australia, China, Argentina
Beryllium	(Be)	4	Lightweight alloys, radiation windows, nuclear reactors	U.S., China
Gallium	(Ga)	31	Low melting-point alloys, high-power high-frequency electronics semiconductors, light emitting diodes (LEDs), solar cells	China, Germany, Kazakhstan, Ukraine
Indium	(In)	49	Liquid crystal displays (LCDs), low melting-point alloys, bearing alloys, transistors, thermistors, photoconductors, rectifiers, mirrors	China, South Korea, Canada
Germanium	(Ge)	32	Fiber optics, infrared optics, solar photovoltaic cells, semiconductors, alloys	China
Antimony	(Sb)	51	Flame retardant, semiconductors, bearing alloys, batteries	China
Tellurium	(Te)	52	Thin-film photovoltaic panels, semiconductors, steel alloys, vulcanizing agent, synthetic fibers	China, Canada, Philippines
Vanadium	(V)	23	Nuclear reactors, springs, carbide stabilizer (alloys), batteries	China, South Africa, Russia
Molybdenum	(Mo)	42	Tempered steel, gun barrels, boiler plates, armor plating, nuclear energy, missile components	China, U.S., Chile
Tantalum	(Ta)	73	Tantalum carbide (hard-metal), Tantalum capacitors	Brazil, Australia, Mozambique, Rwanda
Tungsten	(W)	74	Tungsten carbide (hard-metal), drilling and cutting tools, specialty steels, heat sinks, turbine blades	China
Rhenium	(Re)	75	High-temperature alloys and coatings, jet engines	Chile, U.S., Peru, Poland, Kazakhstan
Palladium	(Pd)	46	Catalytic converters, multi-layer ceramic capacitors (chips), hybrid integrated circuits	South Africa, Russia, Canada, Zimbabwe
Platinum	(Pt)	78	Catalytic converters (diesel)	South Africa, Russia, Canada, U.S.

substitutes for beryllium, and in previous years there was a shortage of high-purity beryllium due to high production costs and health and safety issues. Foreign-sourced beryllium is not of sufficient purity for defense applications.

In 2005, under Title III of the Defense Production Act (P.L. 81-774), DoD invested roughly $90 million in a private-public partnership with domestic beryllium producer Brush Wellman, Inc. (now called Materion Brush Beryllium and Composites) to produce a primary beryllium plant in Ohio.[30] That plant became operational in early 2011, dropping the reported U.S. import dependence from 61 percent in 2010 to 21 percent in 2011. Twelve percent of the annual U.S. beryllium consumption is attributed to defense applications. The USGS reports that the U.S. currently possesses about 65 percent of the world's beryllium reserves and, with the opening of the Materion Brush plant in 2011, accounts for almost 90 percent of world production.[31]

TRANSITION METALS

The group of transition metals contains 38 elements that are grouped together due to their common electron configuration, and are generally hard, malleable, and possess high melting points. They are good electric conductors and are often magnetic. The uses of transition metals are vast, making their use common.

RHENIUM

Rhenium (Re) is a rare metallic element that is important to the defense community because of its contribution to the properties of high-temperature alloys and coatings. The USGS reports that nearly 70 percent of rhenium is used for high-temperature engine turbines common to jet engines, while an additional 20 percent is a key catalyst in refining oil.[32] Rhenium is also used as a promoter in catalysts in gas-to-liquid operations, which may become more important in the future in light of the rapid expansion of shale gas output in the United States and elsewhere.

Rhenium is obtained almost exclusively as a byproduct of the processing of a special type of copper deposit known as a porphyry copper deposit. Specifically, rhenium is obtained from the processing of the mineral molybdenite (a molybdenum ore), which in itself is a copper byproduct. Therefore, rhenium is among the most expensive and volatile metals in the world, and its price fluctuated from $10,000/kg in 2008 to $3,500/kg in March 2013.[33] Currently, the United States is the world's second leading producer of rhenium (after Chile), with about a 12 percent market share. However, because rhenium is a byproduct of a byproduct, its production is limited by the production of molybdenum, which is in turn limited by copper production. In 2012, the U.S. imported nearly seven times its domestic production of rhenium, mainly from Chile and Kazakhstan.[34] Rhenium is part of the DLA Strategic Materials stockpile.

MOLYBDENUM

Molybdenum (Mo) is an important alloying agent that contributes to the hardening and toughness of tempered steels, and is used in steel armor plate, gun barrels, and boiler plates. Almost all ultra-high strength steels contain up to eight percent molybdenum. Molybdenum is used in nuclear energy applications and for missile and aircraft parts. Molybdenum is both mined as a primary ore and recovered as a byproduct of copper. The United States is the second largest producer of molybdenum with about one quarter of the global share, and currently exports about half of its annual output.[35]

VANADIUM

Vanadium (V) is used predominantly as an additive in steel that is then used in nuclear energy applications and in rust-resistant springs and high-speed tools. Ferrovanadium, an alloy of steel, accounts for 95 percent of the vanadium used in the United States. Vanadium is a non-substitutable component of aerospace titanium alloys; however, for many other applications, other metals such as molybdenum, tungsten, manganese, niobium, or titanium may be substituted for vanadium.[36] Small amounts of vanadium are added to iron alloys to improve corrosion resistance; ferrovanadium is mostly used in gears for cars, jet engines, and springs. The type of vanadium used in steel does not face immediate supply constraints. Due to increasing demand for steel in expanding economies, the demand for vanadium is expected to increase.

Three countries —China, South Africa, and Russia—dominate the vanadium market, and together account for more than 96 percent of current global production. The United States depends on imports for 80 percent of its domestic consumption of ferrovanadium; its main import sources are South Korea, Austria, Canada, and the Czech Republic.[37]

Twenty percent of the vanadium market consists of vanadium pentoxide, which is more valuable than ferrovanadium. In 2012, the major exporters of vanadium to the United States were Russia (47 percent), South Africa (32 percent), and China (19 percent). Vanadium pentoxide is used as a catalyst in petroleum refineries, in ceramics, and in super-conductive magnets. Currently, however, vanadium pentoxide is considered suitable for vanadium redox batteries, a new type of advanced rechargeable battery that is able to store renewable energy coming from wind or solar generation. This new type of battery can store more energy more efficiently than Li-ion batteries, with a faster recharge time and a longer lifecycle (see this report's chapter on Li-ion batteries).[38]

Demand for vanadium pentoxide is expected to expand 30 percent in the next three years while supply is tight; 90 percent of the vanadium on the market is not suitable for processing into vanadium pentoxide, and is only appropriate for strengthening steel.[39] Vanadium pentoxide (used in large format batteries) is a byproduct of combusting fossil fuels containing vanadium. The byproducts containing vanadium pentoxide can be in the forms of dust, soot, boiler scale, and fly ash.

TANTALUM

Tantalum (Ta) is used in several alloys due to its thermal and corrosion resistances, ductility, and strength. Many types of tantalum minerals are mined in different parts of the world and possess slightly different properties. In many applications, it cannot be substituted without lessening quality. For example, tantalum carbide is among the most durable materials currently known.[40] The United States has no identified reserves of tantalum and depends on imports for all its tantalum consumption.

Tantalum is found in selected geological regions of the world, namely in the eastern areas of the Democratic Republic of Congo as well as in Australia, Brazil, Canada, and Mozambique. Furthermore, a related mineral, coltan, the industrial name for a columbite–tantalite mineral from which columbium (also known as niobium) and tantalum are extracted,[41] is widely used to manufacture capacitors found in consumer electronics, computers, and automobiles.[42] In the last 10 years, demand for coltan-extracted tantalum has surged, stirring armed conflicts in central Africa as paramilitary groups mine and smuggle the chemical elements in order to finance their own activities. Coltan is the mineral equivalent of "blood diamonds," which received large amounts of publicity and incited a human rights campaign in the late 1990s and early 2000s. Coltan-related conflicts also have destroyed the habitat of lowland gorillas and the livelihood of numerous indigenous communities.

In spite of tantalum's importance to the U.S. economy and national security, the DLA Strategic Materials sold off most of its tantalum mineral, tantalum metal powder, metal ingots, and metal oxides in the 2000s. In 2013, it still holds small quantities of tantalum carbide powder. The latter is extremely hard and brittle, and is commonly used in tool bits for cutting applications or sometimes added to tungsten to create a metal alloy. The United States consumes 120,000 metric tons annually, with no reserves; the United States imports all tantalum from China, Germany, Australia, and Kazakhstan.

Although USGS forecasts that supplies of tantalum are sufficient for projected demand, and significant untapped reserves exist in Brazil and Australia, a third of the current tantalum production originates from politically unstable sub-Saharan African countries.[43]

TUNGSTEN

Tungsten (W) possesses the highest melting point of all metals (3,400 degrees Celsius or 6,150 degrees Fahrenheit) and is nearly as strong as diamond. Additionally, it is an excellent electrical conductor. The most common use is as tungsten carbide, a "hard metal" known for industrial drilling and other cutting tools.[44] Additionally, tungsten carbide and tungsten alloys are used for armaments, heat sinks, turbine blades, and rocket nozzles.[45]

China is the largest producer of tungsten is, accounting for about 80 percent of global production and possessing roughly two-thirds of world tungsten reserves. China is also the world's top consumer of tungsten, and using a majority of the tungsten it produces. The Chinese government actively intervenes in the tungsten industry to limit supply: foreign investment is forbidden; exports are controlled by licenses, taxes, and quotas; overall

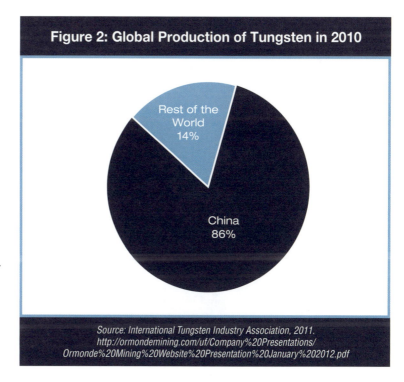

Figure 2: Global Production of Tungsten in 2010

Rest of the World 14%

China 86%

Source: International Tungsten Industry Association, 2011.
http://ormondemining.com/uf/Company%20Presentations/
Ormonde%20Mining%20Website%20Presentation%20January%202012.pdf

production is limited; and exploration and new operations are tightly controlled. In the immediate future, China is expected to be even more protective of its domestic supply, and is likely to attempt to further reduce exports as well as increase tungsten imports.[46]

Accordingly, tungsten prices are expected to increase in light of increasing demand and constricted supply. Historically, the United States and Russia have stockpiled tungsten, although both countries have been disposing of their stockpiles over recent years. The Russian stockpile is thought to be depleted, while the entire U.S. government holding of tungsten has been authorized for disposal.[47]

Although the United States imports a fair amount of tungsten, thanks to improved recycling of scrap consumed by processors and end-users, import reliance dropped from 63 percent in 2010 to 36 percent in 2011.[48] Nevertheless, there is only one domestic source of tungsten concentrates in the United States. The U.S. military cannot function without tungsten, and there are no substitutes for most applications. World demand slackened due to the global financial crisis, but scarcity will push up tungsten prices, especially since strategic manufacturing sectors would be willing to pay inflated prices.[49]

POST-TRANSITION METALS

Post-transition metals are softer than transition metals, with lower melting points, but they have high electronegativity, meaning that they are better at attracting electrons than the transition metals and more readily form polar bonds. They are malleable, ductile, and generally good conductors.

GALLIUM

Gallium (Ga) is not produced in the United States even though it is a critical component of optoelectronic devices, solar cells, light-emitting diode (LED) lights, and photo-detectors. Gallium is essential for creating high-brightness LEDs, and many governments in Asia are committed to introducing widespread LED lighting.[50] Therefore, demand for gallium likely will increase. Moreover, gallium is also a key component for thin film photovoltaic technology, a sector expected to grow by a factor of 9 by 2018; however, falling prices of silicon-based solar cells are limiting the current demand for more expensive gallium-based cells.[51] The primary military application of gallium is in high-power, high-frequency communications, such as those used in missile guidance systems. Gallium semiconductors can function at much higher temperatures than silicon, allowing them to function at a much higher capacity and reliability than more common silicon-based chips.[52] While silicon-based alternatives may be viable for commercial uses, they are not suitable replacements for defense-related applications.

The leading producers of gallium are China, Kazakhstan, and Ukraine. The United States is roughly 99 percent import-dependent on gallium, which is produced as a byproduct of bauxite (aluminum ore) and zinc ores, making it very difficult to accurately calculate gallium reserves. United States bauxite resources generally are not economical to extract, because their high silica content makes domestic production uneconomic and very unlikely.[53] Because gallium is primarily a byproduct of bauxite, and only a small portion of gallium in bauxite is recoverable (approximately 50 parts per million [ppm]), it is uneconomical to recover gallium independently of aluminum. The demand for aluminum will likely continue to dictate the world's supply of gallium.

INDIUM

Indium (In) is used in liquid crystal displays (LCDs) as the compound indium tin oxide, and is a byproduct of zinc ores. Indium is unevenly distributed in the earth's crust, causing the United States to be completely reliant on imports (although lower-grade imported indium is refined domestically). Due to its low abundance in most ores (less than 100 ppm in most zinc ores), recovering indium separately is uneconomical except as the byproduct of refining other ores. Currently, over half the world's indium is produced in China, with another 16 percent coming from South Korea. While there are techniques for reclaiming indium from discarded LCD screens, this option is only economically viable when indium prices are already high.[54]

Indium is used in transistors, thermistors, photoconductors, and low melting point alloys. It can also be used to create corrosion-resistant mirrors.[55] Indium is used in short-wave infrared (SWIR) imaging, including advanced night vision applications. Its advantage over traditional night vision systems is that a single SWIR device can function in both daylight and night, and does not require the extreme cooling that alternative technologies require. Such indium devices are used in Unmanned Aerial Vehicles, such as the Spectre-Finder and Predator. Because this technology does not rely on detecting heat but rather reflected light, it provides crisp images in starlight conditions, allowing for much greater accuracy in identifying targets than the alternative imaging technologies.[56]

METALLOIDS

Metalloids are elements that possess properties of both metals and non-metals. They are generally metallic in appearance, but are often brittle rather than malleable. They often possess good semiconductor qualities, and can serve as good insulators. Chemically, they behave as both metals and non-metals depending on the substance with which they react.

GERMANIUM

Germanium (Ge) is constrained in its availability because it is not found in concentrated deposits. It is relatively rare in the earth's crust (approximately 1.6 ppm), and while certain minerals do contain high levels of germanium, those minerals do not exist in any mineable deposits. Instead, germanium is most often produced as a byproduct of zinc extraction. Significant quantities of germanium are also recoverable from ash that comes from the burning of certain coals in energy production. China is the main producer of germanium, with a 68 percent market share, although significant reserves do exist within the United States. In 2011, the price of germanium nearly doubled as a result of increased Chinese export taxes and the closing of one germanium plant in China due to "environmental concerns."[57] However, germanium recycling has become increasingly common, with roughly 30 percent of consumed germanium coming from recovered scrap (recycled optical devices and window blanks in decommissioned tanks and other military vehicles).[58]

Germanium is used in fiber and infrared optics and in solar photovoltaic cells. Silicon shares many similar semiconducting properties with germanium, and may be a suitable substitute (at the expense of performance).

The estimated value of U.S. germanium consumption in 2012 was only about $55 million. Germanium sales represent an extremely small market. Yet germanium has been considered a critical material, and DLA Strategic Materials holds a small stockpiled inventory in case of sudden shortages. None was released in 2012.[59] The United States has known reserves of germanium though it has not mined them. Certain military applications will not work without germanium, and the metal's price fluctuates wildly because of the policy decisions by the most important mining regions.

The Chinese government restricts supplies by imposing new export controls or closing down germanium mines. These export restrictions are aimed at encouraging more finished production in China and stimulating the growth of an industry that relies on raw germanium such as optical lenses, fiber optics, LEDs, and solar cells. Chinese authorities have also identified germanium as a strategic resource and included it in their stockpile.[60]

ANTIMONY

Antimony (Sb) is used in a variety of applications, including semiconductors and batteries. It is most widely used as a flame-retardant, which accounts for about 36 percent of its use, and for which there is no effective substitute. While antimony sometimes occurs in pure form, it is more common as stibnite (Sb_2S_3, a sulfite), with other heavy metals, and as oxides.

China accounts for about 88 percent of annual antimony production, and over 60 percent of the global antimony reserves. Government officials in the Hunan region (where nearly 60 percent of China's antimony is produced) recently closed many antimony plants, citing health and safety

concerns. As a result, the price of antimony increased by 20 percent between January and September 2011. Additionally, at current production levels, the Chinese supply is projected to be depleted within five years.[61] The U.S. previously stockpiled antimony; however, these stocks were disposed of by 2003.

TELLURIUM

Tellurium (Te) is a relatively uncommon element, and acts as a semiconductor. Tellurium's major use is as an alloying additive in steel to improve machining characteristics. It is also used as a vulcanizing agent for rubber and as a catalyst for synthetic fiber and is important for photovoltaic (solar) cells, which will likely become a major source of solar electricity in the future. These cells are incredibly thin—usually only 1 to 10 micrometers (μm) thick—and can be flexible and highly adaptable to various designs in different applications. Tellurium is also used in creating fiber-optics capable of functioning in harsh environments, which are likely to become increasingly prevalent in military aircraft.

Tellurium is most often produced as a byproduct of copper processing. Tellurium is extremely rare, with its presence in copper concentrates often below 100 ppm.[62] Most imported tellurium comes from China, although tellurium is also produced in the United States, which possesses sizeable reserves (about 15 percent of known global reserves).[63] The metal is commercially profitable to recover only when it is concentrated in residues collected from copper refineries.

EXTRACTION RISK FACTORS

Many of these metals or metal-type elements are in fact byproduct metals of a carrier metal such as zinc, copper, or bauxite. Consequently, many of these metals are uneconomical to produce independent of the production of the carrier metal. Demand for the carrier metal therefore drives the production of these industrial metals, creating the potential for undesirable market conditions including price spikes and shortages. Germanium, gallium, and indium, for example, are all extracted from zinc ores; gallium is also obtainable from the processing of bauxite (aluminum) ore; tellurium, gallium, and molybdenum are recovered as byproducts of copper ores. Rhenium is a special case, as it is produced as a byproduct from molybdenum, which in itself is a byproduct of copper, making it among the most expensive metals in the world.[64]

Many of these elements simply are not found in concentrations high enough to warrant extraction as a primary product and are produced only as the byproduct of other metals. This fact raises problems with both increasing supply and supply availability. For example, it is uneconomical to increase the mining of copper in order to extract more tellurium. In 2009, copper production approached $80 billion, while the production value of tellurium was only about $30 million.[65] Because tellurium's abundance in copper ores is very low (less than 100 ppm), there would have to be a massive increase in copper production to have any impact on the tellurium supply. Given the values of the two markets, and the resultant drop in copper prices if such an expansion were to occur, producers would lose money overall if they attempted to expand the supply

of tellurium. Expanding tellurium production does not appear economically viable despite the fact that tellurium's role in photovoltaic panels that could dramatically reduce the costs of solar energy.

Another example is gallium, which is experiencing a surge in demand due to increased interest in LED lighting. Gallium arsenide (GaAs) is commonly used in high-efficiency, high-brightness LEDs because it has the ability to convert electricity directly into laser light. Many governments, including that of South Korea, are encouraging the adoption of LED lighting in the private sector and mandating it in the public sector, resulting in a rapid increase in gallium demand. According to the USGS, gallium consumption more than doubled between 2009 and 2011, resulting in a price increase of more than 50 percent.[66] However, gallium is mostly extracted as a byproduct of bauxite (aluminum). If demand for bauxite ore declines, then there would also be a reduced supply of gallium, even though the demand for gallium appears to be rapidly increasing.

GEOPOLITICAL RISKS

The United States relies on imports for many of the industrial metals (see Tables 1 and 2), a trend that has grown over the last decade. According to data collected by the USGS, the United States now imports more than 50 percent of 43 key minerals (compared to 29 in 1995). The United States is now totally reliant on importing 19 minerals, compared to 10 in 1995. Thus, import reliance or dependence has increased as the importance of certain minerals has grown.

The concentration of an important commodity among only a small number of sources creates significant potential for supply disruptions. For example, cobalt and tantalum are produced in the Democratic Republic of the Congo. The extraction of these elements has fed political instability, poverty, and human rights violations. In other situations, the presence of raw materials encourages monopolistic practices and price manipulation. For example, South Africa nearly has a monopoly over PGMs; citing concerns over shrinking reserves, China, the dominant producer of antimony, has tightened its production restrictions. As countries become dependent on the extraction and global production of often-scarce elements, they may be tempted to impose extra fees, taxes, and prices in order to exploit their unique position in the global market. They also may be tempted to restrict exports in order to build up a domestic processing and fabricating industry, as China did with the REs market. Even in the best of cases, the United States faces risks if it depends on a few suppliers of critical elements, since a major earthquake, accident, industrial strife, or lack of investments may disrupt supplies.

RARE EARTH ELEMENTS

REEs are necessary for many of the modern world's most advanced technologies: missile guidance systems, flat-screen TVs, cellphones, generators in windmills, and motors in hybrid cars, to name just a few. During the last decade, China has cornered the market on REEs—a group of 17 elements including scandium, yttrium, and 15 lanthanide elements at the bottom of the periodic table (see Table 4). Demand for REEs is expected to continue to increase.

In the short term, REE demand has fluctuated, because the state of the global economy strongly determines the need for REEs. Demand rose again in 2009, after the immediate impact of the global economic crisis had passed. As demand increased, the Chinese authorities cut export quotas, artificially reducing the supply of REEs. This fueled fears of possible shortages and caused stockpiling, driving prices to historically high levels by 2011. In 2012, prices plunged by as much as 90 percent in international markets (see Figure 3).

During the two years of surging prices for REEs, many mining companies and investors decided to go into the business of extracting REE oxides. When prices fell suddenly, mining companies suffered financial setbacks. In fact, the collapse of prices has been devastating for Western mining companies, which were trying to bring online new operations to take advantage of the high prices and reduce the West's dependence on Chinese oxides. Molycorp of the U.S. and Lynas of Australia suffered financial difficulties and ran into operational problems. Both companies have seen their share prices drop by more than half.[67] Many smaller players have also suffered calamitous financial setbacks, and their fate hinges on being able to mine so-called heavy REEs. Not all 17 rare earth elements are equally rare; DoE has identified five of them as "critical." Neodymium, a light REE, and dysprosium, a heavy REE, are used in permanent magnets for wind turbines or electric vehicles. Europium, terbium, and yttrium are heavy REEs, and are used in flat-screen electronics and energy-saving lightbulbs. Demand growth for these REEs will be strong, while mining them will be challenging.

REE mining is unlike any other type of mining. Unlike other metals used in many consumer and defense items, REEs are

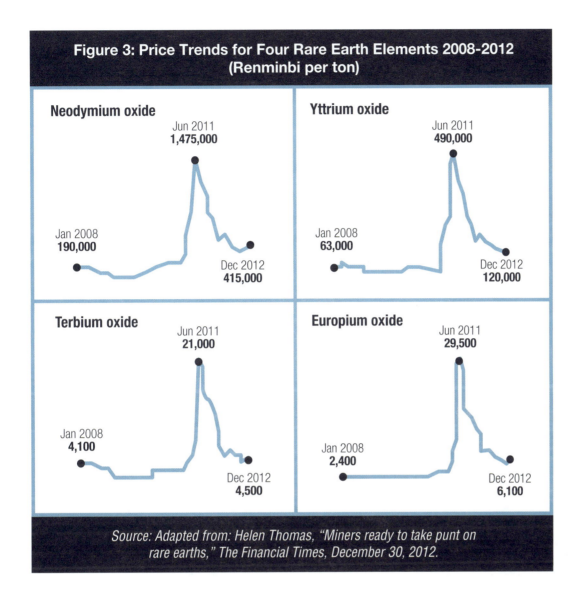

**Figure 3: Price Trends for Four Rare Earth Elements 2008-2012
(Renminbi per ton)**

Neodymium oxide

Jun 2011
1,475,000

Jan 2008
190,000

Dec 2012
415,000

Yttrium oxide

Jun 2011
490,000

Jan 2008
63,000

Dec 2012
120,000

Terbium oxide

Jun 2011
21,000

Jan 2008
4,100

Dec 2012
4,500

Europium oxide

Jun 2011
29,500

Jan 2008
2,400

Dec 2012
6,100

Source: Adapted from: Helen Thomas, "Miners ready to take punt on rare earths," The Financial Times, December 30, 2012.

to some extent abundant though they are hardly ever found in high enough concentrations to make mining them economical. Rather, REEs are mingled with other metals and must be carefully extracted and refined. REEs are often found together; mine operators must identify and isolate the individual oxides. Moreover, each REE oxide possesses different and distinct properties; mine operators must take the customer of their oxides into consideration. Thus, a mine that has a contract to sell neodymium must first refine the oxides and then extract the neodymium

elements. The length of this process makes REE mining costly and complex. First the miner must extract the ore, and then the mine operator must separate the REEs according to atomic weight. The various separation processes differ in complexity because some REEs (such as cerium) are common, while others (such as terbium) require a month of separation before ample oxides can been extracted.[68] Accordingly, mine operators cannot ramp up production quickly in response to changing global demand. Not only is it time-consuming to extract and refine the

Table 4: Rare Earth Elements, their Applications and Uses

Element	Atomic Symbol	Atomic Number	Applications and Uses
Scandium	(Sc)	21	Lightweight alloys
Yttrium	(Y)	39	Lasers, high-temperature superconductors, microwave filters, metal alloys
Lanthanum	(La)	57	High refractive glass, battery-electrodes, fluid-catalytic cracking, hybrid engines, metal alloys
Cerium	(Ce)	58	Chemical oxidizing agent, fluid catalytic cracking, metal alloys
Praseodymium	(Pr)	59	Magnets, lasers, ceramic capacitors
Neodymium	(Nd)	60	Magnets, lasers, neutron capture, hybrid engines, computer components
Promethium	(Pm)	61	Nuclear batteries
Samarium	(Sm)	62	Magnets, lasers, neutron capture, masers
Europium	(Eu)	63	Phosphors, lasers, nuclear magnetic resonance
Gadolinium	(Gd)	64	Magnets, high refractive glass, lasers, x-ray tubes, computer components, neutron capture, magnetic resonance
Terbium	(Tb)	65	Phosphors, magnets
Dysprosium	(Dy)	66	Magnets, lasers, hybrid engines
Holmium	(Ho)	67	Lasers
Erbium	(Er)	68	Lasers, vanadium steel
Thulium	(Tm)	69	Portable x-ray machines
Ytterbium	(Yb)	70	Lasers, chemical reduction
Lutetium	(Lu)	71	PET scanners, high refractive glass, chemical catalyst

Leslie Hook, "Chinese rare earth metals prices soar," The Financial Times, May 26, 2011.
http://www.ft.com/intl/cms/s/0/751cab5a-87b8-11e0-a6de-00144feabdc0.html#axzz25SIRqVhy;
and Department of Energy, Critical Materials Strategy (December 2010).
http://energy.gov/sites/prod/files/edg/news/documents/criticalmaterialsstrategy.pdf

REE oxides, but deposits vary by mine and each separation plant must be tailored to the specific local situation of that particular mine. For this reason, REEs represent some of the most technically challenging mining operations.[69]

Nevertheless, it is worth remembering that REEs are important to many renewable energy technologies. To a large extent, green energy technologies rely on an abundance of REEs. Electric vehicles use large amounts of neodymium and dysprosium (magnets) and lanthanum. Wind

Table 5: Selected Defense Uses of Rare Earth Elements

REE	Defense Use
Lanthanum	Night vision goggles
Neodymium	Laser rangefinders, guidance systems, communications, magnets
Europium	Fluorescents and phosphors in lamps and monitors
Erbium	Amplifiers in fiberoptic data transmission
Samarium	Permanent magnets that are stable at high temperatures, precision-guided munitions, and "white noise" production in stealth technology

Source: Hobart King, "REE - Rare Earth Elements and their Uses."
http://geology.com/articles/rare-earth-elements/

turbines need large quantities of neodymium and praseodymium for their powerful magnets. Energy-efficient lighting, such as LEDs and compact fluorescent bulbs, use RE phosphor powders made from yttrium, europium, and terbium.[70]

In short, the appeal of REEs lies in their ability to perform highly specialized tasks effectively (see Table 4). Europium is needed to create the red phosphor for television and computer monitors; cerium is needed to polish glass. Because they are light-weight and have high magnetic strength, REEs have reduced the size of many electronic components dramatically, and are common in consumer electronics, cars, and many military platforms. Common devices such as flash memory sticks depend on rare earth magnets (REMs), which can contain dysprosium, gadolinium, neodymium, praseodymium, and samarium. These elements are used in nuclear control rods, smart missiles, carbon-arc lamps, miniature magnets, high-strength ceramics and glass, and countless other applications.[71]

In spite of their importance to the overall economy and national security, for most of the past decade, the United States did not have a secure supply of REEs. (The Mountain Pass mine closed in 2002 and re-opened in 2012.) By 2010, Chinese producers moved into the global market for REEs and ended up controlling about 97 percent of world production and refining of REEs (see Chart 2).[72] The situation has changed somewhat since 2012 because U.S. and Australian mining companies, drawn by the high prices, opened or re-opened REE mines. Currently China is estimated to control 90% of global supply of REEs.[73] Since the 1990s, Chinese authorities pursued an explicit policy of controlling a resource they considered "strategic and critical."[74] In the 1990s, Chinese operators (both legal and illegal) flooded international markets with low-priced oxides, ores, and raw materials. Many mining companies in the United States and Australia (a country with a wealth of natural recourses) could not compete against these prices, causing many non-Chinese mining companies to shut down. Subsequently, Chinese

Table 6: United States Usage of Rare Earth Elements (2008)

Usage of Rare Earth Elements	Percent of Usage
Metallurgy & alloys	29 %
Electronics	18 %
Chemical catalysts	14 %
Phosphors for monitors, television, lighting	12 %
Catalytic converters	9 %
Glass polishing	6 %
Permanent magnets	5 %
Petroleum refining	4 %
Other	3 %

Source: Hobart King, "REE - Rare Earth Elements and their Uses."
http://geology.com/articles/rare-earth-elements/

operators have gained control over many different mineral resources while driving out production in advanced economies. In Australia, dozens of mines closed in the early 2000s due to a collapse of prices for many metals. In the United States, the Mountain Pass Mine in California, which is owned by Colorado-based Molycorp, closed in 2002 as production became uneconomical due in large part to Chinese mercantilist practices.

In the 2000s, Chinese authorities decided that, rather than exporting raw materials, it would be preferable if the processing, refining, and fabrication of final product applications would take place in China itself so that Chinese companies could reap the benefits of the added value. In 2007, Beijing instituted a 25 percent export tax on europium, terbium, and dysprosium. In 2010, Chinese authorities implemented further export restrictions on REEs by tightening export quotas.[75] The impact of a series of new measures to restrict the export of REEs meant that foreign REE consumers were paying a third more for REEs than Chinese fabricators. According to the World Trade Organization, Chinese manufacturers of REEs have a distinct price advantage over foreign firms.[76] In response, many foreign refiners and producers of final products that use REEs relocated to China to gain access to REEs and to avoid the export quotas and taxes. Japanese and U.S. companies established a foothold in China and moved production and manufacturing offshore. (In another chapter of this report, we examine permanent magnets and present an extreme case of outsourcing and offshoring that has led to a situation wherein the defense industrial base wholly depends on Chinese processing of REEs and the U.S. economy and defense industrial base must import virtually all of their high-tech magnets.)

China's near monopoly in this strategic sector raised concerns in Washington, D.C., and Tokyo, particularly when China suspended REs shipments to Japan during a diplomatic dispute in 2010. That incident, combined with broader concerns about the reliability of Chinese supply, triggered a surge of investment in RE mines outside China and brought down prices and speculative hoarding of REE oxides. Subsequently, the small REE global market has been depressed. In response, China cut production of REEs at its mines, in an effort to bolster global prices; this production cut has had a huge impact on prices. Current market dynamics do not support high RE prices. Supply is up and demand is down.

Supply is up because non-Chinese companies have aggressively invested in REE mining. Japanese companies have opened rare earths mines and processing in Kazakhstan, India, and Vietnam. The production of elements outside China is predicted to grow tenfold over five years, from 6,000 tons in 2011 to 60,000 tons in 2015.[77] According to industry analysts, as of March 2013, 50 rare earth mineral resources are active, associated with 46 advanced rare earth projects and 43 different companies, located in 31 different regions within 14 different countries. The large and sudden investments in REE mining and processing have brought prices down, especially as global demand has softened.

However, China may ultimately retain its dominant position. The price squeeze is making it unprofitable to continue operations in advanced industrialized countries. Molycorp reopened Mountain Pass Mine when prices skyrocketed. But the mine mostly produces light REs, which are relatively abundant and the least valuable. Australia's Lynas Corp. opened a mine called Mt. Weld, which also produces light rare earth oxides. Both companies have promised to find more valuable heavy REEs. These oxides are more difficult to locate; China possesses them in abundant quantities. Even if mines outside China can locate heavy REs, the issue remains that China is an extremely low-cost producer. It will be difficult for companies in the United States or Australia to compete with Chinese mines when Chinese authorities are lax in enforcing health, safety, and environmental rules. REE mining is notorious for generating massive amounts of toxic waste. Occupational safety rules as well as environmental controls make mining in the United States (and other OECD countries) more expensive than in China. However, the cost differentials between countries may be especially striking when the extraction is accompanied by a comparatively high amount of radioactive tailings, as is the case with REEs

Ultimately, the real issue is not the oxides. Mining and separating the oxides is the first step in using REEs for commercial and defense applications. The real trick lies in converting the oxides into powders, metals, alloys, and magnets. Mining is costly, but the real technological skill involves processing the RE oxides into usable items. That technology has shifted to China, which has sought to build up a "mine-to-magnet" vertical integration. The supply chain starts with oxides and then moves to refining, purification, manufacturing metal alloys, and finally to fabrication of magnets.

The critical technology for manufacturing these magnets is overseas—mostly in China. China captured the market gradually by transferring U.S. technology to China and flooding the market with cheap magnets in the early 2000s. Since then, China has continued to improve its manufacturing expertise and now possesses a depth of engineering skills.

This explains why Molycorp bought a Canada-based REE company, Neo Material Technology, which runs major manufacturing facilities in China. Molycorp cannot process the oxides into fabricated and finished products in the United States.

The U.S. mine ships RE material to China, where REEs such as dysprosium and neodymium are transformed into military-grade magnets.[78]

In the FY2007 National Defense Authorization Act (NDAA), Congress passed reforms to the specialty metals restrictions and created the Strategic Materials Protection Board (SMPB). The SMPB was meant to determine what protections were necessary to ensure the supply of materials for national defense purposes; assess potential risk associated with the non-availability of those materials; and advise policymakers on how to ensure

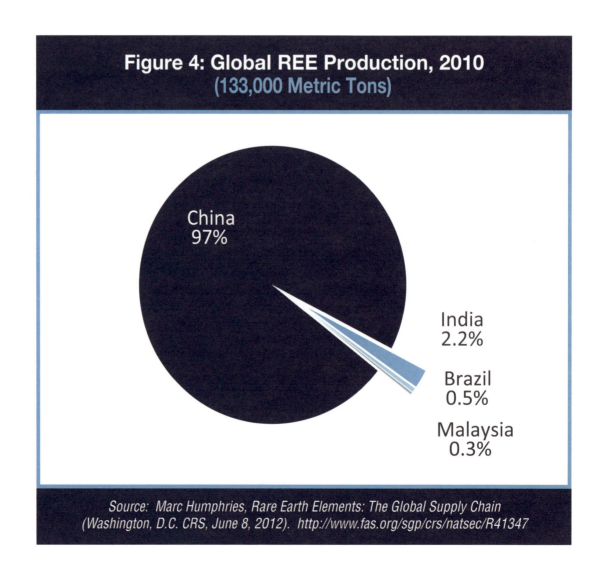

Figure 4: Global REE Production, 2010
(133,000 Metric Tons)

China 97%

India 2.2%

Brazil 0.5%

Malaysia 0.3%

Source: Marc Humphries, Rare Earth Elements: The Global Supply Chain (Washington, D.C. CRS, June 8, 2012). http://www.fas.org/sgp/crs/natsec/R41347

that supply. The SMPB is required to meet at least once every two years, publish recommendations regarding materials critical to national security, and vet the list of specialty metals.

The SMPB met twice in 2008 and issued its report and recommendations in December 2008 and February 2009.[79] The boards concluded that specialty metals were *not* "materials critical to national security," but instead "strategic materials" that warranted monitoring but not domestic source restrictions.[80] Alternatively, the Board recommended relaxing or removing domestic source requirements in an effort to reduce costs and more readily access specialty metals produced abroad.

The FY2010 NDAA required the Government Accountability Office (GAO) to assess the domestic and global availability of REMs, their importance to defense programs, and the potential for the supply of these metals to be restricted. As a result in the April 2010, GAO issued the report "Rare Earth Materials in the Defense Supply Chain" (GAO 10-617R).[81] The report stated that dependence on Chinese suppliers puts future availability of REMs—especially neodymium—at risk. The report also stated that projected domestic supply options would take seven to 15 years before becoming fully operational, primarily due to state and federal regulations. At the time of the GAO report, DoD was still in the process of evaluating defense vulnerabilities, and was scheduled to complete its analysis by September 2010. That report has never been released to the public.

The FY2012 NDAA calls for DLA to submit a plan to DoD to establish a stockpile of REMs, as well as to provide a broader assessment of source reliability. The DLA report, which was scheduled for completion in July 2012, would require a DoD

decision on the plan within 90 days of submission. At present, the DLA maintains a stockpile of 28 materials with a value of about $1.4 billion, but does not currently stockpile any REs.[82] In a significant change that increases the authority of the U.S. government to address stockpile deficiencies, Sec 901(a) of the FY2013 NDAA says that the Deputy Assistant Secretary of Defense for Manufacturing and Industrial Base Policy is now responsible for "[e]nsuring reliable sources of materials critical to national security, such as specialty metals, armor plate, and rare earth elements." DoD issued its *Strategic and Critical Materials 2013 Report on Stockpile Requirements* in March 2013 and identified 23 strategic and critical materials. The report calls for a fund of $1.2 billion to mitigate the shortfall of key materials.

Separate from the NDAA, the 112th Congress introduced at least 13 bills (nine in the House of Representatives, four in the Senate)[83] relating to REs; however, none has yet passed out of the relevant committee. Additionally, the Congressional Research Service has conducted at least three studies focused on REs and specialty metals, while GAO has released one. Broadly speaking, these reports indicate that Congress should demand renewed assessment by DoD of the "strategic materials" categorization in light of recent global supply chain concerns, and suggest policies including stockpiling REs and reinvesting in domestic research and production. These suggestions appear to be conditional on a new assessment of the SMPB/DoD, which appears reluctant to take any further action without an additional mandate from Congress. It does not appear that DoD is likely to alter its opinion expressed in the FY2011 Industrial Capability Report to Congress, which stated that, although securing a non-Chinese source of REs is essential, only minimal provisions (such as prioritizing defense applications over commercial applications) are required.[84]

To some extent, DoD's position dovetails with the interests of large defense contractors who prefer to source the small amount of magnets they need from cheap Chinese suppliers rather than to deal with U.S.-based producers.

In conclusion, although prices have dropped and shortages have disappeared in the short term, the Chinese authorities continue to meddle and intervene in the global market for RE oxides, mostly because they control the global mining of these oxides and seek to take advantage of that position. The long-term Chinese goal is to foster a high-tech RE industry in China while preserving RE reserves.[85]

PLATINUM GROUP METALS

The PGMs (also sometimes called platinum group elements, or PGEs) include iridium (Ir), osmium (Os), palladium (Pd), platinum (Pt), rhodium (Rh), and ruthenium (Ru). PGMs have excellent resistance to heat and serve as catalysts for chemical reactions, contributing to their uniqueness and importance in a variety of applications.

The most prominent application of PGMs is in catalytic converters, which dramatically reduce the pollution from automobiles. Many PGMs, especially palladium, are used as catalysts in fuel cells that find wide applications in the auto industry. Since the global car industry is projected to expand in the next decades (Chinese and Indian consumers), demand for palladium will continue to grow.[86] In addition, palladium is also used in fuel cells in hybrid cars. Thus, the switch to cars emitting fewer pollutants will not necessarily sharply reduce the demand for palladium.

In addition, platinum and palladium are extremely common in most electronic devices, including military hardware. Although the actual per-unit metal content is minute, a huge quantity of palladium is needed to meet the growing demand for electronic goods. Multi-layer ceramic capacitors (MLCC), which regulate the flow of electricity through a circuit, represent the largest demand for palladium from the electronics industry. While the automotive industry mostly consumes palladium as components of catalytic converters, automobiles also contain a large number of hybrid integrated circuits (HIC), which make use of silver-palladium tracks to connect different components of the circuit.[87]

Platinum is reportedly used in some capacity during the fabrication process of more than 20 percent of all manufactured goods.[88] It is malleable, ductile, resistant to corrosion, and possesses a high melting point around 1,770 degrees Celsius (3,215 degrees Fahrenheit). Its uses include electronics and chemical catalysts, in addition to many other applications. Platinum is up to 30 times as rare as gold (another precious metal).

Platinum and palladium supplies are potentially at risk due to their geographic concentration in areas that face political instability. In 2011, global production of platinum was dominated by South Africa (72 percent) and Russia (14 percent). The material is found in large commercial concentration in only a few regions of the world, yet the future of energy, transportation, and the environment relies on platinum. Platinum's catalytic property aids emissions control in transportation and combats pollution. Demand is bound to increase, not only in advanced industrialized countries, but also in emerging markets as governments seek to control emissions and smog. U.S. federal agencies' reports identify platinum as subject to supply risks with enormous consequences for the U.S. defense and the economy at large.[89]

In 2011, South Africa accounted for about 38 percent of palladium production, and Russia 41 percent. In all, South Africa controls more than 95 percent of known PGM reserves.[90] Two North American mines extract palladium, but their share of global new production amounts to only 14 percent.[91] Since the 1980s, the Russian government has held a stockpile of palladium. The actual size of the Russian stockpile has long been a closely guarded secret. But when prices were exorbitantly high in the early 2000s, they sold a large portion of the stockpile, bringing down the price of palladium.

South Africa traditionally has been aligned with the West; its business environment is open to Western foreign direct investments and capital flows. Yet many observers are extremely concerned about the political situation in South Africa and the possibility that its political instability may place future supplies at risk. South Africa copes with many internal tensions and conflicts. For example, different factions within the ruling African National Congress are pressing for a more aggressive policy towards the natural resource sector in order to extract greater revenues to accelerate economic development and foster wider redistribution.

Additionally, the South African government has failed to invest in society's infrastructure; as a result, many public sectors are starved of capital. Also, the current vulnerabilities in the mining sector may create a window of opportunity for more determined outside forces to gain control over a slice of the South African mineral wealth. The South African Mining Charter requires mining companies to be at least 26 percent owned by historically disadvantaged South Africans.[92] After two decades, the black empowerment objectives have not fundamentally changed the ownership structure of the mining industry, except for some smaller junior mines. These mines are scrambling for capital infusions, which may come from Chinese investors, which means that Chinese companies are moving into the PGM sector by propping up junior mining companies in South Africa. Another issue is that labor relations in some of the largest mines are fraught with conflict and tension. In the summer of 2012, a standoff between management and miners resulted in the deaths of dozens of miners and a shutdown of platinum mines. Strikes and labor unrest subsequently spread to other mines, pushing up prices of platinum and gold.[93]

As industrial strife and stoppages reduced the supply of platinum to its lowest level in a decade, the sluggish global economy and a rebound in scrap supply have kept prices within its historic range. Platinum sales from South Africa dropped by 12.5 percent in 2012, yet platinum's price fell from a high of $2,290/oz in 2008 to $1,605/oz in March 2013.

The risk is that the depressed prices will deter investments in ailing South African mines and therefore generate future supply constraints. Low prices for platinum and other PGMs have exacerbated the plight of the South African mining industry, which needs to make enormous investments to upgrade existing facilities and improve productivity.[94]

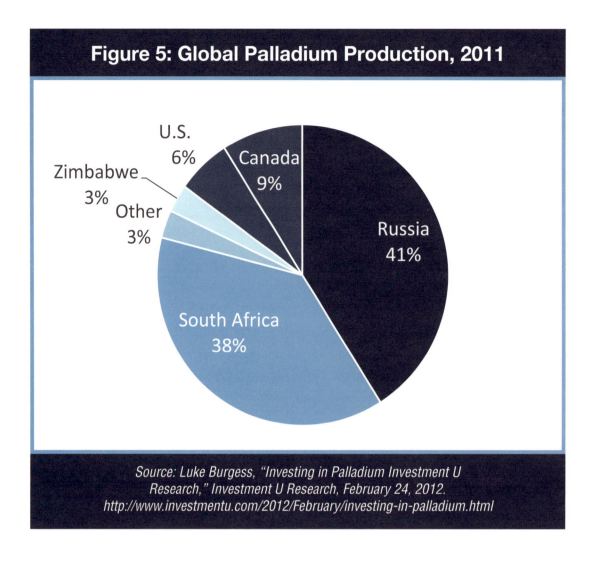

Figure 5: Global Palladium Production, 2011

Russia 41%

South Africa 38%

Canada 9%

U.S. 6%

Zimbabwe 3%

Other 3%

Source: Luke Burgess, "Investing in Palladium Investment U Research," Investment U Research, February 24, 2012. http://www.investmentu.com/2012/February/investing-in-palladium.html

While major South African mining companies face an uncertain and difficult future, Chinese investors have entered the market to assist junior mines in South Africa—a move that matches its larger strategy in sub-Saharan Africa. Concerned about supply risks to its the Chinese economy and determined to build up its military capabilities, Chinese authorities identify access to raw materials as one of their major foreign policy goals. To prevent any supply disruptions, China has been very active in sub-Saharan Africa, which is one of the regions of largely untapped metals and minerals. In turn, China's investments and presence

is welcomed in some African countries. Chinese authorities also do not exert pressure on African governments about human rights, transparency, political freedom, internal politics, environmental standards, or ethical trading practices.[95] The entry by Chinese investors or state holdings into the South African PGM sector should be a source of concern, especially as the established mining sector struggles with low productivity and underinvestment.

For these reasons, most OECD countries perceive PGMs as one of the groups of specialty metals with the single highest risk factor. First, there are no obvious substitutes for palladium and platinum, yet they are indispensable for the global production of vehicles, engines, and computer storage devices. Moreover, supply risks are high because of the political conditions in South Africa, which pull the South African government in conflicting directions, resulting in disappointing mining performance. Labor disputes add another layer of uncertainty, as discontent among workers about working conditions and pay creates a volatile atmosphere. The financial situation in some smaller start-up mines is often delicate, and provides Chinese operators with the means to gain control over sectors of the mining operations. Finally, many mines require major upgrades, and the overall transportation, power, and public service infrastructures in South Africa are in steady decline.[96]

The other country with substantial deposits of PGMs is Russia. Mining in Russia is a risky business and many mines have failed to attract private sector capital. With the fall of communism, state-owned mines were privatized and distributed to a handful of individuals. Because commodity prices were low, capital was sent overseas rather than reinvested in the mines, resulting in the decline of the Russian mining sector.[97]

Today, while greater attention is devoted to the mining sector, Russia is perceived as an unpredictable place for investments. Its economic and political environment is stable, but the mining sector is subject to arbitrary non-transparent decisions and immense bureaucratic hurdles. Obtaining a permit to explore a region is daunting because of the many technical and administrative rules. Once a company has secured an exploration license and identified a resource, it must apply for a mining license, which requires extensive paperwork as well as approvals from different levels of governments and authorities. The whole process may take years and discourages investment and expansion. Foreigners are also dissuaded by various laws that privilege domestic operators over foreign investors. The Russian state has issued laws protecting "strategic" assets, including raw materials.[98]

MITIGATING THE RISKS

The metals and chemical elements discussed in this chapter are a diverse group and require a differentiated approach, but the following recommendations will mitigate risks for most of them.

Increase the exploration of alternative sources for the elements and thereby secure a diversification of the supply chain. Deposits of specialty metals are found in smaller concentrations in various parts of the United States. For example, northeast Minnesota is thought to possess deposits of underground copper, nickel, platinum, palladium, and gold. While it seems unlikely that this region can meet all U.S. needs, mining these deposits would lessen the reliance on imports from unstable parts of the world and also reduce the impact of any future supply restrictions.

The United States should continue the search for substitute and synthetic materials to replace REEs and REMs. Even if mining companies find more geological concentrations of exotic elements, in reality at some point the United States will run out of easily accessible resources. Manmade composites would be the long-term solution to increased dependence on the scarcer elements of the periodic table.

Recycling must be improved, strengthened, and increased. Manufacturers and producers should use extracted materials in ways that facilitate recycling and re-use. The more that is recycled, the less the economy will be dependent on imports.

A new system of stockpiling or inventory should be designed to mitigate the impact of possible supply disruptions. The DLA currently stores 28 commodities valued at over $1.4 billion. Although the stockpile contains quantities of PGMs, it does not hold REEs, and it does not appear to be properly coordinated with other agencies. To operate more efficiently, DLA Strategic Materials should adopt a sensible and proactive plan to acquire materials when prices are weak and coordinate with downstream users. Congress has recently taken steps that will enable U.S. stockpiling efforts to be more proactive; however, sustained, high-level attention will be necessary.

The United States should continue to adequately fund the USGS, which collects and analyzes data, without which it would be very difficult to pursue a mitigation strategy in the first place. USGS is a critical agency in gathering and disseminating information on the state of affairs of our natural resources. Past budget cuts have caused the USGS to struggle to meet one of its principal objectives: to inform the nation of the status of its geological resources and warn of the potential for emerging supply constraints.

Enforce greater interagency coordination, which is critical to mapping out a proper long-term strategy for managing our specialty metals supply chain. DoE, DoD, and the White House Office of Science and Technology Policy all have issued reports on how to address the critical materials agenda. There should be greater coordination and collaboration in establishing a common approach to addressing the risks of supply constraints of critical and strategic materials. In addition, since other advanced industrialized countries face very similar challenges, it would also be helpful to foster greater international cooperation and coordination among the European Union, Japan, Australia, and Canada, including possible collaboration on topics such as resource mapping, substitutes, and recycling.

U.S. foreign and security policy has paid limited attention to sub-Saharan Africa, which possesses some of the world's richest concentrations of key minerals. China has been very active in Africa to ensure that it has a presence in countries with large concentrations of strategic minerals. Because the continent supplies many of the most strategic minerals, U.S. foreign, trade, and security policy should focus on ensuring continued access to African mineral deposits.

CONCLUSION

Many minerals already were labeled as critical and strategic in the early 1980s. Advanced technologies upon which our economy and national security depend are themselves heavily dependent on specialty metals and minerals. Nevertheless, over time the United States has become more dependent on imports of key minerals from countries with unstable political systems, corrupt leadership, or opaque business environments. Moreover, the countries themselves (notably, China) have taken a more aggressive posture towards mineral resources and now compete with Western mining operators for extraction control.

The United States is not the only Western country that has increasingly ignored the economics of mineral extraction. Many electronic devices, green technology, and advanced weapon systems rely on a host of exotic chemical elements. An overarching strategy linking DoD with other government and industry stakeholders is imperative to address potential shortages before they impact U.S. national security.

ENDNOTES

1   National Research Council, Committee on Critical Mineral Impacts of the U.S. Economy, Committee on Earth Resources, *Minerals, Critical Minerals, and the U.S. Economy* (Washington, D.C.: National Academies Press, 2008), 39.

2   U.S. Department of Defense, Office of Manufacturing and Industrial Base Policy, *Reconfiguration of the National Defense Stockpile Report to Congress* (April 2009). http://www.acq.osd.mil/mibp/docs/nds_reconfiguration_report_to_congress.pdf.

3   U.S. Geological Survey, *Mineral Commodities Summary 2013* (Reston, Virginia: 2013), 5. http://minerals.usgs.gov/minerals/pubs/mcs/2013/mcs2013.pdf.

4   According to the American Resources Policy Network, the absence of a working understanding of why a mineral may be critical or strategic has greatly contributed to the delay in securing access to key minerals. Daniel McGroarty and Sandra Wirtz, "Reviewing Risk: Critical Metals & National Security," *American Resources Policy Network* (June 6, 2012). http://americanresources.org/wp-content/uploads/2012/09/ARPN_Fall_Quarterly_Report_WEB.pdf.

5   "Minerals Make National Security," National Mining Association. http://mineralsmakelife.org/assets/images/content/resources/MineralsMakeNationalSecurityFactSheet.pdf.

6   Roderick G. Eggert, Mineral and Energy Economics Program at Colorado School of Mines, quoted in David J. Hanson, "Critical Materials Problem Continues Debate over Use of and Substitutions for Rare Earth Elements Points out a Need for Much More Research," *Chemical & Engineering News* 89 (October 24, 2011), 28.

7   Rare earth elements (REEs) are a group of 17 chemical elements that occur together in the periodic table. These metals have many similar properties, which explains why they often are found together in geologic deposits. However, the group is divided into two categories according to their atomic weight on the periodic table: light and heavy REEs. Many light REEs are in fact abundant and of less value than the heavy ones; heavy REEs are of higher value in spite of a smaller market for them. See Chapter 3 on Magnets for a further discussion of light and heavy REEs.

8   U.S. Geological Survey (UGS), *Principal Rare Earth Elements Deposits of the United States: A Summary of Domestic Deposits and a Global Perspective* Keith R. Long et al., (Reston, VA: USGS & U.S. Department of the Interior, Scientific Investigations Report 2010-5220). http://pubs.usgs.gov/sir/2010/5220/downloads/SIR10-5220.pdf.

9   Brian Sylvester. "Rare Earth Metals—Not Your Father's Mining Biz: Byron King," *The Gold Report*, January 13, 2012. http://www.theaureport.com/pub/na/12440.

10 Elisa Alonso, Frank R. Field, and Randolph E. Kirchain, "Platinum Availability for Future Automotive Technologies," *Environmental Science & Technology* 46, no. 12 (2012), 986-1012.

11 British Geological Survey (BGS), *Risk List 2011*. http://www.bgs.ac.uk/downloads/start.cfm?id=2063. See also, BGS, *Critical Raw Materials*. http://www.bgs.ac.uk/mineralsuk/statistics/criticalRawMaterials.html.

12 Other elements with substantial supply risks are bismuth (China), carbon graphite (China), germanium (China), indium (China), iodine (Chile), rhenium (Chile), strontium (China), and thorium (India). British Geological Survey, *Risk List 2011*. http://www.bgs.ac.uk/downloads/start.cfm?id=2063.

13 Gerrit Wiesmann, "Merkel Strikes Kazakh Rare Earth Accord," *The Financial Times*, February 8, 2012.

14 "European Union Critical Raw Materials Analysis the European Commission Raw Materials Supply Group," *Executive Summary by Swiss Metal Assets* (October 1, 2011). http://www.swissmetalassets.com/wp-content/uploads/2011/10/EU-Critical-Metals-Summary-Oct-3-2011.pdf.

15 U.S. Geological Survey, *Rare Earths Statistics and Information 2012* (Reston, VA: USGS, 2012). http://minerals.usgs.gov/minerals/pubs/commodity/rare_earths/.

16 The Mountain Pass Mine in California was one of the largest producers of REE until it closed in 2002 as a result of falling RE prices as China stepped up production and heightened U.S. environmental regulations and concerns regarding pollution from the extraction and refining processes. A series of environmental disasters between 1984 and 1998 released thousands of gallons of toxic sludge into the California desert, prompting a major lawsuit. The mine reopened in 2012. Russell McLendon, "What Are Rare Earth Metals?" *Mother Nature Network*, June 22, 2011. http://www.mnn.com/earth-matters/translating-uncle-sam/stories/what-are-rare-earth-metals.

17 Bryan Tilt, *The Struggle for Sustainability in Rural China: Environmental Values and Civil Society* (New York, NY: Columbia University Press, 2010).

Shunsuke Managi and Shinji Kaneko, *Chinese Economic Development and the Environment*, (Northampton, MA: Edward Elgar, 2008).

18 "Wish You Were Mine," *The Economist*, February 11, 2012. http://www.economist.com/node/21547285.

19 "Wish You Were Mine," *The Economist*, February 11, 2012. http://www.economist.com/node/21547285.

Alex MacDonald, "Australia Gives Africa a Lesson in Resource Nationalism," *The Wall Street Journal*, February 13, 2012. http://blogs.wsj.com/dealjournalaustralia/2012/02/13/australia-gives-africa-a-lesson-in-resource-nationalism/.

Andrew England, "Zuma Seeks to Allay Fears over Nationalization," *The Financial Times*, February 10, 2012. http://www.ft.com/intl/cms/s/0/cfc3f05a-5403-11e1-bacb-00144feabdc0.html.

20 Brian Sylvester, "Richard Karn: 50 Specialty Metals Under Supply Threat," *The Critical Metals Report*, June 14, 2011. http://www.theaureport.com/pub/na/9867.

21 National Research Council, *Minerals, Critical Minerals, and the U.S. Economy* (Washington, D.C.: National Academies Press, 2008).

22 National Research Council, *Minerals, Critical Minerals, and the U.S. Economy* (Washington, D.C.: National Academies Press, 2008), 165.

23 National Research Council, *Minerals, Critical Minerals, and the U.S. Economy* (Washington, D.C.: National Academies Press, 2008), 58.

24 U.S. Geological Survey, *Mineral Commodity Summaries, Lithium* (Reston, VA: January 2012). http://minerals.usgs.gov/minerals/pubs/commodity/lithium/mcs-2012-lithi.pdf.

25 There is one active lithium mine in Nevada. The mine's production capacity was expanded in 2012, and there is a small amount of recycled lithium. Nevertheless, the U.S. imports more than 70 percent of its lithium. U.S. Geological Survey, *Mineral Commodities Summaries 2013* (Reston, VA: January 2013). http://minerals.usgs.gov/minerals/pubs/mcs/2013/mcs2013.pdf.

The American Physical Society and Materials Research Society, *The Energy Critical Elements: Securing Materials for Emerging Technologies* (Washington, D.C.: February 18, 2011). http://www.aps.org/policy/reports/popa-reports/loader.cfm?csModule=security/getfile&PageID=236337.

26 David J. Hanson, "Critical Materials Problem Continues Debate over Use of and Substitutions for Rare Earth Elements Points out a Need for Much More Research," *Chemical & Engineering News* 89 (October 24, 2011), 29.

27 U.S. Geological Survey, *Mineral Commodity Summaries, Lithium* (Reston, VA: January 2012). http://minerals.usgs.gov/minerals/pubs/commodity/lithium/mcs-2012-lithi.pdf.

28 Thomas Cherico Wanger, "The Lithium Future—Resources, Recycling, and the Environment," *Conservation Letters* 4 (2011), 202-206.

29 Thomas Goonan, "Lithium Use in Batteries," *U.S. Geological Survey Circular* 1371 (Reston, VA: USGS, 2012). http://pubs.usgs.gov/circ/1371/.

30 U.S. Department of Defense, Office of Manufacturing and Industrial Base Policy, *Annual Industrial Capabilities Report to Congress* (2011). http://www.acq.osd.mil/mibp/docs/annual_ind_cap_rpt_to_congress-2011.pdf.

31 U.S. Geological Survey, *Mineral Commodity Summaries, Beryllium* (Reston, VA: January 2012). http://minerals.usgs.gov/ minerals/pubs/commodity/beryllium/mcs-2012-beryl.pdf.

32 U.S. Geological Survey, *Mineral Commodity Summaries, Rhenium* (Reston, VA: January 2012). http://minerals.usgs.gov/ minerals/pubs/commodity/rhenium/mcs-2012-rheni.pdf.

33 Live price of Rhenium Price accessed on March 10, 2013. http://www.taxfreegold.co.uk/rheniumpricesusdollars.html.

34 U.S. Geological Survey, *Mineral Commodity Summaries, Rhenium* (Reston, VA: January 2012). http://minerals.usgs.gov/minerals/pubs/commodity/rhenium/mcs-2012-rheni.pdf.

35 U.S. Geological Survey, *Mineral Commodity Summaries, Molybdenum* (Reston, VA: January 2012). http://minerals.usgs.gov/minerals/pubs/commodity/molybdenum/mcs-2012-molyb.pdf.

36 U.S. Geological Survey, *Mineral Commodity Summaries, Vanadium* (Reston, VA: January 2012). http://minerals.usgs.gov/minerals/pubs/commodity/vanadium/mcs-2012-vanad.pdf.

37 U.S. Geological Survey, *Mineral Commodity Summaries 2013* (Reston, VA: January 2013).

38 Brian Sylvester, "Solving Critical Rare Earth Metal Shortages: Dr. Michael Berry and Chris Berry," *The Critical Metals Report*, January 10, 2012. http://www.theaureport.com/pub/na/12217.

39 Steve Gagnon, "Vanadium," *Jefferson Lab*. http://education.jlab.org/itselemental/ele023.html.

Nick Hodge, "Vanadium: The Best Thing Since Lithium," *Energy & Capital*, May 25, 2011. http://www.energyandcapital.com/articles/the-best-thing-since-lithium/1531.

40 "Tantalum Facts: Chemical and Physical Properties," Anne Marie Helmenstine for About.com. http://chemistry.about.com/od/elementfacts/a/tantalum.htm.

41 Columbite is an older name, no longer in use except in the context of columbite-tantalite. Columbite comes from columbium, which was renamed niobium in the 1950s.

42 U.S. Geological Survey, *Mineral Commodity Summaries, Tantalum* (Reston, VA: January 2012). http://minerals.usgs.gov/minerals/pubs/commodity/niobium/mcs-2012-tanta.pdf.

43 Both the Securities Exchange Commision and OECD are contemplating new rules to force corporations to conduct business ethically in central Africa. The Dodd-Frank Act has banned the export of minerals from regions where mining appears to be intertwined with ongoing civil strife. U.S. Geological Survey, *Mineral Commodity Summaries, Tantalum* (Reston, VA: January 2012). http://minerals.usgs.gov/minerals/pubs/commodity/niobium/mcs-2012-tanta.pdf.

Michael Nest, *Coltan* (New York: Polity Press 2011).

44 "About Tungsten," Ormonde Mining, PLC. . http://www.ormondemining.com/en/investors/about_tungsten/about_tungsten_details.

45 "Tungsten: Statistics and Information," last modified March 11, 2013, U.S. Geological Survey.

http://minerals.usgs.gov/minerals/pubs/commodity/tungsten/index.html.

46 U.S. Geological Survey, *Mineral Commodity Summaries, Tungsten* (Reston, VA: January 2012). http://minerals.usgs.gov/minerals/pubs/commodity/tungsten/mcs-2012-tungs.pdf.

47 U.S. Under Secretary of Defense for Acquisition, Technology, and Logistics, *Strategic and Critical Materials Operations Report to Congress*, January 2011. https://www.dnsc.dla.mil/Uploads/Materials/dladnsc2_9-13-2011_15-9-40_FY10%20Ops%20Report%20-%2005-06-2011.pdf.

48 U.S. Geological Survey, *Mineral Commodity Summaries, Tungsten* (Reston, VA: January 2012). http://minerals.usgs.gov/minerals/pubs/commodity/tungsten/mcs-2012-tungs.pdf.

49 Jonathan Marshall, "Opium, Tungsten, and the Search for National Security, 1940–52," *Journal of Policy History*, 3 (October 1991), 89-116.

Mildred Gwen Andrews, *Tungsten: The Story of an Indispensable Metal*, (Washington, D.C.: Tungsten Institute, 1955).

50 For example, South Korea plans to convert 100 percent of public sector lighting to LED by 2020, and aims to reach 60 percent LED lighting nationwide.

51 U.S. Geological Survey, *Mineral Commodity Summaries, Gallium* (Reston, VA: January 2012). http://minerals.usgs.gov/minerals/pubs/commodity/gallium/mcs-2012-galli.pdf.

52 Keith Gurnett and Tom Adams, "Taming the Gallium Arsenic Dicing Process," *Military & Aerospace Electronics*, December 1, 2006. http://www.militaryaerospace.com/articles/print/volume-17/issue-12/news/taming-the-gallium-arsenide-dicing-process.html.

53 U.S. Geological Survey, *Mineral Commodity Summaries, Gallium* (Reston, VA: January 2012). http://minerals.usgs.gov/minerals/pubs/commodity/gallium/mcs-2012-galli.pdf.

54 U.S. Geological Survey, *Mineral Commodity Summaries, Indium* (Reston, VA: January 2012). http://minerals.usgs.gov/minerals/pubs/commodity/indium/mcs-2012-indiu.pdf.

55 "Indium." Chemicool Periodic Table. *Chemicool.com*, November 9, 2012. http://www.chemicool.com/elements/indium.html.

56 "Technology: Why SWIR? What is The Value of Shortwave Infrared?" Sensors Unlimited, Inc., Goodrich Corporation. http://www.sensorsinc.com/whyswir.html.

57 U.S. Geological Survey, *Mineral Commodity Summaries, Germanium* (Reston, VA: January 2012). http://minerals.usgs.gov/minerals/pubs/commodity/germanium/mcs-2012-germa.pdf.

58 U.S. Geological Survey, *Mineral Commodity Summaries, 2013* (Reston, VA: January 2013). http://minerals.usgs.gov/minerals/pubs/commodity/germanium/mcs-2012-germa.pdf.

59 For example, the 2012 DLA made only 3000 kg available for purchase of a 16000 kg total holding. See Defense Logistics Agency, *Strategic Materials, Annual Material Plan (FY2012)* (2012). https://www.dnsc.dla.mil/Uploads/NewsRelease/bberuete_10-6-2011_11-9-18_3028%20FY12%20AMP.pdf.

60 U.S. Geological Survey, *Mineral Commodity Summaries, Germanium* (Reston, VA: January 2012).

61 U.S. Geological Survey, *Material Commodity Summaries, Antimony* (Reston, VA: January 2012). http://minerals.usgs.gov/minerals/pubs/commodity/antimony/mcs-2012-antim.pdf.

62 U.S. Geological Survey, *Byproduct Mineral Commodities used for the Production of Photovoltaic Cells*, by Donald Bleiwasy. (Reston, VA: 2010). http://pubs.usgs.gov/circ/1365/Circ1365.pdf.

63 U.S. Geological Survey, *Material Commodity Summaries, Tellurium* (Reston, VA: January 2012). http://minerals.usgs.gov/minerals/pubs/commodity/selenium/mcs-2012-tellu.pdf.

64 Live price of rhenium, Tax Free Gold, accessed February 17, 2012. http://www.taxfreegold.co.uk/rheniumpricesusdollars.html.

65 American Physical Society. *Energy Critical Elements: Securing Materials for Emerging Technologies*. http://www.aps.org/policy/reports/popa-reports/loader.cfm?csModule=security/getfile&PageID=236337.

66 U.S. Geological Survey, *Mineral Commodity Summaries, Gallium* (Reston, VA: January 2012). http://minerals.usgs.gov/minerals/pubs/commodity/gallium/mcs-2012-galli.pdf.

67 Molycorp's share price dropped below its IPO price of $14/share. It rose to $76/share on April 28, 2011, and then dropped as low as $5.75/share in November 2012. It traded at $6.50/share in March 2013.

68 Kia Ghorashi, Lucinda Gibbs, Polly Hand, and Amber Luong, *Rare Earth Elements: Strategies to Ensure Domestic Supply*. (Stanford, CA: Bay Area Council and the Breakthrough Institute, March 10, 2011). http://publicpolicy.stanford.edu/system/files/RareEarthElements.pdf.

69 Kia Ghorashi, Lucinda Gibbs, Polly Hand, and Amber Luong, *Rare Earth Elements: Strategies to Ensure Domestic Supply*. (Stanford, CA: Bay Area Council and the Breakthrough Institute, March 10, 2011). http://publicpolicy.stanford.edu/system/files/RareEarthElements.pdf.

70 United States Department of Energy, *Critical Materials Strategy* (Washington, D.C.: DOE/PI-0009, December 2011). http://energy.gov/sites/prod/files/DOE_CMS2011_FINAL_Full.pdf.

71 Russell McLendon. "What Are Rare Earth Metals?" *Mother Nature Network*, June 22, 2011. http://www.mnn.com/earth-matters/translating-uncle-sam/stories/what-are-rare-earth-metals.

72 Marc Humphries, *Rare Earth Elements: The Global Supply Chain* (Washington, D.C.: Congressional Research Service 7-5700, September 6, 2011).

73 ResearchInChina, China Rare Earth Industry Report, 2012-2015 (Beijing: ResearchInChina, April 2013).

74 John Tkacik, Jr., Magnequench: CFIUS and China's Thirst for U.S. Defense Technology (Washington, D.C.: The Heritage Foundation, May 2, 2008). http://www.heritage.org/research/reports/2008/05/magnequench-cfius-and-chinas-thirst-for-us-defense-technology. See also Chapter 3 on Magnets.

75 Patti Waldmeir, "China Ups Rare Earth Reserve," *The Financial Times*, February 12, 2010. http://www.ft.com/intl/cms/s/0/fe51dd3c-1775-11df-87f6-00144feab49a.html.

76 Michael Martina, "China Says Ready to Defend its Rare Earth Policies," *Reuters*, February 1, 2012. http://www.reuters.com/article/2012/02/01/us-china-rare-earths-idUSTRE8101OT20120201.

77 Leslie Hook, "China to Subsidise Rare Earths Producers," *The Financial Times*, November 22, 2012. http://www.ft.com/intl/cms/s/0/b3332e0a-348c-11e2-8986-00144feabdc0.html#axzz2NBcKoeAw.

78 Lou Kilze, "Pentagon 'naive' on rare earth outlook, several experts say," *Pittsburgh Tribune-Review* April 10, 2012. http://triblive.com/x/pittsburghtrib/news/s_7905799.html#axzz2NCMazsVs.

79 U.S. Department of Defense, "Report of Meeting, Department of Defense, Strategic Materials Protection Board, December 12, 2008" (December 2008). http://www.acq.osd.mil/mibp/docs/report_from_2nd_mtg_of_smpb_12-2008.pdf.

"Analysis of National Security Issues Associated With Specialty Metals," *Federal Register* 4, no. 34 (February 23, 2009), 8061-64. http://edocket.access.gpo.gov/2009/pdf/E9-3708.pdf.

80 U.S. Department of Defense, "Report of Meeting, Department of Defense, Strategic Materials Protection Board, December 12, 2008" (December 2008). http://www.acq.osd.mil/mibp/docs/report_from_2nd_mtg_of_smpb_12-2008.pdf.

81 The GAO report is available online at http://www.gao.gov/new.items/d10617r.pdf.

82 However, in March 2013, DoD suggested adding heavy rare earth elements to the stockpile in the amount of $120.43 million. DoD would have to buy these elements from China, because there is no domestic producer of heavy REEs. Dorothy Kosich, "U.S. DoD, Congress Worry on Rare Earths Stockpiles, Supplies," *MineWeb*, March 25 2013. http://www.mineweb.com/mineweb/content/en/mineweb-industrial-metals-minerals-old?oid=183283&sn=Detail.

U.S. Department of Defense, Under Secretary of Defense for Acquisition, Technology, and Logistics, "Strategic and Critical Materials Operations Report to Congress" (Washington, D.C.: January 2011). https://www.dnsc.dla.mil/Uploads/Materials/dladnsc2_9-13-2011_15-9-40_FY10%20Ops%20Report%20-%2005-06-2011.pdf.

83 H.R. 618, H.R.952, H.R.1314, H.R.1367, H.R.1388, H.R.1875, H.R.2011, H.R.2184, H.R.22847, S. 734, S. 1113, S. 1270, S. 1351.

84 U.S. Department of Defense, Office of Under Secretary of Defense for Acquisition, Technology, and Logistics, "Annual Industrial Capabilities Report to Congress" (Washington, D.C.: September 2011), 10-12. http://www.acq.osd.mil/mibp/docs/annual_ind_cap_rpt_to_congress-2011.pdf.

85 Rebecca Keenan, "China May Double Rare Earth Exports as Demand Rebounds," *Bloomberg News*, February 27, 2012. http://www.bloomberg.com/news/2012-02-26/china-may-double-rare-earth-exports-as-overseas-demand-rebounds-on-price.html.

86 Daniel Sperling and Deborah Gordon, *Two Billion Cars: Driving Toward Sustainability* (New York, NY: Oxford University Press, 2009).

87 "Electronic Uses of Palladium," Stillwater Palladium, http://www.stillwaterpalladium.com/electronics.html.

88 "Gold Depository," WHYY, http://www.whyy.org/tv12/secrets/gold.html.

89 Elisa Alonso, Frank R. Field, and Randolph E. Kirchain, "Platinum Availability for Future Automotive Technologies;" *Environmental Science & Technology* 46, no. 12 (2012), 986-1012.

Daniel McGroarty and Sandra Wirtz, "Reviewing Risk: Critical Metals & National Security," *American Resources Policy Network* (June 6, 2012). http://americanresources.org/wp-content/uploads/2012/09/ARPN_Fall_Quarterly_Report_WEB.pdf.

90 U.S. Geological Survey, *Material Commodity Summaries, Platinum Group Metals* (Reston, VA: January 2012). http://minerals.usgs.gov/minerals/pubs/commodity/platinum/mcs-2012-plati.pdf.

91 The two mines are Stillwater Complex in Montana and North American Palladium in the Thunder Bay District of Ontario. Luke Burgess, "Investing in Palladium," *Investment U Research*, February 24, 2012. http://www.investmentu.com/2012/February/investing-in-palladium.html.

Luke Burgess, "The Definitive Guide to Palladium Investing in 2012," *Investment U Research*, February 28, 2012.

http://www.investmentu.com/2012/February/palladium-investing.html.

92 Peter Leon, "Business Should Be Driving Empowerment of Workers," *The Financial Times*, February 8, 2012. http://www.ft.com/intl/cms/s/0/dfe79f60-4ce0-11e1-8741-00144feabdc0.html#axzz25SIRqVhy.

93 Andrew England, "Marikana Mine Bus Crews Face Threats," *The Financial Times* (August 27, 2012). http://www.ft.com/intl/cms/s/0/d405cd72-f058-11e1-b7b2-00144feabdc0.html#axzz25SIRqVhy.

Andrew England, "S Africa Gold Miners Strike as Strife Spreads," *The Financial Times* (August 31, 2012). http://www.ft.com/intl/cms/s/0/1b0f6d3c-f378-11e1-9c6c-00144feabdc0.html#axzz25SIRqVhy.

94 Jack Farchy, "Platinum Shortfall Largest in Decade," *The Financial Times*, November 13, 2012. http://www.ft.com/intl/cms/s/0/99c85de0-2cf9-11e2-beb2-00144feabdc0.html#axzz2NBcKoeAw.

95 Masuma Farooki, "China's structural demand and commodity prices," edited by Christopher M. Dented, *Implications for Africa China and Africa development relations* (New York, NY: Routledge, 2011), 121-142.

Deborah Brautigam, *The Dragon's Gift: The Real Story of China in Africa* (New York, NY: Oxford University Press, 2010).

96 Hein Marais, *South Africa Pushed to the Limit: The Political Economy of Change* (London: Zed Books, 2011).

Bill Freund and Harald Witt, eds., *Development dilemmas in post-apartheid South Africa* (Cottsville, South Africa: University of Kwazulu-Natal Press, 2010).

Janine Aron, Brian Kahn, and Geeta Kingdon, *South African Economic Policy under Democracy* (New York, NY: Oxford University Press, 2009).

97 Global Business Reports, "Mining in Russia: Braving the Bear," *Engineering & Mining Journal*, February 2012, 50-54. http://www.gbreports.com/admin/reports/EMJ-Russia-2012.pdf.

98 Ibid.

CHAPTER 4 • TITANIUM

EXECUTIVE SUMMARY

Titanium alloys are as strong as steel, but are 45 percent lighter. This makes them an ideal structural material for aircraft, certain naval vessels, and other applications. Titanium alloys are present in practically every military and commercial aircraft today. Titanium is non-substitutable, and enables American aerial dominance and force projection.

A handful of producers dominate the global titanium market. They supply nearly all titanium alloys to the aerospace industry, which determines global demand for titanium products. Although the global market is relatively small, compared to that of other metals the titanium market is highly volatile and susceptible to quick shifts in demand. Periods of high demand are often followed by periods of stagnant demand and depressed prices. Unfortunately, because of the technical complexity and capital-intensive nature of titanium fabrication, producers cannot adjust quickly and easily to new global conditions in the titanium market.

The Specialty Metals Clause (SMC) currently mandates domestic procurement of titanium, but some defense contractors have lobbied to weaken or even repeal the SMC. The SMC has, to some extent, smoothed out the volatility domestic titanium producers face by providing them with predictable orders. Defense acquisitions have sheltered domestic producers from the full impact of a sudden downturn in the commercial aerospace industry. However, during periods of high demand, defense contractors and suppliers seek to challenge the SMC's requirements because they would prefer to purchase titanium products abroad in order to avoid backlogs, delays, or other inconveniences.

The main risk facing the titanium industry is that foreign producers may dramatically expand production to undercut U.S. firms and force them out of business. Although defense demand for titanium is significant (and, under the SMC, must be fulfilled by domestic producers), in an average year it represents only 10 percent of total U.S. sales. Increased foreign production—especially by companies with close ties to their respective governments, such as Russian and Chinese producers—can drive down commercial prices and could jeopardize the survival of U.S. firms. So far, the SMC has helped to prevent such a scenario, but efforts to repeal or significantly weaken it could have disastrous consequences for a strategically important industry.

TITANIUM METAL
SPEED AND STRENGTH FOR THE U.S. MILITARY

MANUFACTURING SECURITY

Titanium is an essential component of many national security products

FIGHTER JET **LITTORAL COMBAT SHIP** **LIGHT ARMOR TANK**

DOMESTIC PRODUCTION

The three U.S. titanium producers supply U.S. defense and national security needs

3 DOMESTIC TITANIUM PRODUCERS

DOD DEMAND WAS VITAL FOR CREATING **THE DOMESTIC TITANIUM** INDUSTRY AND DOD DEMAND **IS STILL VITAL TODAY**

SUPPLY VULNERABILITIES

CHINA AND RUSSIA **SUPPLY OVER 40%** OF GLOBAL TITANIUM **METAL PRODUCTION**

VULNERABILITY

The U.S. does not currently supply its own demand for titanium sponge

 THE U.S. IMPORTS **APPROXIMATELY 65%** OF TITANIUM SPONGE

AIR SUPERIORITY

Titanium is a key input for the F-35 fighter

 **22% TITANIUM** BY WEIGHT

PRODUCTION COSTS

Titanium production is highly capital-intensive

SATISFYING UNPREDICTABLE **DEMAND IS DIFFICULT**

MITIGATING RISKS

Smart policy can help sustain U.S. titanium production

ENFORCE SPECIALTY METALS CLAUSE **INVEST IN SPONGE PRODUCTION** **INVEST IN INNOVATION**

MILITARY EQUIPMENT CHART
SELECTED DEFENSE USES OF TITANIUM

DEPARTMENT	WEAPON SYSTEMS	PLATFORMS	OTHER SYSTEMS
ARMY	■ M240L machine gun ■ M777 155mm Field Howitzer	■ M1 Abrams main battle tank ■ M2 Bradley fighting vehicle ■ M113 armored personnel carrier ■ UH-60 Blackhawk helicopter	■ Armor for ground vehicles
MARINE CORPS	■ M777 155mm Field Howitzer	■ SH-60 Seahawk helicopter ■ F-35 Joint Strike fighter ■ V-22 Osprey aircraft	■ Armor for ground vehicles
NAVY	■ Submarines	■ Nimitz-Class nuclear-powered aircraft carrier ■ SH-60 Seahawk helicopter ■ F-35 Joint Strike fighter ■ Independence-class littoral combat ship (LCS)	■ Seawater piping ■ Shipboard switches ■ Terminal boxes ■ Turbines
AIR FORCE	■ AIM-9 Sidewinder air-to-air missile	■ A-10 Thunderbolt II close support aircraft ■ F-15 Eagle fighter ■ F-22 Raptor fighter ■ F-35 Joint Strike fighter ■ HH-60 Pave Hawk helicopter ■ AB-8B Harrier aircraft	■ Aircraft parts ■ Jet engines

INTRODUCTION

Titanium is the preferred metal or alloy used in military airplanes, helicopters, and rockets. Because it is critical in meeting U.S. defense needs, titanium products are covered by the Specialty Metals Clause (SMC). The SMC's stated goal is to protect the U.S. defense industrial base from becoming overly dependent on foreign suppliers, especially in times of conflict, and to encourage U.S. producers to continue to research and invest in the high technology strategic materials that support that base. At the same time, titanium's inclusion in the SMC has fueled protracted struggles in Washington, D.C. Prime contractors and other defense suppliers regularly lobby Congress and the Department of Defense (DoD) to obtain waivers from the SMC that would allow them to purchase titanium products from non-qualifying countries such as Russia, which is the largest producer of titanium and titanium alloys in the world. In response to pressure from contractors and suppliers, DoD has issued numerous exemptions to the SMC. As DoD approves different exemptions, defense contractors subsequently have lobbied to weaken or even repeal the SMC, which would be a disaster for the domestic titanium industry. If the SMC is repealed or weakened, the U.S. defense industrial base would rely too heavily on foreign suppliers for a strategic material necessary to produce modern aerospace and naval platforms and weapons systems.

Titanium has a long history. First discovered in 1790, titanium's main use was as a whitening additive in paints. Although titanium is the sixth most abundant ore (after aluminum, iron, sodium, potassium, and magnesium), it is never found in pure form, and is usually found as a component of the minerals rutile and ilmenite. Titanium is highly reactive, which means that it easily mixes with other elements, such as oxygen or iron. Titanium was only commercialized in the early 1950s because of the many technical barriers involved in extracting and producing a titanium alloy. Once titanium was commercialized, the U.S. federal government and private sector built a new generation of fighter jets that could fly faster, higher, and longer than their Soviet counterparts.

Titanium is regarded as a miracle alloy because it is as strong as steel, but is 45 percent lighter. It is twice as strong as aluminum, but is only 60 percent heavier. It does not corrode and is not affected by salt, oxygen, heat, or chemical applications. By far the most significant end user use of titanium alloys is the commercial aerospace industry, which uses it in applications such as commercial aircraft, military aircraft, space rockets, and missiles. In addition, the automobile industry often relies on titanium alloys to manufacture various engine, brake, and blade components. Military applications represent approximately 10 to 13 percent of the U.S. market for titanium.[1]

A variety of issues have important effects on titanium supply.

First, titanium prices tend to be volatile. Titanium prices often go through sharp cyclical fluctuations, creating boom-bust periods for U.S. producers. These cycles closely follow the cycles in the commercial aerospace industry, especially the build rates. Because the production of titanium metal includes an expensive batch process that requires multiple phases, the high cost and long lead times make it difficult for firms to adjust quickly to turbulent market conditions. The cyclical nature of titanium demand has had adverse effects on U.S. titanium producers, especially when prices suddenly drop.

Second, while DoD demand for titanium represents a small fraction of total U.S. titanium output, that share is critical in guaranteeing the survival of a competitive and innovative domestic sector, especially during recessions or periods of waning demand for titanium. For this reason, in 1973 Congress passed domestic sourcing restrictions for specialty metals (currently found in the SMC [10 U.S.C. 2533b]). The SMC prohibits defense contractors from procuring melted titanium from non-U.S. suppliers. Due to cost and other pressures, however, the Pentagon and defense contractors repeatedly have pushed to weaken or eliminate the domestic sourcing requirement for titanium.

Third, Congress has passed many waivers and exemptions to titanium domestic sourcing requirements, which in turn has made it easier for U.S. companies to blend foreign and domestic titanium. Waivers and exemptions also relieve smaller sub-contractors from having to separate foreign and domestic titanium.

In contrast to U.S. practice, foreign governments recognize the significance of their domestic titanium industry and step in to support their producers during periods of depressed global demand. Along with the United States, Russia is a major titanium producer. The Russian titanium sector is partly state-owned and receives special support from the Kremlin. As such, Russian production faces less pressure to compete for profits.

China also is becoming a major player in the titanium market. The increasing role of China as a supplier may result in greater instability and volatility in this relatively small global market. Many Chinese titanium producers have close ties to the Chinese state or Communist Party-aligned enterprises, and are less constrained by the need to generate short-term earnings for their shareholders. Conceivably, during periods of overcapacity, Chinese firms could sell titanium alloys at prices below costs and thereby exert further downward pressure on global prices. This action would have expensive consequences for American titanium producers. This has happened before with rare earth elements (REEs), in the 1990s, when low prices resulted in the closure of many U.S. and Australian mines. It is clear that governments of other countries recognize the importance of this strategic material and seek to preserve a domestic base.

Key themes discussed in this chapter are:

- Titanium is a strategic material and widely used as a critical input for key U.S. military capabilities, including advanced aerospace applications and weapons systems.

- U.S. defense demand helped to create the titanium industry and should continue to support it today. The SMC has buffered domestic titanium producers from turbulent global markets, dominated by a handful of players.

- Without strong and enforced domestic sourcing requirements for titanium, U.S. production of defense goods that use titanium may become dependent on Chinese and Russian producers.

- Other governments recognize the criticality of their titanium industries and provide support, protection, or other positive incentives.

A NOTE ON CRITICALITY

The combination of high strength and low weight has made titanium the essential input to the modern aerospace sector, and without it many military aircraft would be too heavy to fly. The replacement of steel with titanium alloy enabled U.S. military aircraft to fly faster, higher, and farther than their counterparts, and is a key contributor to U.S. air superiority. Titanium is also used to fabricate the hulls of rockets and missiles, and replaces steel in many applications where steel's weight is restrictive, such as the U.S. Navy's Independence-class Littoral Combat Ship. Restricted access to titanium would prohibit the construction and repair of most U.S. military aircraft and many other advanced weapons systems, diminishing U.S. command over the skies and effectively *incapacitating* U.S. force projection capabilities.

COUNTERFEIT TITANIUM PARTS HIGHLIGHT THE IMPORTANCE OF TITANIUM FOR DEFENSE CAPABILITIES

Counterfeit parts—generally those whose sources knowingly misrepresent the parts' identity or pedigree—have the potential to seriously disrupt the Department of Defense (DoD) supply chain. In an extensive survey of counterfeit parts (GAO-10-389 – March 2010), the Government Accountability Office found instances of substandard titanium in fighter jet engine mounts. Multiple services and government agencies purchased titanium for use on platforms that included F-15 engine mounts and F-22 and C-17 parts. The titanium was substandard and, if it had failed, could have caused casualties and property loss. The supplier has been charged with selling substandard titanium and repeatedly issuing fraudulent certifications that state the titanium passed testing standards. Profit is the primary incentive for counterfeiting, but pressures exerted by DoD for "fair price" as well as tight market conditions create additional incentives for dishonest operators to sell substandard parts.

Titanium faces a high risk of market turbulence as a result of volatile aerospace industry demand and long lead times to start up or close down production. In recent years, demand for titanium has equaled or exceeded supply, resulting in increasing prices. Although the United States is responsible for a significant proportion of global titanium alloy production, its share of titanium sponge production (the raw material that is used to produce the metal in mill forms) is far lower, and has declined significantly over the past decade. This decrease has created a dependence on foreign sources of titanium sponge with Kazakhstan supplying over 50 percent of U.S. imports. This dependence on foreign titanium sponge introduces a *high* dependence on global market supplies and further exposes the U.S. to turbulent pricing.

BACKGROUND

In 2013, the U.S. titanium manufacturing industry consists of three independent titanium mills that compete domestically and supply the U.S. market with different titanium alloys used in the aerospace industry, engine manufacturing, and the automotive sector. TIMET is the largest U.S producer of titanium sponge and melted and rolled products. TIMET also has mills in Europe and extensive facilities to recycle salvaged titanium. In December 2012, Precision Castpart Corp. (headquartered in Portland, Ore.) purchased TIMET, which is now part of a larger aerospace forgings producer.[2] ATI Allvac (Pittsburgh, Penn.) is the second-largest producer and manufacturer of sponge titanium. ATI Allvac supplies the aviation market as well as produces biomedical titanium products (e.g., hip replacement parts). RTI International Metals, Inc. (Pittsburgh, Penn.) is the third-largest titanium producer, and has

a strong focus on aircraft-grade titanium alloy sheet production and sales.

The American titanium industry emerged in the early 1950s at a time when the U.S. government was looking for an aircraft that could fly higher and faster than any other military aircraft. This need eventually led to the development of the SR-71 Blackbird in the 1960s by Lockheed.[3] The aircraft had to be lightweight yet resilient enough to carry a huge amount of fuel in order to travel long distances without refueling. Due to its strength and light weight, titanium was proposed as the solution. DoD invested substantial amounts of capital into fast-tracking the development of a commercial titanium sector.

The Specialty Metals Clause prohibits defense contractors from procuring melted titanium from non-U.S. suppliers. Due to cost and other pressures, however, the Pentagon and defense contractors have repeatedly tried to weaken or eliminate the domestic sourcing requirement for titanium.

Over the subsequent decades, titanium became viewed largely as an aerospace metal used primarily by the military (because of its high cost). Today, titanium is used in many commercial applications, ranging from aviation to sports equipment to artificial hips. It is also used in art and architecture; for example, titanium sheathes large portions of the Guggenheim Museum in Bilbao, Spain.

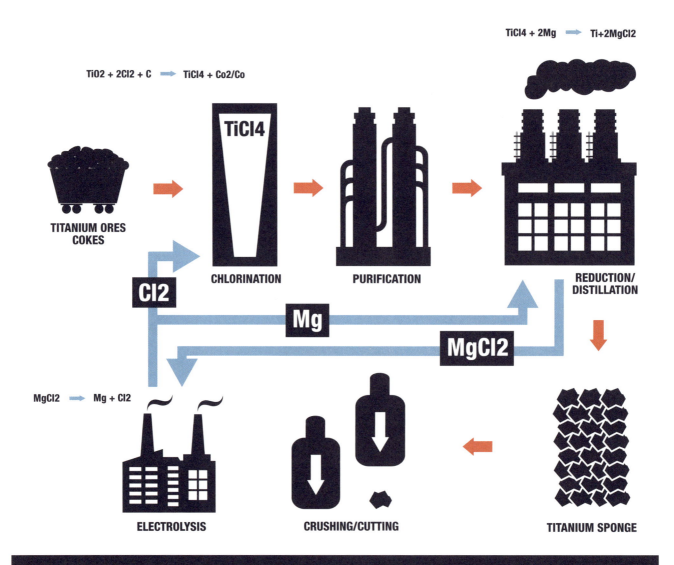

Figure 1. Producing Titanium (The Kroll Process)

$TiO_2 + 2Cl_2 + C \rightarrow TiCl_4 + Co_2/Co$

$TiCl_4 + 2Mg \rightarrow Ti + 2MgCl_2$

TiCl4

TITANIUM ORES COKES

CHLORINATION

PURIFICATION

REDUCTION/ DISTILLATION

Cl2

Mg

MgCl2

$MgCl_2 \rightarrow Mg + Cl_2$

ELECTROLYSIS

CRUSHING/CUTTING

TITANIUM SPONGE

Adapted from: Titaniumexposed.com, "Titanium Industries-One Metal, a Thousand Possibilities," 2013.

Although it is used in many different applications, titanium is still expensive to refine, process, and fabricate. It is five times more expensive than aluminum to refine and 10 times more expensive than aluminum to form into ingots and fabricate into finished products.[4] The processing time is elongated because titanium scrap or sponge must first be obtained before titanium can be turned into ingots (see Figure 1). The primary and secondary fabrication processes pose separate challenges, imposing a need for large capital investments as well as lengthy fabrication time (see Figure 1 on the Kroll process).

Sponge (the raw material of titanium) is the first stage of titanium ore extraction

Table 1: World Sponge Metal Producers and Main Exporters to the U.S.

Area	2011 Sponge Production (tons)	Percentage of Principal Importers to the U.S. (2007-2009)
U.S.	--------	
China	60,000	5%
Japan	56,000	37%
Kazakhstan	20,700	51%
Russia	40,000	
Ukraine	9,000	
World	186,000	

Source: U.S. Geological Survey, Mineral Commodity Summaries, January 2012. US sponge production in 2011 was not published because of concerns about releasing proprietary information.

and requires chlorinating titanium ores to obtain titanium tetrachloride. The Kroll process relies on fractional distillation, the final product of which is a porous mass of titanium metal mixed with byproducts, known as titanium sponge. These byproducts must be reduced with magnesium or sodium to form pure sponge, which is then sold globally. High-quality sponge is bought by the aerospace industry, and low-quality sponge is used in sporting equipment. Sponge fabrication is a chemical process, and since titanium sponge does not fall under the SMC, foreign-produced sponge can be used in U.S. defense goods.

The United States relies heavily on imported titanium sponge, which accounts for approximately 60 to 65 percent of its sponge titanium needs. Japan and Kazakhstan export the most titanium sponge to the United States. China and the Ukraine export less of the material to the United States (see Table 1). Imported sponge is subject to a trade tariff of 15 percent, and so is more expensive than domestically produced sponge.[5]

The second major phase of titanium production is the making of ingots, each of which can weigh several metric tons. Mixing and then melting titanium sponge with titanium scrap creates ingots, which often already include other metals and can be sold as alloys. This is the first fabrication process, and the production of ingots falls under the SMC. This process is

Figure 2: Four Stages of Titanium Production

Stage 1	Stage 2	Stage 3	Stage 4
Sand	Sponge	Ingots	Forged Bar

time-consuming because the sponge and scrap are melted multiple times in order to remove impurities and improve homogeneity. There are approximately 50 different grades of titanium, although only four grades are unalloyed, and therefore pure enough for commercial use as unalloyed titanium. All the other grades are alloys, by definition mixed with other metals to achieve certain properties and tailored to specific applications. Elements combined with titanium to achieve specific properties are aluminum, molybdenum, cobalt, zirconium, tin, and vanadium. The aerospace industry prefers titanium alloy Ti-6Al-4V, which is composed of six percent aluminum and four percent vanadium.[6]

The secondary fabrication process consists of manufacturing titanium parts and components from titanium mill products. The titanium mill forges and applies hot- or cold-rolling to produce plates, sheets, billets, and bars. Again, this part of the production phase poses its own technical trials, because titanium metal is very susceptible to oxidation. All welding must

be done in an inert atmosphere, such as argon or helium, in order to shield it from oxygen and nitrogen. The final phase is the fabrication of finished titanium products. Secondary fabrication consists of forging, extrusion, machining, and casting (see Figure 2). The type of fabrication depends on the final item and its application. Here, too, companies active in the titanium industry face technical complications. In contrast to steel or aluminum, titanium fabrication is expensive because its hardness grinds down tools and its reactivity slows down the machining process. In addition, it tends to generate substantial waste, which raises the price of the secondary fabrication process.

While there is a market for recycled titanium scrap, not all scrap can be reused. Each alloy must be recycled separately and segregated from other scrap titanium. Any dirt, oil, or lubricants render scrap titanium unusable until it is washed with soap. Nevertheless, in 2010, the U.S. titanium industry used around 29,000 tons of titanium scrap, most of which went to the steel industry to create ferrotitanium, which is widely used in stainless steel. Only 1,000 tons were used for super-alloys, purchased by the aerospace industry, compared to 10,000 tons by the steel industry.[7]

Titanium was first isolated and analyzed in the early 1800s, but it was not until the 1950s that the metal became commercially available. Most mined titanium is turned into titanium dioxide (TiO_2, or titanium concentrate), which yields white TiO_2 pigment. Because of its high refractive index[8] and resulting light-scattering ability, TiO_2 is the predominant white pigment for paints, paper, plastics, rubber, and various other materials.[9] Only about five percent of the world's annual production of titanium minerals becomes titanium metal. In 2012, the value of the market of titanium sponge was $388 million ($11.75/kg), while the market for titanium dioxide came to about $3.9 billion.[10] The leading producers of titanium concentrates include Australia, Canada, China, India, Norway, South Africa, and Ukraine.

In 2010, an estimated 75 percent of domestically produced titanium metal was used in aerospace applications.[11] The remaining 25 percent went into armor, chemical processing, marine, medical, power generation, sporting goods, and other non-aviation applications. Titanium's resistance to chlorine and acid makes it an important material in chemical processing. It is used for the various pumps, valves, and heat exchangers on the chemical production line. The oil refining industry employs titanium materials for condenser tubes because of corrosion resistance. This property also makes titanium useful for equipment used in the desalinization process.[12]

VULNERABILITIES TO THE TITANIUM SUPPLY CHAIN

While the commercial market for titanium is much bigger than defense-related production, Congress recognized titanium as a strategic material in the 1970s because certain key military weapons and other devices were wholly dependent on titanium alloys. This dependence generated concern about the risk of unexpected supply disruptions.

The world's supply of titanium is dominated by a handful of producers, and there is currently no substitute for titanium in most military and aerospace applications. The price of titanium is determined in large part by the health and vigor of the global aerospace industry (especially in the United States and Europe). In turn, commercial aviation is influenced by the business cycle, consumer spending, and economic growth. When the economy is buoyant, business and leisure travel goes up. When a recession looms, airlines cut back on orders for new airplanes. Titanium prices are therefore a measure of the health and outlook of the world economy. For domestic titanium producers, a price collapse creates havoc because they cannot quickly adjust production, and

Table 2: Production Share of World (Metal) Titanium Market, 2009

Producing Country	Percentage
United States	33.9%
China	21.4%
Russia	20.9%
Japan	17.1%
Germany	3.5%
Others	0.5%

Source: James Chater, The Shape Of Things To Come: Titanium Loses Its Sparkle But Retains Its Allure, Stainless Steel World (November 2009).

ultimately end up with large inventories. In fact, even sudden high prices are difficult to manage, because a titanium processing plant takes years to build and requires a capital investment of $300-$400 million.[13] The defense market (as opposed to the commercial market) historically has been more stable, which has helped buffer U.S. producers against the impact of a price collapse.

Although the defense portion of the domestic titanium market is comparatively modest in size, the military equipment chart at the beginning of this chapter shows that titanium and titanium alloys are indispensable across a wide range of military applications. A prolonged downturn in titanium availability would have major

ramifications for the U.S. military, and therefore would pose a threat to national security.

In spite of its widespread use in many aerospace applications, missile systems, and naval vessels, DoD must also reckon with the volatility of the titanium market. In 2011, the U.S. Army Aviation and Missile Research, Development & Engineering Center (AMRDEC) Industrial Base Group undertook a study to understand how to protect itself against the price fluctuations and availability of titanium alloys.[14]

What prompted the AMRDEC study was a rise in titanium prices, which began in 2004. As demand for titanium products rose after 2004, there was a shortage of titanium alloys. Three different U.S. Army helicopters (CH-47 Chinook, UH-60 Blackhawk, and AH-64 Apache helicopters) require over 150 different titanium parts. In this time period, titanium capacity simply was insufficient to produce spare parts in a timely fashion. For example, CH-47 aft blades were the number one CH-47 backordered item during 2006 and 2007. The U.S. Army undertook an analysis of the different pieces of the titanium supply chain. They found that titanium capacity and demand were influencing titanium lead-time and price, resulting in a significant delay in procuring necessary parts and leading to substantial spare part backorders. Ultimately, the shortage of spare parts compromised U.S. military readiness and affected DoD's warfighting capacity. No one predicted the sudden increase in demand for titanium, which resulted in a shortage of essential helicopter parts.[15] (Prices did not decrease until 2008, after many mills had expanded production capacity, and after the onset of the global economic crisis.)

VULNERABILITY: BUSINESS CYCLES AND TITANIUM PRICES

Pricing in the global titanium market corresponds to the classic model of an oligopoly. There are only a handful of international titanium producers, including three U.S. companies and one major Russian producer (VSMPO-Avisma [Verkhnaya Salda Metallurgical Production Association]), that are able to deliver high quality titanium for the aerospace industry. VSMPO-Avisma is the world's largest titanium metal producer, holding approximately 30 percent of the global market share in titanium shipments in 2007.[16] The U.S. company TIMET claimed 18 percent of the world's titanium industry and an eight percent market share of world titanium sponge production in 2005.

Table 2 shows the distribution of production in the world titanium market. China is the largest titanium-producing nation, but most of what it produces is for domestic use, and at this point China is unable to supply material that meets standards for high-end aerospace usage. TIMET is the largest U.S. titanium producer, followed closely by ATI. RTI is considerably smaller. VSMPO-Avisma is partially controlled by the Russian state-owned weapons exporter and aerospace manufacturer, Russian Technologies State Corporation, which has close ties with Russian President Vladimir Putin and his United Russia party.[17]

In short, titanium is a small market, much smaller than steel. While the global steel market approached 1,353.3 million metric tons in 2010, the global titanium market was approximately 180 thousand metric tons. Of course, the value of the world titanium market is high relative to its size because of the expense of processing and fabricating titanium metal. But the size of the titanium market means that it is subject to sporadic supply and demand shocks, which cause considerable headaches for both titanium consumers and producers. Titanium prices are usually set through long-term contracts between titanium mills and downstream consumers such as the aerospace industry. Long-term contracts guarantee stable revenues for producers and stable prices for the consumer. There are periods when prices are depressed, and there are periods when prices suddenly rise. Besides pre-arranged contracts, there is also a small spot market to accommodate excess demand or supply.

From 2005 to 2006, the titanium market experienced unusually tight conditions because there was a sudden acute shortage of sponge. Titanium users had to rely on the spot market to purchase additional sponge—at extremely high prices. The spot market price for sponge had historically hovered around $7/kg and rose abruptly to nearly $30/kg by the end of 2006 (see Figure 4).

The U.S levies a tariff on wrought titanium (titanium bars, sheets, etc.) and sponge. Until the early 2000s, the tariffs on sponge and finished titanium were not much of an issue. The end of the Cold War meant

a decline in military spending on military aircraft, and a decline in global demand for titanium.

Once prices skyrocketed in 2005, both the tariffs on titanium and U.S. domestic sourcing requirements provoked intense political feuding, centered on repealing these laws so that prices of titanium would fall. TIMET and a few very small niche producers possessed their own sponge

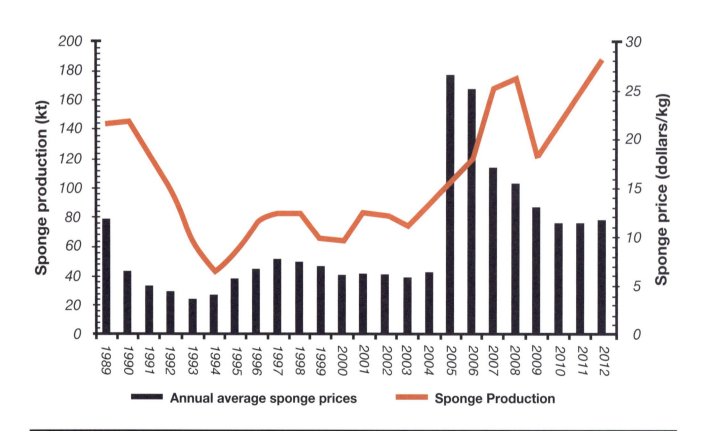

Figure 3: Titanium Sponge Prices and Production (1989 – 2012)

Source: metalprices.com "titanium" and Somi Seong, Obaid Younossi, Benjamin W. Goldsmith, Titanium Industrial Base, Price Trends, and Technology Initiatives (Santa Monica, CA: Rand Corporation-Project Air Force, 2009). http://www.rand.org/pubs/monographs/2009/RAND_MG789.pdf; Mineral Commodity Summaries, U.S. Geological Survey, January 2013, p. 172-173. http://minerals.usgs.gov/minerals/pubs/commodity/titanium/mcs-2013-titan.pdf.

production capabilities, while other tita-nium producers (and downstream users) had to rely on imported sponge. Whereas vertically integrated mills earned healthy profits when prices rose, intermediate pro-ducers scrambled to purchase sponge or ingots at exorbitantly high prices.

Figures 3 and 4 trace the price increases of sponge and ingots during the 2000s. The titanium market was relatively stable until after 2004, when demand exceeded supplies and sponge capacity. It took approximately two years before sponge production caught up with increased demand. Both figures also underscore the cyclical nature of the global titanium mar-ket. Prices dropped steeply from 2008 to 2009 as a result of the collapse of confi-dence and limited prospects for economic growth due to the global economic crisis.

The years 2005 to 2009 were exceptional, as prices skyrocketed and then abruptly fell. Nevertheless, this period can be viewed as a harbinger of future threats to the U.S. titanium sector and end-users of titanium alloys. It is therefore worthwhile to explore why prices rose so quickly, and question what these price fluctuations meant for the global titanium market.

INTERNATIONAL EVENTS AND DEMAND FOR TITANIUM

After the implosion of the Soviet Union in the early 1990s, U.S. defense spend-ing was cut, and orders for sophisticated fighter jets, tankers, naval ships, and other military equipment declined. In the chaos of post-Soviet Russia, the titanium industry was "sold" to private investors, who sub-sequently flooded the market with cheap titanium products and sponge, depressing global prices. U.S. producers struggled in a period of slack demand and a flood of cheap titanium. Only one U.S. firm, TIMET,

continued to produce sponge. (ATI revived sponge production in 2010 and built a new titanium sponge facility in Rowley, Utah, that achieved qualification for most criti-cal applications in 2012). A soft market in the 1990s resulted in an overall decline of sponge production, and world capacity fell by 22 percent between 1997 and 2004. In the United States, sponge production capacity dropped 70 percent between 1995 and 2004.[18]

In 2004, when it looked as if the market for titanium sponge might begin to recover, large producers were reluctant to reinvest in sponge production, because capacity expansion requires large capital invest-ment and involves long lead times. (New factory construction requires an invest-ment of between $300 and $400 million, and construction takes around three years.) Thus, the handful of large-scale titanium producers had to be convinced that the new prices, driven by strong demand, were not a temporary fluke.

What drove prices up after 2004? A con-fluence of discrete developments drove up demand for titanium while supply did not increase to the same extent. Rising fuel prices persuaded the civilian airline indus-try of the need to conserve fuel, which resulted in the development of lighter (and therefore more fuel-efficient) aircraft. Two of these planes, the Boeing 787 Dreamliner and the Airbus A380, use a significant amount of titanium.[19] This change led to a wave of new orders for these and other more fuel-efficient wide-body passenger aircraft.

At the same time, scrap titanium was in short supply, and the Defense Logistics Agency (DLA) had depleted its sponge stockpile in 2005, just when the global market experienced a shortage of sponge and scrap. The original justification for

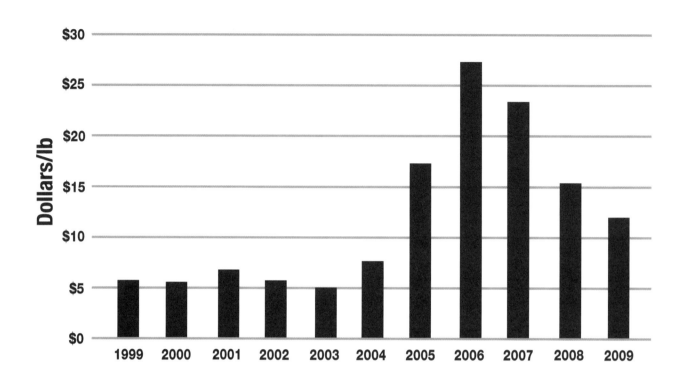

Figure 4: Titanium Ingot Prices (1996 - 2009)

Source: American Metals Market.

stockpiling titanium sponge was to guarantee adequate supplies during national emergencies and to support the domestic titanium industry. As geopolitical and security pressures lessened, in 1997 Congress approved of the disposal of the inventory.[20] What depleted the stockpile completely by 2005, however, was that major titanium producers had aggressively used the sponge stockpile as a substitute for scrap and ferrotitanium in the years prior. When prices began to rise, there was no inventory left at DLA.[21]

The other factor contributing to rising prices and supply shortages was China's emergence. Titanium is widely used in construction and industrial equipment, and China had embarked on a large-scale building boom.

Long-term contracts among stable suppliers and customers determine titanium prices. But the shortage of titanium sponge forced some downstream industries to purchase sponge on the small spot market, which met the extra demand but at extra high prices. Firms paid outsized prices for titanium sponge on the spot market, prompting calls for an overhaul of how the United States handled strategic materials, including titanium.

U.S. domestic sourcing requirements prohibited the use of titanium melted or smelted in non-NATO members in the production of defense items. Downstream users, with facilities supplying the commercial and defense markets, felt that they were burdened with extra logistics because they had to separate titanium melted domestically from a product melted abroad in a non-NATO country. End-users could import and rely on titanium from Germany, the U.K., France, or Italy, but the SMC prohibited them from using Chinese or Russian titanium in the production of military end-items.

Prices have declined since 2008, as several large producers have increased sponge capacity. China, afraid of running out of sponge, accelerated the expansion of its titanium sponge production dramatically by increasing output by 500 percent in three years (2005 to 2008).[22] While Chinese (and U.S.) producers increased sponge production, in late 2008 the global financial crisis and delays in the production of high-titanium content aircraft (such as the Airbus A380 and Boeing 787) caused a sharp decline in titanium demand. This drop caused a steep decline in prices. At the same time, new sponge plants in the United States and Japan, initiated in the earlier boom years, were coming online. In both 2009 and 2010, titanium sponge capacity exceeded demand, and producers delayed further expansions, idled plants, and (in China) closed smaller uneconomical plants.

Many analysts forecast a buoyant market for titanium metal in coming years. With high oil prices at $100 or more per barrel, the commercial aerospace sector continues to seek improvements in fuel efficiency. Aircraft manufacturers rely on titanium to lower the airframe's weight while preserving its strength. Demand will continue to be strong thanks to the growing need for fuel-efficient aircraft in commercial aviation. In addition, titanium also has corrosion-resistant applications, and is widely used in chemical processing plants, water desalinization units, and nuclear and fossil-fuel power stations. Finally, as Western societies age, there is growing demand for joint replacement devices. In the United States alone, 8 to 10 million joint replacement surgeries take place every year.[23] For all of these reasons, most observers believe that demand for titanium will be strong the next few years.

As global sponge production is sufficient, titanium producers are increasingly looking to build an integrated supply chain by investing in downstream activities such as forging and fabrication. The aerospace industry expects quality control in the form of an integrated production line that melts, machines, and forges titanium-made parts. In response, American mills have turned away from increasing their sponge production to deploy capital in fabrication and machining capacity. Of course, as demand for titanium parts rises, there may once again be a squeeze on sponge capacity, with a significant corresponding price increase. Therefore, the titanium market will likely continue to be volatile, as the increased demand for the metal may exceed global sponge production. This gap will not be a concern for the next year or two, but eventually sponge capacity may not be sufficient to meet increasing production rates of the A350 and the 787 Dreamliner.[24]

POLICY RESPONSES AND MITIGATING THE RISKS

Beginning in the 1950s, DoD realized the strategic importance of titanium and supported commercialization efforts. Throughout the Cold War period, titanium was considered a strategic material by the U.S. government, and the Defense National Stockpile Center (DNSC) maintained a large stockpile of titanium sponge. In 1997 Congress authorized the disposal of the DNSC stockpile which at that time still held 33,000 tons of titanium sponge. After 1997 the stockpile shrank quickly, and by 2005 the supply had been completely sold off.[25]

In addition, in order to preserve and protect the U.S. defense industrial base and to ensure preparedness during periods of adversity and war, Congress passed a set of domestic source restrictions that became known as the Berry Amendment. In 1973, specialty metals, including titanium, were added to the Berry Amendment; however, the FY2007 NDAA separated the SMC from the Berry Amendment. The SMC prohibits DoD from acquiring end-units or components for aircraft; missile and space systems; ships; tank and automotive items; weapon systems; or ammunition, unless these items have been manufactured with specialty metals that have been melted or produced in the United States or other qualifying countries.[26]

In the 1990s, budget constraints and the end of the Cold War put pressure on DoD to reduce defense spending. On the one hand, it was agreed that the United States had to preserve a dynamic titanium industry to meet defense needs. On the other,

it was also decided that DoD should seek savings by procuring Commercial-off-the-Shelf (COTS) articles, products, and end items.

Therefore, the SMC was revised to permit DoD to purchase COTS articles, electronic articles, and items containing small amounts of non-compliant specialty metals. Revisions also allowed producers of commercially derivative defense articles to treat domestic and foreign specialty metals as fungible materials, which meant that commercial and defense articles could be produced on the same production line without the need to trace the small amounts of metal used in each article. The rationale was to promote efficiency throughout the defense supply chain. Domestic titanium producers supported those changes to accommodate the logistical concerns of their principal customers.

Once prices began to rise in 2005, DoD and defense contractors fought for special waivers that would allow original equipment manufacturers to buy titanium from foreign sources other than those countries with which the U.S. has signed Memoranda of Understanding (MOU), guaranteeing equal treatment of their suppliers in defense procurement. These countries—NATO members, as well as Australia, Egypt, Israel, Sweden, and Switzerland—can supply titanium to DoD. The only titanium-producing countries that do not fall under an MOU agreement (non-qualifying countries) are Russia and China.

According to the Berry Amendment, and subsequently the SMC, under certain conditions the Secretary of Defense has the authority to waive the requirement to buy domestically. Among the exceptions are situations when DoD or the military departments determine that there is not satisfactory quality or sufficient quantity of a specialty metal available at reasonable market prices, and that as a result DoD should be permitted to procure specialty metals from wherever they are available. From 2005 to 2007, downstream users claimed that the supply shortages and high prices on the spot market justified such a waiver. Subcontractors at the second, third, and fourth tiers complained that they were unable to comply with domestic sourcing requirements for specialty metals because they provide a mixture of final products, most of which are destined for the commercial, non-military market. Some defense contractors and DoD officials who are focused primarily on short-term cost considerations view domestic sourcing restrictions as costly, and counter to the best interests of the U.S. and national security. In this view, such restrictions undermine free market competition and may disincentivize firms from investing in the latest technology and remaining competitive. However, proponents believe that U.S. domestic sourcing requirements for titanium protect an important part of the defense industrial base that is subject to unpredictable and volatile fluctuations.[27] Moreover, there is some question as to whether purchasers were merely attempting to use the political process to achieve price reductions from their titanium suppliers.

The titanium industry, and in particular the three principal U.S. producers, have resisted repeated efforts to dilute and weaken the SMC by permitting sourcing from suppliers in Russia or China. Russia is a large supplier of titanium, and the Russian-owned VSMPO-Avisma already supplies 40 percent of the global commercial titanium market. VSMPO-Avisma was established by the Soviet regime, and Russian President Boris Yeltsin quickly privatized it in the early 1990s. In 2013, it

is once again state-owned. President Putin re-nationalized the company because the Russian state agencies wished to control access to weapons-grade titanium. In the 1990s, VSMPO-Avisma flooded the world market with sponge, leading to a collapse of prices. At the time, U.S. titanium industry leaders pointed out that it would be very risky for the U.S. defense establishment to depend on imports from Russia, which in the past has manipulated commercial contracts and relationships for political ends.[28] Above all, reliance on imports, whether Russian or not, would mean that the defense industry might not have access to sufficient titanium alloys during periods of growing global demand, since suppliers may deliver titanium metal or ingots to domestic or preferred customers first.

No waivers or exemptions were granted at that point, but in 2008 DoD proposed to amend the SMC to address statutory restrictions on the acquisition of specialty metals not melted or produced in the United States or by MOU. DoD argued that the presence of more suppliers would reduce the domestic price of titanium ingots. By that time, the price crisis had passed; titanium users had decided not to seek changes in the law.

Nonetheless, DoD reopened the question of the SMC and suggested again that it should be modified to open up the defense portion of the titanium market to other global players. Representatives of the titanium industry contested this move and pointed out that domestic procurement restrictions have no bearing on the cost of titanium ingots, because raw titanium comes in sponge form (see Figure 1) and titanium sponge is produced through a chemical (Kroll) process not covered by the domestic source restrictions of the SMC. Sponge is later melted and transformed into alloys.

Titanium sponge is sold at world market prices irrespective of national origin.

Opening the domestic industry to foreign competition from relatively low-cost producers would undermine the ability of U.S. mills to continue to supply the domestic market, and could put the U.S. economy at risk of losing its capacity to manufacture titanium metals and alloys. This loss would have clear and undesirable national security implications.

The risk of having low-cost producers corner the market by selling titanium sponge and ingots at exceptionally low prices is not merely theoretical. China has joined the select group of countries that produce titanium, and Chinese producers could make the global titanium market more competitive by selling large volumes of sponge on the world market. The enormous increase in Chinese capacity has had the effect of stabilizing prices in spite of the fact that the titanium market recovered from its steep fall after 2008. But if Chinese firms produced a vast surplus, nothing would prevent them from dumping cheap titanium sponge and metal onto the world market—thereby destroying the profit margins and possible viability of U.S. producers.[29] In fact, such a scenario could potentially result in a repeat of what happened to the mining and refining of REEs, which disappeared in the United States after the deluge of Chinese-mined REEs at prices that no U.S. company could match.

The objective of the SMC is to offset painful and potentially sudden contractions in global demand and collapsing prices by guaranteeing steady demand from defense contracts. Government titanium purchases for defense items soften the impact of market gyrations and ensure that the capital-intensive domestic titanium sector can survive and continue to compete while investing in innovation. Defense procurements and the SMC serve as a safety net during difficult times. While the commercial market is cyclical and unpredictable, the defense sector is typically more stable. Without the existing legal infrastructure, the U.S. titanium supply could be in serious jeopardy. Even though defense sales represent only 10 percent of U.S. titanium output during normal times, they generate sufficient demand such that, during a severe downward cycle, defense orders help preserve the sector. During a downturn, the defense segment of the market can represent as much as 40 percent of sales, supplying enough orders so that the U.S. titanium industry can survive.[30]

Nevertheless, DoD continues to demonstrate a desire to weaken the SMC because of an understandable interest in cost savings, and because U.S. defense contractors already rely on globalized supply chains for many components.[31] While many parts are in fact already sourced from around the globe, the case for retaining and enforcing the SMC for titanium is compelling, especially in light of chronically turbulent markets. Other governments pursue policies favoring their domestic industry, and when a sector is so critical to U.S. security, it is logical that the U.S. government would take steps to protect its interests. Further weakening of the SMC, by contrast, could lead to the use of U.S. defense dollars to support the expansion of the titanium industries in states such as China and Russia, which has the largest and most competitive producer, VSMPO-Avisma. Repealing or weakening the SMC would mean that U.S. defense industrial base firms would purchase titanium alloys from VSMPO-Avisma for U.S. weapons systems and fighter jets.[32]

U.S. national security could be at risk if a Russian titanium mill, controlled by a Russian state-owned weapon exporting company, became the major supplier of defense-related titanium. Not only would the United State lose access to a reliable source of high quality titanium alloys, it would be dependent on the politically driven decisions of a Russian state-owned company. In addition, the United States would undercut the technological knowledge base to design and manufacture complex alloys.

The following are recommendations to address the risks to titanium in the defense supply chain:

Congress should preserve the Specialty Metals Clause. The SMC is a tool critical to the preservation of a domestic titanium industry exposed to unpredictable global market forces beyond its control. Without the SMC, nothing would prevent contractors and suppliers from procuring titanium products from foreign sources, thereby depriving the domestic sector of the positive and stabilizing impact of steady DoD procurements. In the long term, it would compromise the U.S. defense posture and pull expertise and technologies offshore. The complex rules aimed at restricting the use of non-domestic "specialty metals" in the defense acquisition process have helped to preserve a vibrant domestic titanium sector, and have also protected the U.S. defense industry from becoming overly dependent on foreign supplies.

The U.S. commercial titanium industry and DoD should meet regularly to discuss anticipated forecasts and needs. These meetings should occur at least once a year to review market conditions and possible DoD orders.

Congress and DoD should continue to explore alternative titanium fabrication processes to reduce the costs of creating titanium and titanium alloys. In the past, DoD and Congress have provided seed money to encourage the private sector to adopt new technologies or to invest in developing new technologies. Two novel approaches hold promise in bringing down the price of titanium products. First, there are Titanium Metal Matrix Composites (TiMMCs), which offer material properties that enable aircraft designers to engineer components that are stronger, lighter, and more durable than existing steel and pure titanium components. According to defense officials, these improvements can expand U.S. air superiority margins over opposition forces by increasing lethality for U.S. munitions, increasing survivability for the warfighter, and ultimately increasing mission success rates. Congress has provided Title III funding to support the expansion of the domestic production capacity of TiMMCs.

Second, Congress and DoD also have provided funds to spur interest in developing a new process to create titanium powder. The new technology is based on non-melt technology. (Patented under the name of Armstrong Process, this new technology would replace the Kroll process.) The new technology can produce commercially pure titanium powder directly from titanium tetrachloride (without first producing sponge titanium) by injecting it into a stream of liquid sodium. Alloyed titanium powders can be created in the same process by injecting chlorides of the alloying elements. Several producers are experimenting with new technologies that bypass the Kroll process, and thus bypass the chemical fabrication of titanium sponge.[33] Congress and DoD should continue to support efforts to find alternatives to the current process of turning "black sand" into titanium.

Finally, to reduce dependence on imports, Congress should provide incentives for further expansion of high quality sponge production in the U.S. Titanium sponge is part of a chemical process to extract the raw material of titanium. Currently, the U.S. imports a large volume of sponge from countries such as Kazakhstan and Russia. Together with the recycling of scrap titanium, the United States may be able to greatly decrease its dependence on imports.

CONCLUSION

The largest risk to the titanium supply chain is the condition of the global market, which swings from periods of high demand to periods of overcapacity and falling prices. A wide variety of modern weapons cannot be built without titanium, because there is neither a substitute alloy nor a metal with properties comparable to titanium. The U.S. titanium industry is dynamic, vibrant, and globally competitive. However, the domestic industry only is able to compete and invest in new products and new technologies to fabricate these products thanks to the SMC. The SMC has provided a safety net during periods of turmoil and has encouraged investment in new products and technologies. However, the SMC continues to generate political controversy, pitting the U.S. titanium industry against defense suppliers who are less concerned about who supplies this critical metal. Russia is the largest producer of commercial titanium, and the main Russian producer has close ties to the Russian state. It would greatly compromise U.S. security if DoD were to rely on specialized titanium products from a partially state-owned Russian supplier. In the critical defense titanium sector, the United States needs to ensure that its national security interests are protected.

ENDNOTES

1   This figure is an estimate, because not all U.S. titanium companies publish a breakdown of their sale figures. The estimate is based on sales by TIMET, the largest U.S producer of titanium sponge and melted and rolled products.

Somi Seong, Obaid Younossi, Benjamin W. Goldsmith , *Titanium: Industrial Base, Price Trends, and Technology Initiatives* (Santa Monica: Rand Corporation, 2009), 27. http://www.rand.org/content/dam/rand/pubs/monographs/2009/RAND_MG789.pdf.

Specialty Metals Law: Promoting Competitive Excellence in a Secure Defense Industrial Base, Unpublished manuscript drafted by Allegheny Technologies Inc. Electron Energy Corporation: RTI International Metals Titanium Metals Corporation (TIMET), 9.

2   In addition, there are 11 companies producing different, specialized titanium ingots, and 30 companies producing titanium forgings, rolled products, and castings. Precision Castparts Corp. (PCC) manufactures complex metal components and products. It serves the aerospace, power, and general industrial markets. PCC bought TIMET in order to create an integrated aerospace supply chain. PCC was TIMET's largest customer and is also a major supplier to the aerospace industry (especially to Boeing). Frank Haflich, "The Titanium Industry Undergoes Massive Changes into 2013," *American MetalMarket*, January 2013. www.amm.com/Magazine/3155196/Features/Titanium-industry-undergoes-massive-changes-into-2013.html.

3   The Lockheed SR-71 Blackbird was built using 85 percent titanium and 15 percent composite materials, a revolutionary concept at the time. The airframe was constructed of titanium, which resisted the high temperatures generated by flight at speeds of over Mach 3. At that time, the United States did not yet have ample supply of high-grade titanium, and Lockheed imported the titanium from the then Soviet Union without revealing how it was planning to use it. James. Chater, "High Expectations: The Aerospace Driving Titanium Sales and Innovation," *Stainless Steel World*, November 2010, 43. http://www.kcicms.com/pdf/factfiles/titanium/ssw0911_titanium_kci.pdf?resourceId=121.

4   Brian E. Hurless and Sam Froes, "Lowering the Costs of Titanium," *The AMPTIAC Quarterly* 6, no. 2, 4.

5   The United States imposes a 15 percent tariff on wrought titanium (billet, bar, sheet, strip, plate, and tubing) and sponge. The U.S. government occasionally has granted preferential trade status to certain titanium products imported from particular countries (notably wrought titanium products from Russia, which carried no U.S. import duties from approximately 1993 until 2004). These preferences were removed in 2004 as the result of a petition filed by domestic producers. Titanium's strategic importance was a major factor in the decision to remove these preferential benefits.

6   Aircraft part suppliers make use of commercially pure (unalloyed) titanium grades as well as titanium alloys. The most commonly used alloy, Ti-6Al-4V, is capable of withstanding temperatures up to 400°C and is used in aircraft ducts, airframes, landing gear, and space vehicles. Other grades 5 and 23 (Ti-6Al-4V ELI) offer improved ductility for cryogenic service and are used on airframes and in space vehicles. Grades 5 and 23 are examples of alpha-beta alloys, which offer a compromise between the virtues of alpha alloys (lighter weight, higher creep resistance) and those of beta alloys (higher strength, lower creep resistance). James Chater, "Titanium and Aerospace: Two Interdependent Industries," *Stainless Steel World*, November 2009, 34-5. http://www.kcicms.com/pdf/factfiles/titanium/ssw0911_titanium_kci.pdf?resourceId=121.

7   U.S. Geological Survey, *Mineral Commodity Summaries 2011* (Reston, VA: USGS, 2012), 173. http://files.eesi.org/usgs_commodities_2011.pdf.

8   Refractive index is a measure of the bending of a ray of light when passing from one medium into another, a property of a material that changes the speed of light, computed as the ratio of the speed of light in a vacuum to the speed of light through the material. When light travels at an angle between two different materials, their refractive indices determine the angle of transmission (refraction) of the light beam. "Definition of Refractive Index," *PC Magazine*, http://www.pcmag.com/encyclopedia_term/0,2542,t=refractive+index&i=50348,00.asp.

9   U.S. Geological Survey, *Titanium Statistical Compendium* (Reston, VA: USGS, 2011).

10 U.S. Geological Survey, *Mineral Commodity Summaries 2012* (Reston, VA: USGS, 2013), 172. http://minerals.usgs.gov/minerals/pubs/commodity/titanium/mcs-2013-titan.pdf.

11 Globally, 50 percent of titanium (alloys) goes to non-aerospace industrial equipment sector, consisting of various manufacturing industries, while aerospace accounts for 37 percent and the military six percent of the world market. In the United States, the aerospace industry accounted for 60 to 75 percent of the titanium sponge consumption in the 2000s, and 28 percent went to military armor, marine, medical, power generation, and other industrial sectors. "Global Supply of Titanium is Forecast to Increase," Metall-web.de, October 22, 2010. http://www.metall-web.de/home/wirtschaft/news-detail/news/2/1287698400global-supply-of-titanium-is-forecast-to-increase.

12 Several countries in the Middle East, particularly Saudi Arabia, are seeking to establish manufacturing sectors. This development will require fresh water and power sources. Desalination plants will provide the fresh water to run power for the new manufacturing sector.

13 National Defense University, Industrial College of the Armed Forces, *Strategic Materials Industry* (Washington, DC: National Defense University, Spring 2011). http://www.ndu.edu/icaf/programs/academic/industry/reports/2011/pdf/icaf-is-report-strategic-materials-2011.pdf.

14 Department of Defense, Office of Under Secretary of Defense Acquisition, Technology, and Logistics, *Annual Industrial Capabilities Report To Congress* (September 2011). http://www.acq.osd.mil/mibp/docs/annual_ind_cap_rpt_to_congress-2011.pdf.

15 Ibid.

16 VSMPO-Avisma provides 40 percent of the titanium used by Boeing, 80 percent of that used by Airbus, as well as supplying such companies as Goodrich, Pratt & Whitney, Rolls-Royce, SNECMA Moteurs, and General Electric. Up to 80 percent of VSMPO's titanium output is exported.

17 Russian Technologies is the world's second largest aircraft-maker.

18 Somi Seong, Obaid Younossi, and Benjamin W. Goldsmith. *Titanium Industrial Base, Price Trends, and Technology Initiatives* (Santa Monica, CA: Rand Corporation-Project Air Force, 2009). http://www.rand.org/pubs/monographs/2009/RAND_MG789.pdf.

19 James Chater, "Titanium and Aerospace: Two Interdependent Industries," *Stainless Steel World*, November 2009, 34-5. http://www.kcicms.com/pdf/factfiles/titanium/ssw0911_titanium_kci.pdf?resourceId=121.

20 National Defense Authorization Act for Fiscal Year 1998 Section 3304, "Disposal of Titanium

Sponge in National Defense Stockpile."

See also, *Proposed Reconfiguration of the National Defense Stockpile*: Hearing Before the Readiness Subcommittee of the Committee of Armed Services, House of Representatives, 111th Cong. 7 (2009). http://www.gpo.gov/fdsys/pkg/CHRG-111hhrg52723/pdf/CHRG-111hhrg52723.pdf.

21 Somi Seong, Obaid Younossi, and Benjamin W. Goldsmith. *Titanium Industrial Base, Price Trends, and Technology Initiatives* (Santa Monica, CA: Rand Corporation-Project Air Force, 2009), 39. http://www.rand.org/pubs/monographs/2009/RAND_MG789.pdf.

22 "Wirtschaft, Leichtmetalle," Metall-web.de, October 22, 2010. http://www.metall-web.de/home/wirtschaft/news-detail/news/2/1287698400global-supply-of-titanium-is-forecast-to-increase/.

23 "Titanium Gets a Lift from Evolving Aviation Market," *American Metal Market* magazine, October 2012. http://www.amm.com/Magazine/3096924/Magazine-Archives/Titanium-gets-a-lift-from-evolving-aviation-market.html.

24 "Titanium Producers Look to Diversify Downstream," *American Metal Market* magazine, October 2012. http://www.amm.com/Magazine/3096936/Magazine-Archives/Titanium-producers-look-to-diversify-downstream.html.

25 Somi Seong, Obaid Younossi, and Benjamin W. Goldsmith, *Titanium: Industrial Base, Price Trends, and Technology Initiatives*, (Santa Monica, CA: Rand Corporation-Project Air Force, 2009), 38. http://www.rand.org/pubs/monographs/2009/RAND_MG789.pdf.

26 Specialty metals covered by this provision include certain types of cobalt, nickel, steel, titanium, titanium alloys, zirconium, and zirconium base alloys.

27 For a summary of these views, see Valerie Bailey Grasso, *The Berry Amendment: Requiring Defense Procurement to Come from Domestic Sources* (Washington, D.C.: Congressional Research Service RL31236, January 13, 2012). http://digital.library.unt.edu/ark:/67531/metadc84034/.

28 Jakob Hedenskog and Robert Larsson, "Russian Leverage on the CIS and the Baltic States," *Defence Analysis* FOI-R--2280—SE (Sweden: June 2007).

Marshall Goldman, *Petrostate: Putin, Power, and the New Russia* (New York: Oxford University Press, 2008).

29 National Defense University, Industrial College of the Armed Forces, *Strategic Materials Industry* (Washington, D.C.: National Defense University, Spring 2007), 8. http://www.ndu.edu/icaf/programs/academic/industry/reports/2007/pdf/icaf-is-report-strategic-materials-2007.pdf.

30 *Specialty Metals Law: Promoting Competitive Excellence in a Secure Defense Industrial Base*, Unpublished manuscript drafted by Allegheny Technologies Inc. Electron Energy Corporation: RTI International Metals Titanium Metals Corporation (TIMET), 9.

31 Scott Hamilton, "Outsourcing U.S. Defense: National Security Implications," *National Defense Magazine*, January 2011. http://www.nationaldefensemagazine.org/archive/2011 /January/Pages/OutsourcingUS DefenseNationalSecurityImplications.aspx.

32 Lionel Beehner, "Russia-Iran Arms Trade," Council on Foreign Relations website, November 1, 2006. http://www.cfr.org/iran/russia-iran-arms-trade/p11869.

33 U.S. Geological Survey, *Mineral Commodity Summaries 2011* (Reston, VA: USGS, 2012), 173. http://files.eesi.org/usgs_commodities_2011.pdf.

CHAPTER 5 • HIGH-TECH MAGNETS

EXECUTIVE SUMMARY

High-tech magnets made from rare earth elements (REEs) increasingly are used in advanced weapons systems and military vehicles. High-tech magnets are uniquely able to maintain their magnetic properties in extreme heat and perform other vital functions to enable many high-tech, modern weapons systems. There is currently no domestic neodymium iron boron (NdFeB) magnet producer, and only one domestic Samarium-Cobalt (SmCo) magnet producer. China currently controls 80 percent of global REE production and is still one of the main sites for fabricating high-tech metal alloys using REEs.

Although rare earth magnet (REM) technology was developed in the United States, 60 percent of SmCo magnets and 75 percent of NdFeB magnets are currently fabricated in China. Because SmCo magnets contain cobalt, they fall under the Specialty Metals Clause (SMC), which requires the Department of Defense (DoD) to procure them from domestic sources. This law has preserved the existence of a sole U.S. producer of SmCo magnets. NdFeB magnets do not currently fall under the SMC, and they may be acquired for defense applications from any country. No domestic NdFeB manufacturer remains.

An initial assessment indicates that access to REEs is the principal bottleneck threatening the supply of high-tech magnets. However, given that Molycorp reopened the Mountain Pass Mine—the largest rare earth mine outside China—in California in 2012, the United States seems simply to lack the engineering skill to turn crude REE oxides into metal and fabricated products. U.S. companies do not have the capacity to process the RE (rare earth) oxides into a metal and then make the metal into magnets. Thus, mining REE oxides is only the first step in recovering self-sufficiency, and must be paired with the reintroduction of metal-making engineering knowledge and fabrication skills.

In 2010, the global price of REEs shot up, which motivated miners in the United States and Australia to start mining RE oxides again. However, as prices crashed in 2012, each of the 17 REEs has assumed a different market value. "Light" rare earths are abundant, and their prices have dropped by 90 percent. "Heavy" rare earths are in fact scarce, and it is much harder to extract the oxides and make

HIGH-TECH MAGNETS
A FORCE IN U.S. NATIONAL SECURITY

MANUFACTURING SECURITY

Rare earth magnets are a major component used in many national security and U.S. military end-items and platforms

MILITARY HUMVEE

APACHE HELICOPTER

JOINT STRIKE FIGHTER

GUIDED MISSILE DESTROYER

VIRGINIA-CLASS SUBMARINE

UNDERSTANDING SUPPLY

The U.S. has no domestic production of Neodymium-Iron-Boron magnets

 **0 DOMESTIC** PRODUCTION FACILITIES

SUPPLY VULNERABILITY

China is the leading supplier of high-tech magnets essential to U.S. security

 **CHINA MAKES** MOST OF THE WORLD'S **HIGH-TECH MAGNETS**

SUPPLY TRENDS

Trends in U.S. production of high-tech magnets

DOMESTIC SUPPLY

DOMESTIC DEMAND

HEAVY VS. LIGHT RARE EARTHS

Heavy Rare Earths: Scarce and Hard to Extract
Light Rare Earths: Abundant

 **VS.**
HEAVY **LIGHT**

MITIGATING RISKS

Avoiding uncertainties in the high-tech magnets industry

IMPROVED RECYCLING

EFFECTIVE STOCKPILING

RESUME RARE EARTH MINING

MILITARY EQUIPMENT CHART
SELECTED DEFENSE USES OF HIGH-TECH MAGNETS

DEPARTMENT	WEAPON SYSTEMS	PLATFORMS	OTHER SYSTEMS
ARMY	■ AGM-114 HELLFIRE air-to-surface missile ■ Joint Direct Attack Munition (JDAM) precision guidance kit ■ PAC-3 Anti-ballistic missile ■ M109 Paladin Howitzer	■ AH-64 Apache helicopter ■ M2 Bradley fighting vehicle ■ M1 Abrams main battle tank ■ Stryker fighting vehicle	■ Laser rangefinder ■ Laser target designators ■ Satellite communication
MARINE CORPS	■ AIM-9 Sidewinder air-to-air missile ■ AIM-120 advanced medium-range air-to-air missile (AMRAAM) ■ Harpoon anti-ship missile ■ Joint Direct Attack Munition (JDAM) precision guidance kit	■ F-18 Hornet fighter ■ F-35 Joint Strike fighter ■ M1 Abrams main battle tank	■ Towed decoys
NAVY	■ AIM-9 Sidewinder air-to-air missile ■ AIM-120 advanced medium-range air-to-air missile (AMRAAM) ■ Harpoon anti-ship missile ■ Joint Direct Attack Munition (JDAM) precision guidance kit ■ Trident D5 submarine-launched ballistic missile	■ Arleigh Burke-class destroyer ■ Nimitz-class nuclear-powered aircraft carrier ■ F-18 Hornet fighter ■ F-35 Joint Strike fighter ■ Littoral combat ship (LCS) ■ Unmanned underwater vehicle (UUV) ■ SSN-774 Virginia-class nuclear-powered attack submarine	■ Aegis radar ■ Firefinder anti-rocket/anti-artillery radar ■ Towed decoys ■ Underwater mine detection
AIR FORCE	■ AIM-9 Sidewinder air-to-air missile ■ AIM-120 advanced medium-range air-to-air missile (AMRAAM) ■ Harpoon anti-ship missile ■ AGM-114 HELLFIRE air-to-surface missile ■ Minuteman-III intercontinental ballistic missile	■ B-52 bomber ■ F-15 Eagle fighter ■ F-16 Falcon fighter ■ F-22 Raptor fighter ■ F-35 Joint Strike fighter ■ MQ-1B Predator drone	■ Towed decoys

the product sufficiently pure for industrial use. U.S. companies do not mine heavy REEs and thus the U.S. defense industry is wholly reliant on Chinese suppliers.

An appropriate mitigation strategy would be to support the mining of heavy REEs, build up a stockpile of REEs, and provide incentives for the few remaining domestic magnetic technology manufacturers to acquire and master the technology of fabricating NdFeB magnets. In the meantime, efforts should be made to develop long-term alternatives to high-tech magnets for defense-critical applications while pursuing shorter-term policies to protect supply through magnet recycling and reuse.

INTRODUCTION

U.S. military supremacy depends on a defense industrial base that can constantly innovate and produce a wide variety of advanced military systems. To sustain a technological edge, U.S. companies must have access to sophisticated inputs with unique properties, many of which may be scarce. High-tech magnets are among these critical inputs, and are found in many of the most important U.S. military capabilities.

Unfortunately, the U.S. magnetic material sector is all but defunct, emblematic of the challenges faced by the U.S. defense industrial base as a whole. The defense industrial base lacks domestic magnetic material manufacturing capacity as well as rare-earths-oxide extraction, processing, refining, and fabrication capacity. This decreased capacity resulted from a lack of awareness among policy-makers and defense planners about how small, seemingly commonplace components such as high-tech magnets require valuable

know-how and advanced manufacturing technologies that warranted measures to sustain them.

The following chapter investigates the challenges the United States faces in securing reliable access to high-tech magnets, also known as REMs. Some REMs consist of inputs such as neodymium or samarium and are combined with iron or cobalt to create small, powerful, and heat-resistant magnets.[1] Currently, dozens of commercial retailers sell high-tech magnets; however these magnets are not manufactured in the United States. They are imported, mostly from China, in large blocks, and then cut according to the customer's specifications. Some of the magnets used in advanced military weapons are made to precise specifications, but they too are fabricated abroad.[2] High-tech magnet manufacturing is an example of a capital-intensive industrial sector driven by research and development (R&D) that has progressively shifted operations from more advanced economies like the United States to smaller, lower cost overseas producers. This shift has led to a corresponding loss of domestic talent, jobs, and future innovation possibilities in the United States.

U.S. scientists discovered and were the first to mass-produce REMs. In 1983, scientists at General Motors (GM) and Hitachi (then Sumitomo Special Metals) simultaneously developed NdFeB magnets with exceptional magnetic properties; the two companies eventually agreed to split the intellectual property rights to the discovery. In 1986, GM set up a separate magnet manufacturer, Magnequench, to supply the automobile industry with heat-resistant magnets. In 1995, as GM divested itself of many of its subsidiaries, a consortium of two Chinese groups and a U.S. investment firm purchased Magnequench.[3] At the time, union leaders sought guarantees from GM that the Anderson plant would stay open for 10 years and that GM would attempt to convince the buyers of the plant not to close it down. Suspecting that the consortium would not abide GM's requests, union leaders approached the Committee on Foreign Investment in the United States (CFIUS) and asked its members to include a provision that would guarantee the Anderson plant would stay open for 10 years. CFIUS reviewed the purchase, and extracted a promise from the consortium that they would keep the plant running for at least 10 years.[4] In August 2001, it was announced that the plant would be closed, following claims that it was unprofitable; the production line had been duplicated in China.[5] By 2005, the last U.S.-based NdFeB magnet producer closed its doors. As a result, the U.S. economy and defense industry are mostly dependent on the supply of RE oxides from foreign (Chinese) mines and dependent on foreign (Chinese and Japanese) NdFeB magnet manufacturers. With the disappearance of the U.S. magnet industry and the end of domestic ore extraction, U.S. leadership in patents, innovation, and manufacturing technology in this area also has declined. The closure of domestic magnet producers and mines was partly the result of a successful strategy pursued by China to become the dominant producer of REEs and REMs.

High-tech magnets share many of the risk factors and supply vulnerabilities discussed in the chapters on specialty (minor) metals: reliance on a single geographic source, the risk of supply restrictions, lack of domestic capacity, and an absence of substitute elements.

"In 1992, during his visit to Bayan Obo (China's largest rare earth mine), Chinese leader Deng Xiaoping declared, 'There is oil in the Middle East; there is rare earth in China.' Seven years later Chinese president Jiang Zemin wrote, 'Improve the development and application of rare earth, and change the resource advantage into economic superiority.' In 1996, Chinese authors Wang Minggin and Dou Xuehong, both from the China Rare Earth Information Center at the Baotou Research Institute of Rare Earth in Inner Mongolia, published a paper called "The History of China's Rare Earth Industry." In it they wrote, 'China's abrupt rise in its status as a major producer, consumer, and supplier of rare earths and rare earth products is the most important event of the 1980s in terms of development of rare earths.'"[a]

Key themes discussed in this chapter are:

- Oxides used to create NdFeB magnets are mined mostly in China, a country with which the United States has a challenging relationship. Chinese authorities have sought to control the price of REEs by imposing export quotas or regulating mine production.

- Chinese authorities' involvement in REE mining has given Chinese markets a structural competitive advantage over non-Chinese producers. Chinese mine operators also excel in the extraction, separation, and refining of RE oxides.[6]

- All NdFeB magnets are now fabricated outside the United States. The single largest producer of high-tech magnets is China, accounting for 75 percent of the global market. Japan controls 22 percent, with the remainder of NdFeB magnets coming from Europe/Germany.[7] The U.S. economy risks losing its knowledge base and therefore its ability to innovate and design new applications for high-tech magnets, which require complex and capital-intensive production methods. This regression occurs at a time when the use of magnetic technology promises many new applications, including in the field of nanotechnology, medicine, and green energy.

- Demand for high-tech magnets continues to grow because virtually every contemporary electric motor uses high-powered REMs. Green technology (including hybrid cars and wind turbines) consumes a large amount of REMs. This demand could mean that prices will continue to rise, while the United States may miss out on the next generation of innovative designs for these magnets (since, at this point, the production of these magnets has largely disappeared from the United States.)[8]

- Substitutes for permanent REMs do not exist, and probably won't be developed over the next 10 years. Even if substitutes could be found, it may be extremely costly to re-engineer complex electric engines and motors to adapt to them.

- REMs are used in internal guidance systems for missiles, microwave and communications systems, radars, and motors and generators that power aircraft and ships. REMs are used in actuators for electric propulsion, in space systems, and in pumps and control rod actuators found in nuclear reactors. While the actual amount (and thus the value) of each bit-sized magnetic component is modest, many U.S. military capabilities cannot function without these magnets. Therefore, high-tech magnets are critical for maintaining military capabilities and economic competitiveness.

A NOTE ON CRITICALITY

The ability to field a modern military depends heavily on REMs, which have essential roles in numerous advanced weapons systems across all U.S. military services (see Military Equipment Chart at the beginning of this chapter). The inability to acquire REMs would be *incapacitating* to U.S. military capabilities, removing its ability to construct nearly all military ground vehicles, aircraft, naval vessels, and missile systems. These magnets are also necessary for advanced radar systems, and enable certain stealth technologies. Because there are no substitutes for these magnets, a lack of supply could impede the level of available military technology for decades to come.

Two separate risk factors contribute to an *extreme* vulnerability to this supply chain. First, China produces 80 percent of the global supply of REMs, including neodymium, an important element necessary for high-tech NdFeB magnets. In the 1980s, China decided to ramp up production, driving out competitors and cornering the market. As it struggled to clamp down on illegal mines, it relied on export quotas to restrict global supplies of REEs. In 2010, China decreased its export quota by 40 percent, which created panic in international markets but made it suddenly economical to re-open dormant mines outside China. In response, Japanese companies also increased the recycling of old electronic appliances and used car batteries. Supply of REEs rose while the global financial crisis hurt advanced industrialized economies and softened demand. Currently, prices of REMs have fallen. However, RE minerals contain one or several of a collection of 17 different elements, and world production has focused on the "light" RE oxides. Light REEs include lanthanum (La), cerium (Ce), praseodymium (Pr), neodymium (Nd), and samarium (Sm). They are relatively abundant and easier to mine than heavy REEs. Heavy REEs are less common and therefore more valuable; these include europium (Eu), gadolinium (Gd), terbium (Tb), dysprosium (Dy), holmium (Ho), erbium (Er), thulium (Tm), ytterbium (Yb) and lutetium (Lu). The United States wholly depends on China for the supply of heavy REs. REMs can contain a mix of light and heavy REEs, including neodymium, dysprosium, samarium, praseodymium and terbium. In addition, mines in advanced industrialized countries struggle to compete with Chinese mining operators. Labor costs are lower in China; more importantly, the enforcement of environmental, health, and safety standards are more lax, which makes it more economical to mine there.

Furthermore, the center of the magnet manufacturing sector has shifted from the United States to China and Japan, creating the possibility that even upon regaining a domestic neodymium source, the U.S. industrial base will have lost the knowledge base to reestablish efficient high-tech magnet production. Together, these factors influence an assessment of extreme risk, even as domestic production capacity for REs is restarted.

BACKGROUND AND GENERAL USE

Because of their widespread use, magnets come in different materials, shapes, and strengths, and have varying density, flexibility, magnetic range, heat resistance, and corrosion resistance. Globally, the permanent[9] magnet market came to $7.98 billion in 2010, and it is expected to grow to $17.1 billion by 2020.[10] Between 2009 and 2011, RE metal prices peaked. Since then, average prices for REEs such as cerium,

CONSEQUENCES OF A FUTURE SUPPLY DISRUPTION (a notional though realistic scenario)

In 2022, a decade-long Chinese effort to effectively halt the export of rare earth elements (REE) has taken its toll on the U.S. defense industrial base. Belated efforts by the U.S. government to reconfigure its National Defense Stockpile DLA Strategic Materials stockpile to provide a buffer supply of REEs prove insufficient. Without a guaranteed supply of REEs to produce samarium-cobalt magnets capable of withstanding the extreme temperatures required for military applications, U.S. technological superiority on the battlefield is diminished. This reduced supply of magnets delays research on the successor to the F-35 Joint Strike Fighter, as well as other advanced systems, leaving the future of U.S. air superiority in question.

lanthanum, neodymium and praseodymium declined by 80 percent. Magnetic materials and alloys also dropped in price by 25 percent since late 2011. RE metal prices increased by 80 percent since late 2011.[11] U.S. companies do not have a presence in the metal-making phase of RE processing; alloy fabrication and metal-making take place offshore in Asia.

Many common household items and defense applications employ REMs. They are used in motors, sensors, actuators, acoustic transducers, hard-disk drives, chemical catalysts, catalytic converters, petroleum refining, hybrid engines, and wind-turbines. Since the 1980s, most electric motors have relied on high-tech magnets.

The extraction and processing of RE oxides is complex, beginning with RE ore production, which yields mixed concentrates. A separation process sorts out individual REEs and subsequently refines them into oxides. Oxides are then turned into pure metals. Additives such as iron or cobalt are added to the rare earth metal according to the desired composition. The materials are melted in a vacuum induction furnace and turned into metal alloys. The alloys are cast into ingots, which are pulverized into particles that are protected by nitrogen and argon to prevent oxidation. Finally, the magnetic particles are pressed and subsequently treated by heat in a sintering furnace.[12] Sintering increases the density of the magnets and shrinks their size. Afterwards, a machine cuts, grinds, and shapes the magnets. The final step is to test and magnetize them. After all these different steps, a magnet emerges with a high "anisotropy," meaning that it has powerful magnetic properties, is difficult to demagnetize, and is able to withstand high temperatures. Magnet fabricators need to know in advance what the final application

of the REM will be because the separation and metal-making process differs depending on the end-use of the magnet.

The original discovery of REMs' potential was a major technological breakthrough, but translating the scientific knowledge into a commercial product still involves multiple phases of processing that consume significant capital investments. The most common way to manufacture NdFeB magnets is through classical powder metallurgy, in which metallic powders are melted in a furnace and cast into ingots that are later magnetized.[13] Because of the costs involved, the "mine-to-market" model is highly efficient. This means that the miner is vertically integrated and oversees the entire process of exploration, mining, processing, separating, metal-making, and alloy production. Profits increase as the producer moves up the supply chain. The objective of most mining companies is to establish RE mine-to-market supply chains.[14]

1. Rare earth oxides are refined into pure rare earth metals.

2. Rare earth metals are combined with other elements to create the desired alloy.

3. The elements are melted in an induction furnace and cooled into ingots.

4. Ingots are pulverized into small particles, which are protected from oxidation by nitrogen and argon.

5. A magnetic field is applied as the particles are pressed into the desired form.

6. The shaped magnets are sintered, reducing their size and increasing their density.

Adapted from: Shin-Etsu, Rare Earth Magnets.
http://www.shinetsu-rare-earth-magnet.jp/e/masspro/

The industrial magnets market consists of four separate categories:

Ceramic or ferrite –
Ceramic or ferrite magnets are the most commonly used today. They are relatively strong, but their strength is diminished by their susceptibility to changes in temperature. Approximately 65 percent of worldwide production is based in China.[15]

Aluminum nickel cobalt (AlNiCo) –
AlNiCo magnets are commonly used and possess strong temperature stability; however, they are relatively easy to demagnetize. Moreover, they are not as strong as magnets formed from rare earths.[16] Approximately 50 percent of worldwide production is based in China.[17]

Samarium cobalt (SmCo) –
SmCo magnets were introduced in 1961, and an improved composition appeared in the early 1970s. They are four to five times as strong as ferrite magnets of the same weight. Moreover, because they are stable at high temperatures, they are well suited for military uses, including aircraft and missiles. For that reason, cobalt has been included in the Berry Amendment, and later in the Specialty Metals Clause (SMC). However, these magnets are more brittle, and the price of cobalt has fluctuated wildly. In 2009, SmCo magnet worldwide production was listed at 5,000 tons and worth about $227 million. There is currently one U.S. company (Electron Energy Corporation in Landisville, Pennsylvania) that still produces SmCo magnets, although it is reliant on REEs imported from China.[18] Sixty percent of worldwide production is based in China.[19]

Neodymium iron boron (NdFeB) –
NdFeB magnets first became available in 1984. They are top-of-the-line, and their superior strength allows for the use of smaller and lighter magnets. These magnets consist mainly of iron, which is cheaper and more abundantly available than cobalt. NdFeB magnets are approximately 1.5 times stronger than SmCo magnets. Because NdFeB magnets lack the extreme temperature resistance qualities of their SmCo counterparts, they are not used in applications that need to withstand temperature over 300 degrees Fahrenheit. Owing to their strength, they are used in consumer electronics, computers, and green technology applications such as hybrid cars and wind turbines. In 2012, NdFeB sintered magnet production came to 63,000 metric tons.[20] The market for NdFeB magnets is 20 times larger than that of SmCo permanent magnets. Seventy percent of worldwide production is based in China and 22 percent in Japan.[21]

It is forecast that by 2020 the global market for sintered NdFeB magnets will grow to 126,000 tons, with an estimated value of $11.5 billion, and that SmCo magnet production will reach 3,000 tons, with an estimated value of $275 million. By 2020, China will account for 80 percent of global NdFeB magnet production and 74 percent of SmCo magnet production.[22]

U.S. production capacity of NdFeB magnets has moved offshore while the larger U.S. magnetic material sector has experienced decline. In the 1990s, 14 U.S. magnet producers (five AlNiCo producers, five SmCo producers, and four NdFeB producers) employed roughly 6,000 workers. By 2008, only four companies remained (three AlNiCo producers and one SmCo producer), employing only about 500 workers.[23]

MAGNETS AND DEFENSE APPLICATIONS

Magnets are essential for a wide range of defense and civilian applications. A magnet is any object that has a magnetic field and attracts ferrous objects, including iron, steel, nickel and cobalt. High-tech magnets come in thousands of different strength, shapes, and applications. There is a significant consumer market for NdFeB magnets, which are used in headphones, hard-disk drives, and automobiles; the magnets are also required for a wide range of special defense applications. To meet U.S. Military Specifications (Mil-Spec), U.S. defense contractors must do business with companies licensed as official DoD suppliers. Magnets made to Mil-Spec may be imported from Germany or Japan (both countries have two DoD-licensed companies), or from China, which has six DoD-licensed companies. Because of the SMC, AlNiCo and SmCo magnets must either be domestically melted and poured or bought from qualifying countries with reciprocal defense procurement memorandums of understanding.[24]

High-tech magnets are widely used in electric motors, and many defense applications rely on them. (The flight-control systems of precision-guided munitions rely upon SmCo magnets.) If these magnets were no longer available, more expensive and significantly larger hydraulic systems would be needed to control "smart bombs" such as the Joint Direct Attack Munition (JDAM). These magnets are also critical in the white-noise concealment stealth technology used for helicopter rotors.[25]

MAGNETIC MATERIALS INDUSTRY AND DEFENSE VULNERABILITIES

Less than 15 years ago, the United States was self-reliant and produced RE oxides as well as magnets. In the early 1980s, GM and Hitachi simultaneously developed technology for NdFeB magnets following a conference presentation demonstrating magnetic properties of a rare earth 50:50 lanthanum-terbium and iron-boron mixture. Subsequent laboratory research discovered that NdFeB had superior permanent magnetic properties; GM and Hitachi submitted applications for several patents.[26] The intellectual property rights were split between the two companies, with GM receiving patents for bonded NdFeB magnets while Hitachi assumed control of the sintered magnet patent. By 1986, GM had finalized a production process, and in 1987 the company opened Magnequench at a factory in Anderson, Indiana, to produce NdFeB magnets for the automobile industry. At first, magnets were used for automobile sensors and airbags; eventually, R&D found other applications such as in lasers and hard-drives.

Although NdFeB magnets produced by Magnequench had found important uses in critical defense applications, notably precision-guided weapons systems, GM was forced to sell Magnequench when it underwent restructuring in the early 1990s. In 1995, Magnequench was purchased by a consortium among the U.S.-based Sextant Group and two Chinese firms (San Huan New Material and China National Nonferrous Metals Import and Export Company (CNNMIEC)).[27] Together, the two Chinese firms, both with close ties to the Chinese government, took a

62 percent interest in the now offshore Magnequench.[28]

Because Magnequench manufactured sensitive components for smart bombs and supplied 85 percent of the magnets used in servo-motors for precision-guided munitions, the U.S. government had to approve the company's sale. Although CFIUS approved the sale, union leaders had their reservations, and a representative of the United Auto Workers Local 662 representing the workers at Magnequench reached an agreement with GM to negotiate a deal that would keep the magnet production plant in Anderson open for at least 10 years. The consortium accepted the agreement struck between GM and its unions, and promised to keep the plant open for at least 10 years. However, union leaders suspected that the (Chinese) buyers would renege and would not keep the plant open for 10 years. They were correct. In the sixth year (August 2001), the plant was closed. [29]

In 1997, less than two years after the deal was completed, the Anderson production line was duplicated in China, and the Chinese investors began considering the closing of the magnet production line in Indiana. The Magnequench plant in Anderson had patented a complex process that converted neodymium, iron, and boron into powder in order to manufacture REMs. The engineers and officials in China first waited to make sure that the Chinese facility could operate properly; by 2001, the consortium was ready to shut down the Anderson plant, telling union representatives that they were going bankrupt.[30] When Archibald Cox (head of the Sextant Group, and the new President and CEO of Magnequench) was asked about the potential impact of the closure, he minimized the national security ramifications even while acknowledging that

Magnequench manufactured the magnets for the JDAM guidance kit. Cox also acknowledged that under terms of the CFIUS review of Sextant's purchase he had promised to keep the Anderson plant open for at least 10 years, but countered that the affected workers received a fair deal once Magnequench bought out their contract.[31]

In 2000, Magnequench also had bought out other small magnetic material producers, including UGIMAG, located in Valparaiso, Indiana, and GA Powders, a spin-off of the Idaho National Engineering and Environmental Laboratory (a U.S. national laboratory).[32] Three years later, in 2003, Magnequench announced the planned closure of the Valparaiso plant, with the loss of 225 manufacturing jobs.

By 2003, the climate had changed, and there was much greater awareness of the implications of the closure of Magnequench's advanced magnet factory in Valparaiso, Indiana, which at its peak made approximately 80 percent of the magnets bought by DoD. Then-Senator Evan Bayh (D-IN) and Representative Pete Visclosky (D-IN) submitted a request to the Bush Administration to intervene, because the closure would leave the United States without a significant domestic source of the REMs used in smart bombs.[33] Because Magnequench manufactured magnets for DoD applications, the Bush Administration and CFIUS could have forced the Chinese firms to divest their holdings in this strategic industry, per the 1998 Exon-Florio Amendment to the Defense Production Act. However, the Administration declined to intervene.[34]

Magnequench announced the closure of its U.S. production line because "almost all of the raw materials for Magnequench's powder products come from China, and

90 percent of [their] customer base is in Asia."[35] U.S. REE production declined in the late 1990s due to increased price competition from dozens of unregulated mines in China, which depressed world prices and hurt Western mining operations, and ultimately contributed to the closure of the Molycorp RE mine in 2002. Despite protests and political pressures, Magnequench shut down its second facility in Valparaiso in 2004. According to the company, the Valparaiso plant was no longer economical, in part because Magnequench established its own Chinese operations as an internal competitor to the Valparaiso operations when it transferred equipment and technology to China.[36]

Once Magnequench closed its doors, soon thereafter all domestic production of NdFeB magnets ceased in the United States.[37] In 2005, Hitachi closed the plant in Edmore, Michigan that it had acquired from General Electric during the 1990s. The closure of the magnet plants gained political traction during the 2008 presidential election, when Democratic political candidates used the offshoring of magnet fabrication to point to the loss of U.S. jobs to China under a Republican president. Hillary Clinton accused the Bush Administration of failing to stop the closure of the last magnet factory in Indiana, notwithstanding that the preceding Clinton Administration had approved the original sale of this plant to China in 1995. However, at that time, a handful of U.S. advanced magnet producers still remained, and it was not obvious that the sale of Magnequench to offshore owners would hasten the disappearance of the U.S. magnetic materials sector.

GM's decision to sell Magnequench nearly two decades ago marked the beginning of the end of a defense-related critical industry. The departure of one magnet manufacturer accelerated the erosion of the entire magnetic material sector, something that policy officials and lawmakers likely had not anticipated. Nor did it seem to be fully appreciated at the time that high-tech magnets were required to support cutting-edge military capabilities, including precision-guided weapons such as smart bombs and cruise missiles.

What lessons can we draw from this chain of events? In retrospect, it seems apparent that CFIUS failed to examine the acquisition adequately or ask the right questions about the long-term national security ramifications of the sale. That the two Chinese partners in the consortium, San Huan New Material and CNNMIEC, were state-owned should have raised a red flag with CFIUS. Since the 1980s, Chinese leaders had announced their intention to develop its domestic RE production capacity by acquiring foreign technologies. Moreover, just months before its acquisition of Magnequench, San Huan New Material had been fined $1.5 million by the U.S. Trade Commission for patent infringement and corporate espionage.[38] Additionally, these companies had sought to purchase Magnequench from GM since 1993, before GM announced its willingness to sell.[39]

As the magnetic material industry was sold to foreigners, the single largest RE mine in the United States closed its doors in 2002. The Mountain Pass Mine in California primarily closed down because of Chinese mercantilist activities, but other factors such as environmental regulations also played a role.[40] With the closing of Mountain Pass, the U.S. defense industrial base lost both its main supplier of manufactured magnets and access to the oxide ores essential for all defense electronics. After a high-grade deposit of REE ores was found in Mountain Pass in the 1950s, the United States was the dominant producer of REE raw materials. Molycorp, which operated the mine, had invested millions of dollars in researching the potential uses of REs. Between 1965 and the mid-1980s, the Mountain Pass Mine was the single most significant global source of REs, with reserves upwards of 13 million metric tons.[41] Over time, as more REEs were discovered, R&D on high-tech magnets increased, and magnetic technology was incorporated into many important defense applications.

When U.S.-based rare earth mining and fabrication began to decline in the 1990s, Iowa State University's Rare-earth Information Center (RIC) shut down after 36 years. Until 2002, financial support from industry and Iowa State University stimulated important new research on magnetic technology, and the center was a dominant source of information on REs for government, industry, and academics. One of the results of the erosion of REE mining and of RE permanent magnet fabrication was that U.S. firms filed fewer patents while most of the new applications for REMs came from abroad. In short, U.S. leadership was overtaken by Japan, Korea, and China.[42]

In less than a decade, the permanent magnet market experienced a complete shift in leadership. In 1998, some 90 percent of global magnet production was based in the United States, Europe, and Japan. The United States accounted for 80 percent of rapidly solidifying magnets, with the remainder manufactured in Europe. Also, together with Europe, the United States accounted for between 20 and 30 percent of all fully sintered magnets (magnets formed by heating but not melting metals). Japan led in the production of sintered magnets, commanding a market share between 70 and 80 percent. Since the late 1990s, more than 130 sintered NdFeB magnet manufacturers have emerged in China, where the industry is growing at about 30 percent annually.[43]

POLICY AND POLITICAL ISSUES

Volatile prices over the past decade have forced non-Chinese producers out of the market. RE prices are determined by Chinese state officials who control the world's largest producers of RE oxides. In the 1980s, China began to exploit its vast reserves of RE metals and sold such large quantities that it forced nearly every other mine in the world out of business. By 2001, when China joined the World Trade Organization, China's global market share for RE metals was nearly 97 percent (the equivalent of 15,000 tons). Then in 2010, China decided to restrict its export quota by 40 percent, and panic set in within the global RE market, especially in the electronics industry—the primary end-users of high-tech magnets. In 2011, RE prices went through the roof after the Chinese authorities decided to crack down on illegal RE mines, curb export quotas, and after hoarding by speculators and end-users. Prices of some REEs rose an astonishing 1,000 percent (see this report's chapter on Specialty Metals).

Export licenses are based on tonnage and auctioned twice a year, and the export quota functioned like a tax. Chinese export quotas, including those on neodymium, only apply to the metals in their alloy form; manufactured products that make extensive use of RE metals may be exported without similar restrictions. This distinction prompted Japanese and other foreign corporations to relocate to China to take advantage of lower prices and greater availability (since they are the main fabricators of high-tech magnets). The aim of the Chinese authorities is to encourage "mine–to-market" vertical integration, since the value added is mostly reaped at the higher tiers of the production process. The quotas and other taxes raise prices for non-Chinese manufacturers and make it difficult for them to compete with Chinese manufacturers. Since the 1990s, Chinese authorities have recognized magnet materials and other applications of REs as a strategic resource.[44] China's control over the global supply of RE metals received heightened attention in 2010, when China halted exports of REs to Japan in the wake of a maritime dispute in which the Japanese detained the captain of a Chinese fishing boat that had rammed a Japanese Coast Guard vessel. (The dispute occurred near a set of Islands in the East China Sea that are claimed by both countries, though currently administered by Japan.) When the Japanese refused to release the captain, China cut off exports of REs to Japan.[45] The captain was released three weeks later, defusing the standoff. But the export cutoff caused great concern among other countries that are dependent on Chinese REE, because they feared that China would use its near-monopoly over REs to extort concessions from dependent countries.

In response, Japanese companies aggressively invested in reducing their dependence on imported REEs. They have improved recycling from old electronic appliances and batteries. In addition, the high prices of dysprosium, a heavy RE, convinced some Japanese companies to spray magnetic material on their engines to conserve usage. (Spraying requires less dysprosium than relying on fabricated magnets.) In some cases, Japanese electric motor manufacturers designed around the RE magnets and replaced them with cheaper ferrite magnets. [46]

For its part, U.S. government encouraged Molycorp to reopen the Mountain Pass Mine which it did in 2012. Projections indicated that Mountain Pass would mine 19,050 metric tons annually of RE oxides, and then expand capacity to 40,000 metric tons annually.[47] However, Molycorp has run into a host of difficulties, and its survival as an independent company is at risk. First, technical challenges delayed the opening of the mine. In addition, because of the depressed RE prices, the management scrapped the second phase of expansion and no longer aims to increase production to 40,000 tons annually. More importantly, Molycorp does not possess the capacity to turn oxides into metal alloys. To compensate for this lack of capacity, Molycorp bought Neo-Material Technologies, which has facilities in China, in order to separate and refine the crude oxides.

MITIGATING THE RISKS

As the manufacturing base in magnetic materials has eroded, there is an increased risk that technological knowledge will follow. In this scenario, the United States will lose its ability to compete in global markets and to design the next generation of applications.[48] Without private and public investments in REMs, it remains unclear how DoD will guarantee the viability of this important sector of the defense industrial base. Furthermore, since permanent magnets are widely used in electronics, hard-disk drives, wind turbines, and hybrid cars, demand is expected to grow by 10 to 16 percent annually over the next several years.[49] Both the U.S. economy and defense industrial base are placed at risk if they depend on research, innovation, and manufacturing performed in offshore markets for clients who may have very different strategic interests than the U.S..

We recommend the following actions to mitigate the risks to the U.S. high-tech permanent magnets sector.

RECYCLING

Manufacturers should research ways to reduce RE content in their products, and increase the effectiveness of recycling. Recycling RE substitute materials and efforts to improve product efficiency can play a small but important part in mitigating the risks to the supply chain presented by the offshoring of magnet production.[50]

While recycling post-consumer magnets is possible, the technology is still in its infancy. In Japan, several different research projects and trials are underway to recycle REs from motors and generators. Likewise, recycling processes for the extraction of neodymium magnets from compressors and hard drives are expected to be in use by 2013. Another Japanese company, Shin-Etsu Chemical, is planning to build a RE recycling plant in Vietnam. Shin-Etsu hopes to recycle up to 1,000 tons of REs from discarded appliances annually (Japan uses approximately 26,000 tons of REEs a year).[51]

A report commissioned by the British Department for Transport concluded that the costs of the recycling extraction processes may be prohibitive, and that there are a number of obstacles to be addressed before it will be possible to economically recycle small amounts of magnetic scrap. Permanent magnets are used for hundreds of different applications and are embedded in components of hundreds of different products. For many of these products, removing the RE magnetic materials may be impossible. Furthermore, NdFeB magnets come in many grades with different chemical compositions. This chemistry would be another challenge to address during the post-consumer recycling phase.[52]

Aside from the costs, both local and federal U.S. authorities must create an efficient collection system in order to begin the process of recycling and extraction. Until now, discarded electronic scrap has been exported to China and other developing countries, which dismantle machinery, engines, and electronics but do not have the capacity to extract REEs.

In short, it may be worthwhile to invest in a comprehensive system of recycling, something that European governments as well as the Japanese are considering.[53] It may take several years to set up a collection and recycling scheme, but in the long run, as the supplies and prices of REEs remain uncertain, this effort may pay off.[54] It is possible that individual magnet producers will be able to introduce small-scale in-house recycling, but it is hard to envision how salvage recovery from motors and electronic devices will be economically feasible at a wider or larger scale.

DLA STRATEGIC MATERIALS STOCKPILE

The Defense Logistics Agency should consider adding key high-performance magnet materials, including cobalt, samarium, neodymium, dysprosium and praseodymium, to the DLA Strategic Materials stockpile. The stockpile stores and sells critical raw materials, but it currently includes no REEs.[55] Some REEs have never been classified as strategic minerals; the government only became interested in these REEs' use in the 1990s.

In a February 2010 proposal to the U.S. federal government, the U.S. Magnetic Materials Association sponsored a study that recommended stockpiling a five-year supply of REEs to support the government's defense-critical needs while allowing the private sector to rebuild the domestic supply chain.[56] Taking advantage of combined buying power, a second approach would be to transform the DLA Strategic Materials stockpile into a cross-agency Strategic Material Security Program (SMSP), tasked with acquiring strategic materials for all governmental agencies.[57] The SMSP could underpin affordable prices by buying large quantities of REEs during periods of low prices. Of course, to make this work, the new agency would need considerable funding and continuous access to financing, therefore slowly building an inventory of key REEs. At the same time, it may be worthwhile to consider phasing out the DLA Strategic Materials stockpile and shifting to a reliance on one specialized agency to guarantee that the United States will not face a shortage of critical rare elements.[58]

The FY2013 NDAA goes a long way toward rectifying structural problems with the DLA Strategic Materials stockpile, by allowing greater authority and new tools to diagnose and mitigate supply chain vulnerabilities for critical raw materials, including RE metals. This development is promising, but revival of the U.S. domestic RE metals sector will require much more effort and time. In fact, since the government plans to purchase the stockpiles from U.S. mining operators, Congress will have to make sure that U.S. mines can in fact produce the REEs selected for inclusion in the new strategic stockpile.

RE-OPENING OF CLOSED MINES AND MINING OF RARE EARTH ELEMENTS

U.S. industry and government—working in concert with foreign allies—should collaborate to recapture a larger presence of the RE mining industry. The United States possesses 13 percent of

global REE reserves, and Australia, Russia, and Canada each possess significant reserves.[59] While China is responsible for 80 percent of current RE production, it accounts for only 50 percent of estimated total global reserves. Prior to 1990, the United States was largely self-sufficient in meeting its REE and permanent magnet requirements. In 2012, Molycorp, Inc., re-opened Mountain Pass Mine; the estimates are that output should begin at about 3,000 to 5,000 tons per year, reaching 20,000 tons by 2014. In addition, deposits in North Fork, Idaho, look promising.[60]

Several mines are scheduled to open in Australia, with production starting in 2012. Lynas Corporation's Mount Weld mine is projected to produce 22,000 tons of REEs annually by 2014, and Nolans, owned by Arafura Resources Limited, should produce 20,000 tons of REEs by 2013.[61] Canada and South Africa will also be opening or restarting mines, which are supposed to come into production in 2013. There is even talk of opening REE mines in Jamaica and Greenland. Thus, in the short-term, more REEs will be available on the global market, lessening China's monopoly. However, Chinese mining costs are low, and mines in many other countries will lack competitive advantage against Chinese producers. Additionally, there is another major caveat stemming from the high global demand for permanent magnets for multiple uses. Permanent magnets are a key component of hybrid vehicles and wind-turbines; in fact, the entire world's renewable energy strategy implicitly relies upon REMs.[62] For example, every Toyota Prius requires approximately 1 kilogram (kg) of magnets. NdFeB magnets are important components in many wind-turbine generators, which require an estimated 3 kg of NdFeB magnets for an average-sized generator.[63] In the meantime, magnets are widely and extensively used in the electronics industry, a fact that suggests demand for high-tech magnets will continue to increase. [64] While there seems to be an abundance of light RE metals, heavy ones are genuinely scarce. There is not enough dysprosium to satisfy the growing demand for magnets, and industry will have to come up with an alternative solution.

In the near term, output from Mountain Pass appears to be sufficient to meet the demand for light RE oxides. But the future of Mountain Pass is nonetheless at risk. The mine has been plagued by budget overruns, technical difficulties, and huge losses. In addition, the mine does not yield many heavy REEs (dysprosium, terbium, europium, yttrium), which are more valuable and less abundant. Moreover, Molycorp does not possess the technology to separate crude oxide concentrates into pure RE metals. Instead, it ships its RE ore to China for processing, defeating the purpose of trying to become self-sufficient in REEs. If the United States wants to mine a significant fraction of its REEs domestically, more extensive mining efforts are necessary. There are REE deposits in Idaho, Montana, Colorado, Missouri, Utah and Wyoming. However, after capital is secured it takes anywhere from seven to 15 years before production can begin. No country has such a cumbersome and impenetrable authorization process as the United States, which receives the lowest rating for a "mine-friendly" business environment among advanced industrialized countries. While seven to 15 years is the standard for bringing a mine into production in the United States, Australia and Canada can complete the process in two to four years. Thus, the U.S. federal government and Congress will have to reform the authorization/permit process to expedite the opening mines that hold promising critical of minerals. It makes sense to begin exploring the possibilities of starting mining activities to extract REEs sooner rather than later, since it takes many years before a mine is operational and before the oxides can be separated and processed.[65] Ucore, a Canadian company, is in the process of exploring the opening of a mine in in Alaska's Bokan Mountain that holds a sizable concentration of heavy REEs (dysprosium, terbium and yttrium). DoD has approached the company to conduct a mineralogical and metallurgical study on its land claim in Bokan Mountain. Moreover, Alaska Senators Lisa Murkowski (R-AK) and Mark Begich (D-AK) have joined efforts to expedite the authorization process.[66] DoD's support could fast-track Ucore into production of REE concentrate and the separation of RE oxides.[67] The fate of U.S. national security cannot rest with one single mine (Mountain Pass), which has struggled since reopening with budget overruns, technical delays, and inadequate technological capacity.

ALTERNATIVE MATERIALS

The quest for alternative materials continues, but even George Hadjipanayis, a co-inventor of the NdFeB magnet, is doubtful that the United States can find a better magnet or an improved substitute material that can equal the property of high-tech magnets in weight or strength. Current U.S. Department of Energy (DoE)-funded research investigates the potential of using NdFeB nano-particle structures to generate usable magnetic strength with only a small particle of the metal (one-billionths of a meter). The small-scale structure of these compounds greatly increases the magnetism found in the metal alone, requiring much less metal to achieve the same or better results found in normal magnets and reducing the amount of scarce neodymium required.[68] Nevertheless, it is unlikely that new compounds will soon arrive on the market in sufficient commercial quantities to displace rare earth oxides. While it will eventually be possible to design a permanent nano-composite magnet,[69] it may be years before such magnets can be developed and commercially produced at a competitive. For the foreseeable future, NdFeB magnets remain the most advanced commercially viable magnet.[70]

Second, many applications of REEs are highly specific, with substitute materials either unavailable or usable only at the risk of inadequate performance. Redesigning engines, motors, and electronics to accommodate a substitute for high-tech magnets would be very expensive. Nevertheless, there are several efforts to develop permanent magnets that do not require REEs or which greatly reduce the amount of neodymium and samarium content. In 2008, DoE funded research at Iowa State's Ames Laboratory that replaces pure neodymium in the NdFeB permanent magnet crystal structure with a combination of neodymium, yttrium, and dysprosium.[71] This replacement alters the crystalline structure of the magnet, allowing the magnet to maintain its strength when exposed to heat. Although the standard NdFeB magnet is superior at low temperatures, the new magnet outperforms the original at temperatures exceeding 75 degrees Celsius.[72]

There are various attempts to move away from the reliance on REMs; eventually some of these endeavors will result in success. DoE has provided grants to the University of Delaware and GE Global Research to develop new magnets using nano-particles to preserve the increasingly small supply of RE metals typically used in the strongest magnets made today. These new magnets are also stronger and lighter than traditional magnets and should increase efficiency as well as conserve the

dwindling supply of neodymium, dysprosium, and terbium. NovaTorque,[73] a California startup, has developed electric motors using low-cost ferrite magnets that the company claims outperform those made out of neodymium.

POLITICAL RESPONSE TO RARE EARTH CRISES

Congress and federal agencies should invest proactively in programs to address the present RE crisis and the future demand for REM. Elected officials appear to have been caught by surprise by the tension surrounding REM. It was not until late 2009, when it became painfully clear to U.S. policy-makers that REEs were critical to U.S. national security, that Congress turned to DoE to formulate a policy. At that time, neither DoE nor DoD was equipped to address the potential shortage of REEs, lacking the funding, staff, and expertise to engage in REE policy. The U.S. Office of Science and Technology Policy is in charge of the National Critical Materials Council (NCMC), which coordinates the federal government's policies for critical materials. However, following the cutbacks and staff re-organization of the 1990s, the NCMC barely functioned, and its responsibilities were divided and delegated to other agencies.

In recent years, seven departments and agencies (the Government Accountability Office [GAO], the Congressional Research Service [CRS], the Institute for Defense Analyses [IDA], DLA, DoD, NRC, and DoE) have issued reports and made statements on REEs and RE metals and magnets. The statements and recommendations often diverge, and definitions of the problem also vary. As a result, the policy approach has been haphazard and often minimalist.[74]

In a similar vein, in the last few years, Congress has moved ahead and proposed several new bills to address the problems of a shortage of REs. For example, the House passed bill H.R. 6160, the Rare Earths and Critical Materials Revitalization Act, which proposed funding to incentivize private investment in domestic RE mining and refining.[75] Meanwhile, national concern about the REE situation continues to grow. Congress has proposed several other bills focused on the U.S. reliance on Chinese decisions and prices. Several proposals are on the table, notably H.R. 1388, the Rare Earths Supply Chain Technology and Resources Transformation Act, sponsored by Rep. Mike Coffman (R-CO); H.R. 2011, the National Strategic and Critical Minerals Policy Act, sponsored by Rep. Doug Lamborn (R-CO); its amended version that includes H.R. 1314, the Resource Assessment of Rare Earths Act, sponsored by Rep. Hank Johnson (D-GA); and S.1113, the Critical Minerals Policy Act, sponsored by Sen. Lisa Murkowski (R-AK). H.R. 4402, the National Strategic and Critical Minerals Production Act of 2012, sponsored by Rep. Mark Amodei (R-NV), would require a more efficient development of domestic sources of REEs,, and calls for easing the permit and authorization process for mineral exploration and mine development.

In partial response to supply concerns, DoE's Advanced Research Projects Agency (ARPA-E) introduced the Rare Earth Alternatives in Critical Technologies program in 2011. The program distributes $22 million in research grants among 14 different projects aimed at reducing or eliminating the RE requirement for permanent magnet motors, while retaining output and efficiency on an equal-cost basis.[76] DoD has agreed, after some hesitation, that it will invest in stockpiling nine metals, four of them REEs. Considering the current federal budget climate, DoD announced

that it would put aside about $1.2 billion to purchase the identified RE metals at risk. DoD would acquire these metals from domestic mining companies, because it will not import them from China. The problem with this strategy is that $1.2 billion is a miniscule amount in light of the fact that it may cost $1 billion alone to explore and open a mine while it may take seven to 15 years before a mine operator can obtain all the necessary permits. As one market observer notes, it is questionable whether the Pentagon fully understands the precarious stage of U.S. mining and the extent of U.S. dependence on foreign REEs.[77]

CONCLUSION

The disappearance of the domestic high-tech permanent magnet sector is a sorry tale of missed opportunities, indifference, and neglect. While U.S. officials were focused on free trade and free capital flows, Chinese authorities seized the opportunity to acquire U.S. technological know-how and advanced manufacturing techniques. The Chinese understood the key role played by REMs in missile defense, aviation, and other military capabilities. Although REMs are just small parts of sophisticated and multimillion dollar weapon systems, without these tiny magnets, guided missile systems, military vehicles, naval vessels, and submarines could not operate. How could our leadership and military establishment have failed to understand that the U.S. needs to preserve and retain this small though key component of the defense industrial base? We may never know the full answer, but we should draw the correct lesson from this example: to nurture our technological advantages and carefully preserve important sectors of our defense industrial base from the actions of foreign powers whose strategic interests may diverge from ours.

ENDNOTES

a   Cindy Hurst, "The Rare Earth Dilemma: China's Market Dominance," *The Cutting Edge*, November 22, 2010. http://www.thecuttingedgenews.com/index.php?article=21786&pageid=&pagename.

1   Very strong permanent magnets also contain tiny amounts of dysprosium, which augments the strength of the magnet while also increasing its resistance to corrosion. Dysprosium is one of the most critical and scarce rare earth elements. See Robert Beauford, "Rare Earth Elements and the US," *Rare Earth Elements Educational Resources Home Page* (2010). http://www.rareearthelements.us/.

2   Valerie Bailey Grasso, *Rare Earth Elements in National Defense: Background, Oversight Issues, and Options for Congress* (Washington, DC: Congressional Research Service, R41744, April 25, 2012). http://www.fas.org/sgp/crs/natsec/R41744.pdf.

3   Cindy Hurst, "The Rare Earth Dilemma: China's Market Dominance," *The Cutting Edge*, November 22, 2010. http://www.thecuttingedgenews.com/index.php?article=21786&pageid=&pagename=.

4   CFIUS does not address job losses and union concerns when vetting a proposal. Thus, it is unclear if CFIUS actually imposed a stipulation on the new owners to keep the plant in the United States, or if the committee informally requested that the owners protect the hundreds of jobs at the Anderson plant. John Tkacik, Jr., *Magnequench: CFIUS and China's Thirst for U.S. Defense Technology* (Washington, DC: The Heritage Foundation, May 2, 2008). http://www.heritage.org/research/reports/2008/05/magnequench-cfius-and-chinas-thirst-for-us-defense-technology.

5   John Tkacik, Jr., *Magnequench: CFIUS and China's Thirst for U.S. Defense Technology* (Washington, DC: The Heritage Foundation, May 2, 2008). http://www.heritage.org/research/reports/2008/05/magnequench-cfius-and-chinas-thirst-for-us-defense-technology.

6   National Defense University, Industrial College of the Armed Forces, *Strategic Materials Industry 2010* (Washington, DC: National Defense University, Spring 2010). http://www.ndu.edu/icaf/programs/academic/industry/reports/2010/pdf/icaf-is-report-strategic-mat-2010.pdf.

7   U.S. Magnet Materials Association, *The U.S. Magnet Materials Story: Past, Present, and Future.* http://www.electronenergy.com/media/USMMA%20Presentation%20General%2012-07.pdf.

8   Richard McCormack, "China's control of high-tech magnet industry raises U.S. alarms," *Manufacturing & Technology News* 16 (September 30, 2009), 1-3.

National Defense University, Industrial College of the Armed Forces, *Strategic Materials Industry 2010* (Washington, DC: National Defense University, Spring 2010). http://www.ndu.edu/icaf/programs/academic/industry/reports/2010/pdf/icaf-is-report-strategic-mat-2010.pdf.

9   Permanent magnets differ from temporary magnets in that their atomic structures are permanently aligned. While permanent magnets are more difficult to magnetize, their magnetic charge is not disrupted by external conditions such as heat. In temporary magnets, heat can cause the atoms to realign, which makes the magnet lose its charge. "What Is the Difference Between a Permanent Magnet and a Temporary Magnet?" Timothy Boyer, http://www.ehow.com/info_8180685_difference-permanent-magnet-temporary-magnet.html#ixzz1ypZw90Mz.

10 Ed Richardson, *Overview of the Current State of U.S. Permanent Magnet Production*, ALMA Symposium: Neodymium Magnet Workshop held on January 14, 2012. http://www.almainternational.org/assets/Documents/WinterSymposia/Panels/neodymiumWS2012/richardson-alma%20neo%20wkshp-01.14.12.pdf.

11 Seeking Alpha Contributor "Shock Exchange," "Molycorp: 'The Pain' Is Here" (January 10, 2013). http://seekingalpha.com/article/1106271.

Mani, "Molycorp: Remain On The Sidelines As Mountain Pass Ramp Up Will Take Time," iStockAnalyst, January 14, 2013. http://www.istockanalyst.com/finance/story/6234738/molycorp-remain-on-the-sidelines-as-mountain-pass-ramp-up-will-take-time.

12 Sintered magnets are the result of treating the compressed powder of the original alloy material. Sintering involves the compaction of fine alloy powder in a die and then using heat to fuse the powder into a solid material. While the sintered magnets are solid, their physical properties are similar to ceramics. They are easily broken and chipped. This fragility is one of the problems when considering the possibilities of recycling post-consumer waste to extract magnetic material.

13 There is another, less common process called the Bonded Magnet Process. Here, a thin ribbon of NdFeB is melted, spun, and pulverized into particles; blended with a polymer; and then either compressed or injection-molded into bonded magnets. In this fashion, it can be shaped in many intricate forms. Approximately 5,500 tons of NdFeB bonded magnets are produced each year.

14 Marc Humphries, *Rare Earth Elements: The Global Supply Chain* (Washington, D.C.: Congressional Research Service R41347, September 6, 2011).

Michel Nestour, *Technology Minerals: The Rare Earths Race Is On!* (London: Ernst and Young, April 2011). http://www.ey.com/Publication/vwLUAssets/Tech_Minerals_Rare_Earth_Paper_FINAL/$FILE/Tech_Minerals_Rare_Earth_Paper_FINAL.pdf.

15 Dan Vukovich, *Overview of the World's Magnet Supply*, (Alliance LLC, October 2004). http://www.allianceorg.com/pdfs/OverviewontheWorldofMagnets.pdf.

16 Geno Jezek, "Types of Magnets," 2006. http://www.howmagnetswork.com/types.html.

17 Dan Vukovich, *Overview of the World's Magnet Supply*, (Alliance LLC, October 2004). http://www.allianceorg.com/pdfs/OverviewontheWorldofMagnets.pdf.

18 Electron Energy Corporation of Landisville, Pennsylvania, produces samarium cobalt (SmCo) permanent magnets. This company still depends on the import of small amounts of gadolinium, dysprosium and terbium, which are only available only from China at the moment. Marc Humphries, *Rare Earth Elements: The Global Supply Chain* (Washington, D.C.: Congressional Research Service R41347, September 6, 2011).

19 Walter T. Benecki, Terry K. Clagett, and Stanley R. Trout, *Permanent Magnets 2010-2020* (Highland Beach, FL: Walter T. Benecki LLC, December 2010).

20 Walter T. Benecki, *The Permanent Magnet Market – 2015*. Talk delivered at Magnetics 2013 Conference, Orlando, Florida, held on February 7 and 8, 2013. http://www.waltbenecki.com/uploads/ Magnetics_2013_ Benecki Presentation.pdf.

21 There is also a small production of bonded NdFeB (5,000 tons) with a dollar value of $395 million in 2009. U.S. Magnet Materials Association, *The U.S. Magnet Materials Story: Past, Present, and Future*. http://www.electronenergy.com/media/USMMA%20 Presentation%20General%2012-07.pdf.

22 Malcolm Southwood and Paul Gray, "Rare Earths: Too Late To The Party?" *Goldman Sachs Thematic Commodities* (4 May 2011). http://www.fullermoney.com/content/2011-05-12/RareEarths.pdf.

Peter Dent, Interview, January 25, 2012.

23 Richard McCormack, "China's control of high-tech magnet industry raises U.S. alarms," *Manufacturing & Technology News* (September 30, 2009), 1.

Peter Dent, "High Performance Magnet Materials: Risky Supply Chain," *Advanced Materials & Processes* (August 2009), 27-30.

U.S. Magnet Materials Association, Presentation, (2008). http://www.usmagneticmaterials.com/documents/USMMA-Story.pdf.

24 Under DFARS 255.003 (10), qualifying countries are Australia, Austria, Belgium, Canada, Denmark, Egypt, Finland, France, Germany, Greece, Israel, Italy, Luxembourg, the Netherlands, Norway, Portugal, Spain, Sweden, Switzerland, Turkey, and the United Kingdom.

25 Justin C. Davey, "Enduring Attraction: America's Dependence On and Need to Secure Its Supply of Permanent Magnets," *Joint Forces Quarterly* 63, no. 4 (2011): 76-83.

26 One of the ironies is that in the early 1980s, neodymium was more abundant and cheaper than samarium. In addition, cobalt was costly to refine, and most cobalt came from mines in unstable regions of Africa. Thus, neodymium was considered a good substitute for samarium. Jim Witkin, "A Push to Make Motors With Fewer Rare Earths," *The New York Times*, April 20, 2012. http://www.nytimes.com/2012/04/22/automobiles/a-push-to-make-motors-with-fewer-rare-earths.html?pagewanted=all.

27 Cindy Hurst, *China's Rare Earth Elements Industry: What Can the West Learn?* (Institute for the Analysis of Global Security, March 2010). http://fmso.leavenworth.army.mil/documents/rareearth.pdf

28 The senior Chinese investor (a son-in-law of the former Chinese "paramount leader" Deng Xiaoping) took over as the company's new chairman. Both Chinese companies were closely tied to the Chinese Academy of Science and Technology for Development (CASTED), an agency established by the Chinese Ministry of Science and Technology. Another smaller partner in the consortium was George Soros of Soros Fund Management. Soros was bought out in 1999. John Tkacik, Jr., *Magnequench: CFIUS and China's Thirst for U.S. Defense Technology* (Washington, DC: The Heritage Foundation, May 2, 2008). http://www.heritage.org/research/reports/2008/05/magnequench-cfius-and-chinas-thirst-for-us-defense-technology.

Steven Thomma, "Clinton Blasts Bush for Project Bill OK'd," *McClatchy Newspapers*, April 30, 2008. http://www.mcclatchydc.com/2008/04/30/35337/clinton-blasts-bush-for-project.html#storylink=cpy.

29 Union representatives asked the government and GM to obtain an agreement from the new owners that jobs would remain in Anderson, Indiana. Many accounts of the sale mention that CFIUS insisted that the Sextant Group retain jobs and equipment in Indiana; however, this raised the question of CFIUS' ability to enforce such a stipulation in light of CFIUS' statutory mandate to vet proposals on the basis of national security concerns. Scott L. Wheeler, "Missile Technology Sent to China," *INSIGHT Magazine*, February 18, 2003. http://www.freerepublic.com/focus/f-news/836496/posts.

30 Jeffrey St. Clair, "The Saga of Magnequench," *Bloomington Alternative*, April 23, 2006. http://www.bloomingtonalternative.com/node/7950.

Cindy Hurst, "The Rare Earth Dilemma: China's Market Dominance," *The Metal's Edge*, November 22, 2010. http://www.thecuttingedgenews.com/index.php?article=21786&pageid=&pagename=.

31 Scott L. Wheeler, "Missile Technology Sent to China," *INSIGHT Magazine*, February 18, 2003. http://www.freerepublic.com/focus/f-news/836496/posts.

32 Ibid.

33 Steven Thomma, "Clinton blasts Bush for project Bill OK'd," *McClatchy Newspapers*, April 30, 2008. http://www.mcclatchydc.com/2008/04/30/35337/clinton-blasts-bush-for-project.html#storylink=cpy.

34 Scott L. Wheeler, "Missile Technology Sent to China," *INSIGHT Magazine*, February 18, 2003. http://www.freerepublic.com/focus/f-news/836496/posts.

35 John Tkacik, Jr., *Magnequench: CFIUS and China's Thirst for U.S. Defense Technology* (Washington, DC: The Heritage Foundation, WebMemo 1913, May 2, 2008). www.heritage.org/Research/AsiaandthePacific /wm1913.cfm.

36 Scott L. Wheeler, "Missile Technology Sent to China," *INSIGHT Magazine*, February 18, 2003.

37 Susan Erler, "Production Halted, Doors Closing at Magnequench Plant," *NWI Times*, September 30, 2003. http://nwitimes.com/articles/2003/09/30/business/business/9f898fd15142a5dc86256db00077cc88.txt.

38 Jeffrey St. Clair, "Outsourcing US Guided Missile Technology," *Counterpunch* (October 25-27, 2003). http://www.counterpunch.org/2003/10/25/outsourcing-us-guided-missile-technology/.

39 Ibid.

John Tkacik, Jr., *Magnequench: CFIUS and China's Thirst for U.S. Defense Technology* (Washington, DC: The Heritage Foundation, WebMemo 1913, May 2, 2008). www.heritage.org/Research/AsiaandthePacific /wm1913. cfm.

40 Lisa Margonelli, "Clean Energy's Dirty Little Secret," *The Atlantic*, May 2009, http://www. theatlantic.com/magazine/archive/2009/05/ clean-energys-dirty-little-secret/307377/.

41 Brian J. Fifarek, Francisco M. Veloso, and Cliff I. Davidson, "Offshoring Technology Innovation: A Case Study of Rare Earth Technology," *Journal of Operations Management* 26 (2008): 222-238.

42 Brian J. Fifarek, Francisco M. Veloso, and Cliff I. Davidson, "Offshoring Technology Innovation: A Case Study of Rare Earth Technology," *Journal of Operations Management* 26 (2008).

43 Cindy Hurst, "*China's Rare Earth Elements Industry: What Can the West Learn?*" (Washington, DC: Institute for the Analysis of Global Security (March 2010). http:// fmso.leavenworth.army.mil/documents/rareearth.pdf.

Geno Jezek, "How Magnets Work." http://www. howmagnetswork.com/rare_earth.html.

44 Leslie Hook, "China rare earth metals soar as Beijing cuts sales," *The Financial Times* (May 27 2011), 23. http://www.ft.com/intl/cms/s/0/751cab5a-87b8-11e0-a6de-00144feabdc0.html.

Javier Blas, "In their element," *The Financial Times* (January 29, 2010), 7. http://www.ft.com/intl/cms/s/0/ de71da64-0c74-11df-a941-00144feabdc0.html

Leslie Hook and Jonathan Soble, "China's Rare Earth Stranglehold in Spotlight," *The Financial Times*, March 13, 2012.

45 "China's Territorial Disputes," *The New York Times*, September 27, 2010.

Mure Dickie, "Japan Reacts Over Rare Earths Ban," *The Financial Times*, September 28, 2010. http:// www.ft.com/intl/cms/s/0/4781ea2e-cb2b-11df-95c0-00144feab49a.html.

Hiroko Tabuchi, "Block on Minerals Called Threat to Japan's Economy," *The New York Times*, September 28, 2010. http://www.nytimes.com/2010/09/29/business/ global/29rare.html.

46 "Companies Seek to Reduce Reliance on Chinese Rare Earths," *The Asahi Shimbun*, September 25, 2012. http://ajw.asahi.com/article/asia/china/AJ201209250059.

Justin Pugsley, "Innovations from Japan do not bode well for primary rare earths demand," Metal-Pages Blog, March 5, 2013. http://blog.metal-pages.com/2013/03/05/ innovations-from-japan-do-not-bode-well-for-primary-rare-earths-demand/.

47 Molycorp Minerals, LLC – Form S-4 (2012). http:// sec.passfail.com/filing/s-4/0001489137-12-000013/ cik-1562261/.

See also, Mani, "Molycorp: Remain On The Sidelines As Mountain Pass Ramp Up Will Take Time," *IStockAnalyst* (January 14, 2013). http://www.istockanalyst.com/ finance/story/6234738/molycorp-remain-on-the-sidelines-as-mountain-pass-ramp-up-will-take-time.

48 Richard McCormack, "China's Control of High-Tech Magnet Industry Raises U.S. Alarms," *Manufacturing & Technology News,* 16 (September 30, 2009), 1-3

National Defense University, Industrial College of the Armed Forces, *Strategic Materials Industry 2010* (Washington, DC: National Defense University, Spring 2010). http://www.ndu.edu/icaf/programs/academic/ industry/reports/2010/pdf/icaf-is-report-strategic-mat-2010.pdf.

49 Marc Humphries, *Rare Earth Elements: The Global Supply Chain* (Washington, D.C.: Congressional Research Service R41347, September 6, 2011), 4.

50 Chris Bryant, "Industry Explores Rare Earth Options," *The Financial Times*, November 2, 2011. http:// www.ft.com/intl/cms/s/0/c70e916e-f4c1-11e0-a286-00144feab49a.html#axzz25ccLcBWs.

51 "China Rare Earths Quota Spurs Japan*." The Financial Times*, March 13, 2012.

52 Oakdene Hollins Research & Consulting, "Lanthanide Resources and Alternatives, A Report for Department for Transport and Department for Business, Innovation and Skills" (May 2010). http://www.oakdenehollins.co.uk/ metals-mining.php.

T. Anand, B. Mishra, D. Apelian and B. Blanpain, "The Case for Recycling of Rare Earth Metals — A CR3 Communication," *JOM: The Journal of The Minerals, Metals & Materials Society* 63, no. 6 (June 2011): 8-9. http://www.wpi.edu/Images/CMS/MPI-CR3/Case_of_ Recycling.pdf.

53 DG Enterprise and Industry - The Ad-hoc Working Group on Defining Critical Raw Materials, "Critical Raw Materials for the EU". (Brussels: European Commission, June 2010). http://ec.europa.eu/enterprise/policies/ raw-materials/files/docs/report-b_en.pdf.

European Commission, "A Resource-Efficient Europe: Flagship Initiative under the Europe 2020 Strategy" COM (2011) 21.

German Federal Government, Ministry of Industry and Technology and Science. http://www.bmwi.de/ BMWi/Navigation/Wirtschaft/Industrie/Ressourcen/ rohstoffpolitik,did=383302.html.

54 Öko-Institut, "Study on Rare Earths and Their Recycling," Darmstadt (January 2011).

55 Defense National Stockpile Center is a branch of the U.S. Defense Logistics Agency and stores and sells raw materials. Materials it holds include aluminum oxide, beryllium, chromium, cobalt, diamonds, ferrochromium, ferromanganese, iodine, iridium, mica, niobium, platinum group metals, talc, tantalum, thorium, tin, tungsten and zinc. It has 12 locations where it stores raw materials.

56 "USMMA Calls for Rare Earth Strategic Reserve," *Business Wire*, February 23, 2011. http://www. businesswire.com/news/home/20110223006331/en/ USMMA-Calls-Rare-Earth-Strategic-Reserve.

57 Defense National Stockpile Center, *Reconfiguration of the National Defense Stockpile* (Washington, DC: Defense National Stockpile Center, April 2009). http:// www.acq.osd.mil/mibp/docs/nds_reconfiguration_ report_to_congress.pdf.

Justin C. Davey, "Enduring Attraction: America's Dependence On and Need to Secure Its Supply of Permanent Magnets," *Joint Forces Quarterly* 63, no. 4 (2011).

58 The Department of Energy listed dysprosium as the most critical element resource, in terms of both importance to clean energy technology and in terms of vulnerability of supply. Demand for dysprosium is expected to exceed current global production in the short term. United States Department of Energy, *Critical Materials Strategy* (Washington, D.C. DOE/PI-0009, December 2010). http://energy.gov/sites/prod/files/DOE_CMS2011_FINAL_Full.pdf.

59 Marc Humphries, *Rare Earth Elements: The Global Supply Chain* (Washington, D.C.: Congressional Research Service R41347, September 6, 2011).

60 UK House of Parliament, Office of Science and Technology, "Rare Earth Metals," POSTNOTE 368 (January 2011). www.parliament.uk/briefing-papers/POST-PN-368.pdf.

61 Ibid.

Peter Dent, "Rare Earth Future," *Magnetics Technology International* (2011), 56.

Peter Dent, *Rare Earth Materials Supply Chain* (November 11, 2011).

62 U.S. Department of Energy, *Critical Materials Strategy* (December 2011).

Cindy Hurst, "The Rare Earth Dilemma: Trading OPEC for China," *The Metal's Edge*, November 8, 2010. http://www.thecuttingedgenews.com/index.php?article=21774.

63 W. D. Jones, "The Rare-Earth Bottleneck," *IEEE Spectrum* 47 (2010), 80.

64 Dysprosium can be substituted for a small amount of neodymium in NdFeB magnets without affecting quality.

65 In March 2012, Molycorp, Inc., decided to acquire one of the world's leading processor of RE oxides (Canada-based Neo Material Technologies, which merged with Magnequench in 2005) for $1.3 billion. Molycorp is planning to export oxides to be processed at the plant acquired from Neo Material Technology, in China. Molycorp's decision defeats the whole purpose of restarting RE mining in the U.S. It reinforces China's role as the fabricator of specialized magnets, and undermines the revival of an U.S. industry that designs and fabricates high-tech magnets used in consumer electronics and sophisticated weapons. In light of its investment of substantial amounts of money and time to re-open its rare earth mine, Molycorp's plans are troubling and feed into China's long-term strategic goal of encouraging domestic vertical integration and capturing foreign technology. Ed Crooks, "Molycorp to start China rare earth exports," The Financial Times, March 9, 2012. http://www.ft.com/intl/cms/s/0/96d95eee-6a09-11e1-a26e-00144feabdc0.html.

Larry Bell, "China's Rare Earth Metals Monopoly Needn't Put An Electronics Stranglehold On America," Forbes, April 15, 2012. http://www.forbes.com/sites/larrybell/2012/04/15/chinas-rare-earth-metals-monopoly-neednt-put-an-electronics-stranglehold-on-america/.

Ed Crooks, "Molycorp to Start China Rare Earth Exports," The Financial Times, March 9, 2012. http://www.ft.com/intl/cms/s/0/96d95eee-6a09-11e1-a26e-00144feabdc0.html.

66 "U.S. and Alaska State Senators Introduce Legislation to Advance Bokan," Ucore press release, February 5, 2013. http://ucore.com/u-s-and-alaska-state-senators-introduce-legislation-to-advance-bokan.

67 Ucore has been selected as a DoD partner because of its mining claim in southern Alaska, and because it has the completely revolutionary technology (solid-phase extraction [SPE] metallurgy) to extract the valuable oxides.

68 Janice Karin, *Nanoparticle Magnets Conserve Rare Earth Metals* (July 26, 2011). http://thefutureofthings.com/news/11220/nanoparticle-magnets-conserve-rare-earth-metals.html.

69 Jim Witkin, "A Push to Make Motors With Fewer Rare Earths," *The New York Times*, April 20, 2012. http://www.nytimes.com/2012/04/22/automobiles/a-push-to-make-motors-with-fewer-rare-earths.html?pagewanted=all.

70 George Hadjipanayis and Alexander Gabay, "Magnet Makeover*,"* *IEEE SPECTRUM* 48 (August 2011), 32-36.

71 D. Bauer, et.al. *U.S. Department of Energy Critical Materials Strategy* (December 2010).

Jeremy Hsu, "Scientists Race to Engineer a New Magnet for Electronics," *Live Sciences*, April 10, 2010. http://www.livescience.com/8187-scientists-race-engineer-magnet-electronics.html.

72 "Industry News: Metals, Polymers, Ceramics," *Advanced Materials and Processes* (March 2008). http://www.asminternational.org/pdf/innovations/innovations_metpolcer.pdf.

73 NovaTorque is located in Fremont, CA (94538). NovaTorque, http://www.novatorque.com/

74 Daniel McGroarty and Sandra Wirtz, *Reviewing Risk: Critical Metals & National Security* (Washington, D.C.: American Resources Policy Network, June 6, 2012). http://americanresources.org/wp-content/uploads/2012/06/ARPN_Quarterly_Report_WEB.pdf.

75 Kia Ghorashi, Lucinda Gibbs, Polly Hand, and Amber Luong, *Rare Earth Elements: Strategies to Ensure Domestic Supply*, (Stanford, CA: Bay Area Council and the Breakthrough Institute, March 10, 2011). http://publicpolicy.stanford.edu/system/files/RareEarthElements.pdf.

76 Department of Energy Advanced Research Projects Agency-Energy, *Rare Earth Alternatives in Critical Technologies* (REACT). http://arpa-e.energy.gov/ProgramsProjects/REACT.aspx.

Jim Witkin, "A Push to Make Motors With Fewer Rare Earths," *The New York Times*, April 20, 2012. http://www.nytimes.com/2012/04/22/automobiles/a-push-to-make-motors-with-fewer-rare-earths.html?pagewanted=all.

77 Daniel McGroarty, "The Pentagon's Metals Gap," *RealClearPolitics*, March 6, 2013. http://www.realclearworld.com/articles/2013/03/06/sequestration_pentagons_metals_gap_100593-full.html.

CHAPTER 6 • FASTENERS

EXECUTIVE SUMMARY

Fasteners hold larger component parts together, and are used in a broad range of military and civilian end-items. Fasteners are almost always found in the lower tiers of the supply chain as component parts of sub-assemblies. They typically receive little attention; however, they are essential components of virtually every piece of military equipment, including aircraft, ships, and vehicles. Many weapons systems, such as stealth aircraft, need specialty fasteners that require significant investments in design. Because fasteners literally hold military equipment together, the failure of a fastener or the corruption of fasteners in the defense supply chain can disrupt equipment supply chains and lead to mission failure or combat casualties. Because fasteners are found throughout the defense supply chain, the integrity of the fasteners supply chain is important for American national security. Fasteners are prone to counterfeiting because of the complexity and opacity of defense supply chains, and because fasteners enter these supply chains at the lower tiers. Prime contractors of military equipment are often several steps removed from the companies that fabricate fasteners, making it difficult to trace faulty fasteners to their origins and hold the counterfeiters accountable. This distance increases the risk that counterfeit fasteners will enter the military supply chain undetected.

Increased competition from foreign fasteners manufacturers, in conjunction with defense spending cuts, has led to a higher prevalence of foreign-manufactured fasteners in defense supply chains. Though the 1990 Fastener Quality Act (FQA) imposed some standards, these standards were only partially implemented, and the burden of enforcement typically falls on the end-user. Additionally, efforts to standardize the industry globally created avenues for circumventing quality controls, resulting in fraudulent certifications and, ultimately, substandard fasteners.

Government and industry collaboration is critical to the success of efforts to protect the fasteners supply chain. Recent studies (2010, 2012) by the U.S. Government Accountability Office (GAO), have documented the presence of substandard and counterfeit fasteners in the defense supply chain, but have concluded that there is not enough

FASTENERS
BUILDING BLOCKS OF U.S. DEFENSE EQUIPMENT

MANUFACTURING SECURITY

Fasteners are an essential component across all military items

FROM THE SMALLEST WEAPON TO THE LARGEST DOD PLATFORM

FASTENER FAILURE

The risks associated with fastener failure include:

DELAYS IN DELIVERY **OF CRITICAL SYSTEMS**

DECLINE IN CRITICAL **COMBAT READINESS**

COMBAT CASUALTIES

MISSION FAILURE

VULNERABILITY

Dangerous counterfeit fasteners are difficult to trace back to the original manufacturer

TRACING FASTENERS

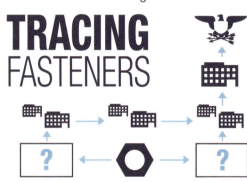

U.S. POLICY MUST **ASSURE QUALITY**

WITHOUT CREATING **A BURDEN FOR U.S. INDUSTRY**

MITIGATING RISKS

Avoiding dangerous uncertainty in fasteners supply chains

TIGHTEN ANTI-COUNTERFEITING

REVIEW QUALITY ASSURANCE

U.S. MADE AND MELTED

UNDERSTAND SUPPLY CHAINS

MILITARY EQUIPMENT CHART
SELECTED DEFENSE USES OF FASTENERS

DEPARTMENT	WEAPON SYSTEMS	PLATFORMS
ARMY	Most Army Weapon Systems Including: ■ M16 Assault Rifle ■ M110 Sniper Rifle ■ M1014 Combat Shotgun ■ M252 81mm Mortar ■ M198 Howitzer ■ High Mobility Artillery Rocket System	All U.S. Army Platforms Including: ■ M1 Abrams Main Battle Tank ■ HMMWV (Humvee) ■ UH-60 Blackhawk Helicopter
MARINE CORPS	Most Marine Corps Weapon Systems Including: ■ M16 Assault Rifle ■ M110 Sniper Rifle ■ M1014 Combat Shotgun ■ M252 81 MM Mortar ■ M198 Howitzer ■ High Mobility Artillery Rocket System	All U.S. Marine Corps Platforms Including: ■ AAV-7A1 Assault Amphibious Vehicle ■ M1 Abrams Main Battle Tank
NAVY	Most Navy Weapon Systems Including: ■ 57 Mk110 Naval Gun System	All U.S. Naval Platforms Including: ■ Nimitz-Class Nuclear-Powered Aircraft Carrier ■ Littoral Combat Ship (LCS) ■ Virginia-Class Nuclear-Powered Attack Submarine
AIR FORCE	Most Air Force Weapon Systems Including: ■ Joint Direct Attack Munition (JDAM) Smart Guidance Kit	All U.S. Air Force Platforms Including: ■ HH-60 Pave Hawk Helicopter ■ F-15 Eagle Fighter ■ F-22 Raptor Fighter ■ F-35 Joint Strike Fighter ■ C-130 Military Transport Aircraft

evidence about the scope of the problem. Understanding of the scope of the problem is necessary for industry and government to review the adequacy of fasteners quality assurance measures, including improved traceability to fasteners manufacturers. Addressing risks to the fasteners supply chain requires reliable data on fastener failure, which in turn requires better procedures for documenting and preventing the entry of substandard and counterfeit fasteners into the supply chain, as well as a better understanding of the lower tiers of the defense industrial base.

INTRODUCTION

"For want of a nail, the shoe was lost. For want of a shoe, the horse was lost."

Buying commercial-off-the-shelf (COTS) defense items saves money, leverages innovation, and enhances supply chain efficiency, with benefits to cost, innovation, and efficiency. The fasteners sector is no exception, particularly with the acceleration of globalization. However, embracing the use of COTS items in the supply chain has caused vulnerabilities, including lack of traceability, introduction of counterfeit or substandard parts, and potential weakening of industry standards. Each of these problems could lead to supply chain disruption and equipment failure. Although the U.S. domestic fasteners industry reliably produces the high-performance, specialty fasteners required for U.S. defense use, the fasteners supply chain is potentially compromised by these vulnerabilities, constituting a threat to an industry that is vital to U.S. national security.

Fasteners hold larger component parts together, and are used throughout a broad range of military and civilian end-items. Fasteners are almost always found in the lower tiers of the supply chain as component parts of sub-assemblies. The U.S. domestic defense fastener industry is a relatively small part of the U.S. domestic fasteners industry, and an even smaller part of the global fasteners industry. Original equipment manufacturers (OEMs) obtain fasteners directly from manufacturers (either foreign or domestic) and from distributors (again, either foreign or domestic). In turn, manufacturers sell to distributors who sell to the Department of Defense (DoD). Domestic fastener distributors typically provide fasteners to OEMs within the same geographical region to take advantage of supply chain efficiencies.

Fasteners are critical to the defense industrial base. The fasteners industry is comprised of three segments: industrial (a very broad category that includes nuts, bolts, and screws), automotive, and aerospace. Industrial fasteners and automotive fasteners are also components of defense end-items, but these fasteners are almost always found in the commercial market, so the defense market is rarely unique for these segments.

Defense critical issues primarily tend to reside within—but are not confined to—the aerospace fastener segment of the industry.

Most defense-unique fasteners are aerospace fasteners made to military specifications (Mil-Spec) rather than COTS. These fasteners often must meet special standards for defense aerospace use and are manufactured using exotic materials and complex design geometry. (For example, stealth aircraft fasteners must themselves be "stealth" quality.) Military aerospace fasteners often have years of research and development (R&D) behind them. Because of their special characteristics, these aerospace fasteners are critical to the performance of many defense end-items. On the other hand, fasteners used for military aerospace purposes are not necessarily defense-unique; indeed, most aerospace fasteners are civilian-military dual-use. U.S. aerospace industries (especially military aerospace industries) have an interest in preserving the health of the aerospace fasteners sector because military aerospace systems with counterfeit or substandard fasteners risk mission failure, which has national security consequences. Therefore, military aerospace fasteners supplied to OEMs of military airframe, engine, and flight components, and in turn to DoD, must comply with defense procurement agency constraints.

The U.S. Government Accountability Office (GAO) is concerned about the risk of counterfeit and substandard fasteners in the defense supply chain. A GAO report issued in March 2010 addressed to U.S. Senators Sherrod Brown and Evan Bayh (in their capacities as Chairmen of Subcommittees on the Committee on Banking, Housing, and Urban Affairs), stated that "Almost anything is at risk of being counterfeited, from fasteners used on aircraft to electronics used on missile guidance systems, and materials used in body armor and engine mounts."[1] Moreover, the report noted that because DoD procures parts from a "large network of suppliers in an increasingly global supply chain, there can be limited

Almost anything is at risk of being counterfeited, from fasteners used on aircraft to electronics used on missile guidance systems.[a]

visibility into these sources and greater risk of procuring counterfeit parts."[2] Then in March 2012, a GAO report of an investigation conducted in cooperation with the Defense Logistics Agency (DLA) concluded that there were pervasive risks to the supply chain from introduction of counterfeit parts. Although the 2012 report noted that the GAO sampling was not broad enough to estimate the extent to which parts are being counterfeited, it stated that DLA views the problems seriously and pledges steps to remove counterfeit parts from their supply chain, including continuing to require suppliers "to provide documentation as a means of assuring the part's origin can be traced back to its original manufacturer."[3]

Industry experts have also addressed the counterfeit fasteners issue, emphasizing the need for awareness and countermeasures. In its March 2011 report entitled *Counterfeit Parts: Increasing Awareness and Developing Countermeasures,* the Aerospace Industries Association (AIA) noted the uncertainty associated with predicting the time and place of counterfeits' entry into the aerospace supply chain. The report called for more collaboration between industry and government to increase diligence and active control measures to prevent counterfeit parts from entering the supply chain.[4]

This chapter will investigate the vulnerabilities in the U.S. defense industrial base as they pertain to the U.S. domestic fasteners industry, and demonstrate how these vulnerabilities pose risks to U.S. national security. This chapter will describe how these vulnerabilities have persisted over the past two decades, as defense procurement shifted from military-specification-based procurement (Mil-Spec) to COTS. More importantly, as globalization has changed the characteristics of the defense supply chain in ways that are essential to industrial health and potentially irreversible, this chapter will highlight the issues that must be addressed if we are to prevent coincident insidious effects. This chapter will conclude with recommendations of how best to address the vulnerabilities, solutions that depend upon collaboration between government and throughout the commercial and defense sectors of the industry.

Key themes discussed in this chapter are:

- The fasteners supply chain is at risk from the introduction of counterfeit and substandard fasteners, but insufficient data is available to determine the scope of the problem.

- Fasteners industry standards are key to ensuring quality products, but current amendments to those standards potentially weaken them.

- The risks to defense readiness and mission success can be effectively mitigated with increased supply chain visibility, thorough implementation of industry control measures, tightening of the Fasteners Quality Act (FQA), and "made and melted in the U.S.A." requirements for defense critical fasteners.

A NOTE ON CRITICALITY

The uncertainty surrounding the quality of fasteners used in the defense industrial base results in part from an increasing rate of counterfeit and substandard parts being introduced at lower tiers of the supply chain. The increase in counterfeit fasteners seems to correspond with the decline of domestic production; however, there is little reliable information about the scale of the counterfeit problem, because it is difficult to identify substandard and counterfeit fasteners until after they fail, and because it is difficult to track counterfeits within the defense supply chain.[5] Because fasteners almost always are found among the lowest tiers of the supply chain, fasteners production is often far removed from the OEMs who use them. This separation results in a lack of traceability and low accountability. This places the fasteners supply chain in jeopardy, and presents a *high* risk originating from this node of the defense industrial base.

These risks could have an *incapacitating* impact on U.S. defense capabilities. As a worst-case scenario, a single batch of substandard, counterfeit fasteners could lead to the catastrophic failure of key defense weapons systems, such as military aircraft. Because of the low-tier location of fasteners in the aircraft production line, and because of the uncertainty about the extent of the counterfeiting problem, failure of one aircraft due to substandard fasteners would suggest an increased potential for similar failure in other aircraft. Any system making use of a fastener from that batch could potentially be compromised. Although it is more likely that the problem would probably be isolated to specific weapons systems, we cannot rule out the possibility that naval vessels,

FASTENERS THAT FAIL CAUSE MISSIONS TO FAIL (a notional though realistic scenario)

Elements of B Troop deploy along the Iran/Afghanistan border to interdict illicit trade. Due to the unit's remote location and the threat of insurgent ambushes, B Troop requires aerial resupply. Resupplies are unavailable, however, because substandard main rotor blade fasteners have caused maintenance delays for rotary aircraft. The commander decides that rations, ammunition, and medical supplies will be airdropped by a Kandahar-based C-130 in order to sustain combat operations.

The C-130 drops five pallets, releasing them at 1,200 feet above ground level. Although three pallets land safely, fastener failures in the parachute release systems cause one pallet containing ammunition and another carrying water to land incorrectly. B Troop manages to salvage some of these supplies. Insurgent forces attack B Troop that evening. Although it repels the attacks, B Troop is now critical in water and ammunition. Given these shortages, the commander decides to move the unit back to the Forward Operating Base until resupply can be re-established. This relocation allows insurgents armed with Man-Portable Air Defense Systems to move through the area unopposed, posing an increased threat to fixed-wing transport aircraft landing in Kabul and Bagram.

recently refitted Mine-Resistant Ambush-Protected Vehicles (MRAPs), or military aircraft are exposed to the vulnerability of counterfeit or non-conforming fasteners.

BACKGROUND

As component parts of component parts, and located in the lower tiers of the supply chain, the fasteners market flies under the radar.

According to the Industrial Fasteners Institute (IFI), the trade association for the U.S. fasteners industry's manufacturers, a fastener is "a mechanical device for holding two or more bodies in definite position with respect to each other. A high percentage of fasteners have threads as part of their design, but unthreaded items such as rivets, clevis pins, machine pins, etc., are considered fasteners as well."[6] The fasteners industry includes manufacturers

of metal bolts; nuts; screws; rivets; washers; formed and threaded wire goods; and special industrial fasteners. Using primarily carbon and steel alloys, fastener manufacturers employ a variety of cold-forming and rolling processes to produce simple parts with greater strength.[7]

The modern fasteners industry originated in the early stages of industrialization, and this timing enabled machine-based innovations in fastener production.[8] The invention of the bar lathe, which allowed the manufacturing of accurate and duplicable threads, initiated an era of standardization. That progress has evolved into the contemporary standards applied to modern fasteners: reliability, precision, and interchangeability.[9]

The U.S. machine-produced fasteners industry began in the mid-1830s. Following the Civil War, (during which the fasteners industry benefited from the production of

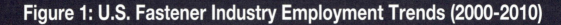

Figure 1: U.S. Fastener Industry Employment Trends (2000-2010)

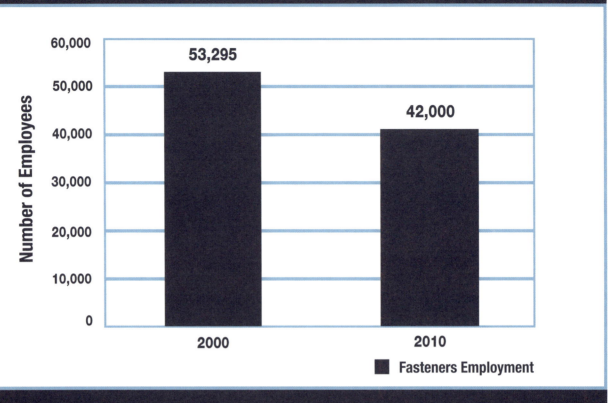

Sources: *Reference for Business, "Bolts, Nuts, Screws, Rivets, and Washers SIC 3452," Encyclopedia of Business, 2nd Edition, (Advameg, Inc., data accessed March 26, 2012) http://www.referenceforbusiness.com/industries/Fabricated-Metal/Bolts-Nuts-Screws-Rivets-Washers.html; Industrial Fasteners Institute, 2010 Abbreviated Annual Report, p. 8. http://www.indfast.org/assets/pdf/2010-Abbreviated-Annual-Report.pdf*

firearms, machinery, and railroad equipment), the U.S. fasteners industry shifted from the Northeast to the Midwest to take advantage of burgeoning transportation networks and iron and steel production facilities.[10]

By 1969, the U.S. fastener industry had reached its production peak, with 450 companies operating 600 plants, employing more than 50,000 workers, and manufacturing more than two billion fasteners per year. However, by 1984, the U.S. fastener industry had shrunk to 250 manufacturers operating 350 plants and employing 35,000 people as a result

of severe challenges from foreign competition and dramatic changes in OEM requirements.

The biggest challenge came from foreign fasteners producers, who took advantage of inexpensive offshore labor and material costs. U.S. domestic fasteners manufacturers were seriously threatened by this influx of cheap, foreign-made fasteners. Domestic manufacturers went from supplying 80 percent of American bolts, nuts, and large screws in 1969 to just 44 percent in 1984. The U.S. fasteners industry began to rebound in the mid-1980s. Fasteners manufacturers allied themselves with companies in need of technically sophisticated products rather than simple standardized commodities, and the falling value of the U.S. dollar drove up the prices of foreign products. During the same period, OEMs (especially automobile manufacturers) pressed fastener manufacturers to develop specialized products at lower costs. The production of these items sustained many companies, but drove smaller, less technologically advanced companies out of the industry.

Fasteners manufacturers have traditionally clustered around OEMs in the automotive, defense, and aerospace industries. The modern American fasteners industry is now largely concentrated in the auto-producing states of the upper Midwest and the defense and aerospace-oriented regions of Southern California. However, automobile industry slumps during the 1980s posed severe challenges to fasteners manufacturers, and defense downsizing in the 1990s posed an equally significant threat to the aerospace industry.[11]

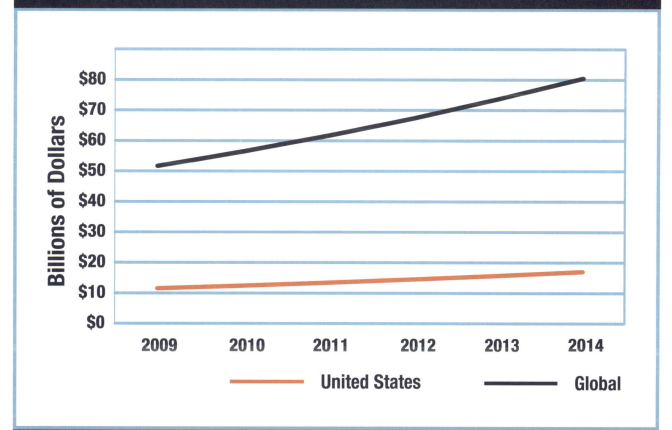

Figure 2: Demand for Fasteners in Revenues (2009-2014)

(Note: Based on current fastener consumption and industry forecasts of an 8 percent annual increase in U.S. demand and an 8.5 percent annual increase in global demand)

Source: IFI, 2010 Abbreviated Annual Report, p. 8-9. http://www.indfast.org/assets/pdf/2010-Abbreviated-Annual-Report.pdf

Since 1990, demands for stronger, lighter, and easier-to-use products have led to improved technology in the fasteners industry. Buyers demanded innovative and diverse fasteners, such as self-locking, self-cinching, or self-sealing screws bolts, nuts, and threaded inserts. Fasteners manufacturers also developed more environmentally friendly products (such as fasteners that maintain lubricity without the use of plating materials such as cadmium, a suspected carcinogen).[12]

As a result of several large fasteners industry mergers and acquisitions in the 1990s and 2000s,[13] the U.S. domestic fasteners industry experienced another decline in the number of manufacturing entities and number of workers employed. In 2010, IFI reported that there were nearly 840 fasteners manufacturing entities (a decline from 937 in 2004), of which about 350 were believed to be substantial manufacturing entities. Meanwhile, U.S. fasteners industry employment declined from 53,295 in 2000 to 42,000 in 2010.

On the other hand, demand for fasteners continues to grow. In 2010, U.S. demand for fasteners was projected to grow eight percent per year. World fasteners demand is projected to grow at an average of 8.5 percent per year through 2014 (with China itself creating 25 percent of that demand). This projected growth would equate to $80.5 billion in fasteners consumption in 2014.[14]

IFI data from 2010 shows domestic fasteners industry production revenues in the $10.5 to $11 billion range. In 2010, the industry operated at just over 64 percent of capacity. IFI's 2009 data shows about 50 upstream industries supplying the fasteners industry, while product flowed to 352 identified downstream fasteners-using industries. U.S. consumption of fasteners is estimated to be in the $11.5 to $12 billion range, with $3.5 billion in imports and $2.5 billion in exports. Three countries accounted for 68 percent of fastener imports to the United States: Taiwan (31 percent), China (23 percent), and Japan (14 percent). Sixty-eight percent of U.S. fasteners exports went to either Canada or Mexico.[15]

An IFI assessment of the segments of the U.S. domestic fasteners industry, including aerospace and defense aerospace, found them to be healthy. The U.S. domestic fasteners industry expects to continue to respond reliably to demands, including from the defense industry. According to an IFI spokesperson, there is no current risk to domestic aerospace fastener production capacity.[16] In 2010, IFI projected that the aerospace fasteners sector, having experienced sustained growth during the previous decade of nine percent annually, would remain a healthy and profitable segment of the industry for the foreseeable future. U.S. fasteners manufacturers for the defense aerospace sector, whose specifications usually call for high-end products for critical applications, appear to be well-positioned to meet increased demand for aerospace fasteners.

The projected decrease in defense spending is not expected to depress demand for fasteners market-wide, as defense is a relatively small part of the entire domestic fasteners industry. However, the cumulative effect of the decline in defense spending, especially as military aircraft procurements and programs are cut, likely will have a dampening effect on the military aerospace industry at large. Defense cuts are expected to affect military aerospace R&D in turn, as the military aerospace industry likely will decline in relation to the civilian aerospace industry. This could potentially lead to negative consequences for the manufacturers of defense-unique aerospace fasteners and could increase industry response time for newly identified defense-critical requirements.

The United States has a competitive edge in the aerospace sector because of the sheer size and technological superiority of the U.S. aerospace industry versus the global aerospace industry. However, this advantage does not necessarily translate to a competitive edge at the lower levels of the supply chain, because both OEMs and distributors are incentivized to lower costs by outsourcing supply to cheaper foreign manufacturers. Additionally, distributors and foreign manufacturers may be incentivized to reduce costs by exercising their option for exemption from quality assurance measures such as the FQA.

STANDARDS – THE FASTENER QUALITY ACT (FQA)

The FQA of 1990 was passed in response to complaints in the 1980s about poor quality and fraud on the part of foreign fasteners manufacturers, and was meant to ensure that the domestic fasteners industry did not have to compete against substandard foreign parts.[17] Fasteners standards are the cornerstone of the fasteners industry and are the basis for production, procurement, and use during the engineering process on any complex equipment. Fasteners must conform to (and should be routinely tested against) standards for hardness, surface hardness, load-bearing strength, and axial and wedge tensile strength.[18] However, quality assurance systems that are meant to assure conformity break down when the standards are inadequately applied. Most important,

substandard quality fasteners can potentially cause equipment failure, which can mean mission failure for defense systems.

Fasteners industry standards are similar to most voluntary standards and also must be consistently applied. Instances of non-compliance with fastener quality standards date back to the 1980s, with widespread reports of unethical manufacturers and distributors misrepresenting standards,[19] and "bogus bolts" leading to equipment failure and loss of life.[20] Enforcement problems were exacerbated because the burden of enforcement was (and remains) on the end-user of the fastener. Although the supplier (either the manufacturer or distributor) is responsible for failure to comply with the voluntary standard, the customer (end-user) is responsible for enforcing compliance by requesting appropriate conformance records from the supplier.

In 1986, media reporting of "bogus bolts" that failed to withstand high loads, leading to equipment failure and, in one case, loss of life, led IFI and other industry advocates to urge Congressional investigation.[21] The investigation determined that millions of fasteners (among the billions on the market) were mismarked, substandard, or counterfeit.[22] Among other measures taken in response to the investigation, one notorious foreign fastener supplier, Voi-Shan, was suspended from doing business with the government for "routinely falsified manufacturing reports and test results from January 1980 through February 1989."[23]

In 1988, a U.S. House subcommittee report concluded that "the failure of substandard and often counterfeit fasteners has killed people, reduced our defense readiness, and cost both the American taxpayer and the American industry untold millions in breakdowns, downtime, reconstruction, and other unnecessary inefficiencies."[24] Moreover, the report noted that the substandard and counterfeit fasteners at fault were largely foreign-made. With strong support from the domestic fasteners industry, Congress passed Public Law 101-592, the FQA of 1990, which provided for the "testing, certification, and distribution of certain fasteners used in commerce within the United States."[25]

> "The Act protects the public safety by: (1) requiring that certain fasteners sold in commerce conform to the specifications to which they are represented to be manufactured, (2) providing for accreditation of laboratories engaged in fastener testing, and (3) requiring inspection, testing and certification in accordance with standardized methods."[26]

The investigation called into question the quality of foreign-made fasteners and affirmed the quality of United States-made fasteners. As a result, the FQA implemented stricter control on foreign-produced fasteners. One consequence of the FQA's was a soaring demand for U.S.-made fasteners because of their reliable quality control records.[27]

Despite passage, the FQA was only partially implemented. A contentious, decade-long process culminated in significant amendments to the FQA, which enabled opt-outs to the act's provisions. The first reason for delay in implementation was the government's reluctance to interfere with the $6 billion domestic fasteners industry. Further, the government wanted to give the fasteners industry time to establish approved testing facilities. (In response, more than 400 testing facilities were operational by 1999.[28]) A second reason for delay was that, since the FQA's passage

in 1990, the U.S. fasteners industry lobbied hard for amendments to the FQA that would ease the regulatory burden on the industry.[29] Despite their strong support for the FQA of 1990, by the mid-1990s the predominant position of the U.S. fasteners industry was that full implementation of the FQA would unduly burden the industry. Many in the U.S. fasteners industry believed that they could assure a safe, high-quality produce simply through the industry's own internal policies, self-policing, and record-keeping. "Powerful forces, including the American Automobile Manufacturers Association—the nation's largest user of industrial fasteners—and the General Aviation Manufacturers Association want to defang the Fastener Quality Act," wrote David Sharp in *Engineering News Record.*[30]

A fasteners industry advisory group that studied the 1990 FQA recommended, among other changes, that the FQA be amended to allow the sale, rather than the destruction, of fasteners with minor non-conformance from standards and specifications. Long-standing industry practice permitted the sale of fasteners with minor non-conformance issues that did not affect form, fit, or function, provided the purchaser was informed of the nonconformance and agreed to purchase them. Moreover, the advisory committee recommended that the FQA be amended to allow commingling of more than two fasteners lots, despite the difficulties associated with traceability created by commingling. The fasteners industry held the position that requiring full lot traceability by manufacturers, importers, and distributors placed an unnecessary economic burden on the industry, and accordingly recommended that the FQA be amended to allow limited commingling.[31]

Figure 3: FQA Requirements Applicability Decision Chart

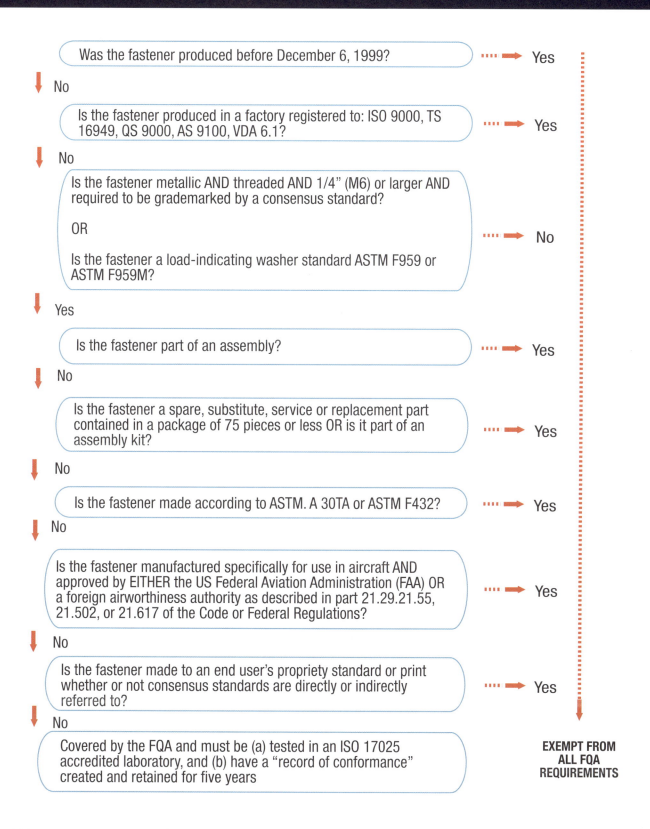

Was the fastener produced before December 6, 1999? ····▶ Yes

No

Is the fastener produced in a factory registered to: ISO 9000, TS 16949, QS 9000, AS 9100, VDA 6.1? ····▶ Yes

No

Is the fastener metallic AND threaded AND 1/4" (M6) or larger AND required to be grademarked by a consensus standard?

OR

Is the fastener a load-indicating washer standard ASTM F959 or ASTM F959M? ····▶ No

Yes

Is the fastener part of an assembly? ····▶ Yes

No

Is the fastener a spare, substitute, service or replacement part contained in a package of 75 pieces or less OR is it part of an assembly kit? ····▶ Yes

No

Is the fastener made according to ASTM. A 30TA or ASTM F432? ····▶ Yes

No

Is the fastener manufactured specifically for use in aircraft AND approved by EITHER the US Federal Aviation Administration (FAA) OR a foreign airworthiness authority as described in part 21.29.21.55, 21.502, or 21.617 of the Code or Federal Regulations? ····▶ Yes

No

Is the fastener made to an end user's propriety standard or print whether or not consensus standards are directly or indirectly referred to? ····▶ Yes

No

Covered by the FQA and must be (a) tested in an ISO 17025 accredited laboratory, and (b) have a "record of conformance" created and retained for five years

EXEMPT FROM ALL FQA REQUIREMENTS

Source: Industrial Fasteners Institute, 10 of the Most Frequently Asked Questions About the Final FQA (1999). p. 5.
http://www.indfast.org/assets/pdf/FQA%20%20Bulletin%20%2010%20Questions%20about%20Final%20FQA%20%20080401.pdf

Less than a month from the deadline for full implementation of the FQA in 1999, then President Bill Clinton signed into law a series of amendments to the FQA, supported by the American fasteners industry (including the Fastener Industry Coalition, the National Fasteners Distribution Association, and IFI), which "makes the legislation more focused and less burdensome."[32] Supporters of the amendments argued that the amended FQA shifted the focus from government-mandated regulations to more "preventative measures," and recognized the industry's initiatives to improve quality control. (For example, in the 1990s, the American fasteners industry instituted state-of-the-art manufacturing procedures such as end-of-line quality control assessments.) Accordingly, the amended FQA accepted the industry position on both nonconformance and limited commingling. In addition, the amended FQA sharply reduced the government testing requirements, limited coverage to high-strength fasteners, and allowed companies to transmit and store records and reports electronically.[33] Indeed, the combination of more robust quality control at both the distributor and manufacturer levels and the incorporation of the industry positions on nonconformance and commingling led some in the industry to proclaim that the FQA itself was redundant.[34]

Figure 3 describes the array of exemptions from the FQA, notable among which are exemptions if the production facility is registered to International Organization for Standardization (ISO) 9000, an international certification of quality control standards, or Aerospace Standard (AS) 9100 (applicable to aerospace fasteners manufacturers).

VULNERABILITIES TO THE FASTENERS SUPPLY CHAIN

Counterfeit threaded fasteners have been discovered in cranes, elevators, and forklifts; in aircraft engines and attachments (wings, tail and landing gear); and in vehicles as components of engines, brakes, and steering mechanisms. Counterfeit fasteners have also been found in nuclear facilities, in valves, compressors, and vessels used to contain radioactive fluids.[35] Moreover, the aforementioned 2010 GAO report also documented substandard and counterfeit fasteners in the following defense equipment:

- Aviation braking: Self-locking nuts, purchased from an unauthorized source, were found to be cracking.

- Air Force helicopters: Rotor retaining nuts, used to hold the rotor to the helicopter mast, were reported to be substandard. Failure of rotor retaining nuts causes the helicopter to crash. "The Air Force reported that a supplier willfully supplied a substandard rotor retaining nut, but the supplier maintained its innocence and claimed that it was unaware that the part it procured was a counterfeit part."

- Army helicopters: "The Army reported that a bolt, intended for use in helicopters, was counterfeit. The problem was detected when Army officials recognized the serial number on the part and identified it as a defective part that had been cut in half for destruction. An X-ray test confirmed the bolt had been welded back together."

- Navy aircraft: DLA procured hook point bolts, "used to stop aircraft when they land on aircraft carriers. Failure of the part could result in loss of life or aircraft. A supplier rubbed serial numbers off hooks that were too thin to use, welded additional material onto the hooks, and reused them. This problem was detected when premature part failure triggered an investigation and the welded material showed up in X-rays."[36]

Despite this evidence, the report noted that insufficient information is available to indicate the scope of the problem of counterfeit and substandard fasteners.[37] The 2010 GAO study of counterfeit parts in defense supply chains confirmed DoD's concern about the scope of the problem and acknowledged a lack of specific measures to address counterfeit parts. The report recommended that DoD conduct further efforts to collect data.[38]

The 2010 GAO study of counterfeit parts (including fasteners) determined:

- The prevalence of counterfeit parts in DoD's supply chain is unknown.

- DoD is in the early stages of gathering information on the counterfeit parts problem.

- DoD's existing practices are limited in protecting its supply chain against counterfeit parts.

- DoD relies on existing procurement and quality control practices that are not specifically designed to address counterfeit parts.

- Some DoD components and contractors have taken initial steps to address counterfeit parts.

- A number of commercial initiatives exist to mitigate the risk of counterfeit parts in supply chains.

- As DoD collects data and learns more about the nature and extent of counterfeit parts in its supply chain, additional actions may be needed to help better focus its risk-mitigation strategies.

What is certain is that counterfeit fasteners in the defense supply chain potentially jeopardize the performance, reliability, and safety of a range of aerospace, space, and other defense products. Counterfeit and substandard fasteners impact safety, security, and our nation's ability to adequately defend itself. Consequences could range from the inability to successfully deliver critical combat airdrop resupply (thereby potentially causing maneuver elements' failure to meet assigned missions), to the failure of a critical engine part in the midst of combat, to the failure of critical components of nuclear reactors aboard U.S. Navy vessels in port.

Although the FQA (as amended in 1999) clarified the standards expected of the domestic fasteners industry, additional amendments have watered it down. Exemptions from the FQA's provisions are widespread, enabling the introduction of counterfeit or substandard fasteners into the supply chain.[39] Because the industry standards essentially rely on self-policing, fasteners manufacturers can essentially opt-out at will, by registering their factories with ISO 9000 or AS 9100 certificates (or other exemption criteria). Moreover, an ISO certificate may be falsely obtained,[40] allowing exemption from FQA testing and certification requirements without incorporating the ISO's quality control certifications. Some in the U.S. fasteners industry are skeptical of the integrity of ISO 9000 procedures in offshore fasteners factories, as the ISO system inspectors are themselves certified by suspect foreign certification processes.[41]

> "Counterfeit parts—generally the misrepresentation of parts' identity or pedigree—can seriously disrupt the Department of Defense supply chain, harm weapon systems integrity, and endanger troops' lives."[b]

According to AIA, problems with counterfeit parts (including but not limited to fasteners) begins when "defense contractors and the government are obliged to purchase both electronic and non-electronic parts and materials to support fielded and new systems from independent distributors/ brokers."[42] As a result, manufacturers lose revenue. The auto industry alone estimates up to $12 billion per year is lost in global revenue due to counterfeit parts.[43] Distributors who buy fasteners for supply to OEMs rely on manufacturers to ensure that their products conform to standards.[44] However, because the FQA as currently amended relies on self-policing of the standards, less scrupulous participants in the market easily can evade the rules in order to turn more profit.

As the 2010 GAO report concluded, one problem with counterfeits in the defense fasteners market is the lack of a DoD definition of the term "counterfeit." The GAO study, also noted that many DoD logistics officials are uncertain how to define counterfeit parts.[45] Unsurprisingly, the report indicated that a common definition would be useful, and noted that "in the absence of a department-wide definition of counterfeit parts, some DoD entities have developed their own."[46] AIA's Counterfeit Parts Integrated Project Team defines counterfeit as "a product produced or altered to resemble a product without authority or

right to do so, with the intent to mislead or defraud by presenting the imitation as original or genuine."[47] However, each separate DoD program currently retains discretion on the definition and scope of "counterfeit," as well as on whether to use the standard.

Another major problem with identifying fasteners deficiencies is the difficulty in collecting failure data. The 2010 GAO report also notes the difficulty in documenting instances of counterfeit parts, in part because databases used by DoD to report deficient parts did not even include data fields for users to track suspected counterfeit parts. No code exists on maintenance records to capture a deficiency as counterfeit; instead, users must indicate suspected counterfeits in narrative descriptions. This inefficiency complicates the aggregation of suspected counterfeit parts, because not only is there a lack of common terminology in the narratives, but also automated searches for such instances are themselves much more difficult, if not impossible, to identify.[48]

There is a particular need for fasteners failure statistics on U.S. military aircraft to include documentation of the specifics of such a failure (e.g., due to corrosion, stress failure, etc.). Fastener replacement and repairs are included in other systems maintenance actions. A 1989 U.S. Air Force study concluded that substandard fasteners were the second largest problem area for aircraft maintenance personnel, with defective tools considered the only area that was worse.[49] The same study found that fasteners failure accounted for more than 40 percent of the structural failures on U.S. Air Force aircraft. Corrosion was the single most common cause of fasteners failure. (Of note, a contemporary U.S. Navy study showed that the Navy had the same deficiencies in tracking fasteners' failures.) Minor maintenance, such as tightening screws on panels during post-flight checks, was not routinely documented. Instead, fastener tightening was documented as a post-flight check, which includes other items not related to fasteners.[50]

Part of the difficulty in identifying instances of counterfeit and substandard fasteners lies in the integrity of the ISO 9000 family of international standards themselves.[51] ISO 9000 conformity means that a supplier has implemented a systematic approach to quality management, and is therefore an indirect measure of conformity of the product itself. Those who conduct inspections for ISO 9000 conformity are certified to serve as inspectors by national-level bodies. Variations in national-level certification processes raise doubts as to their integrity. ISO does not certify manufacturers, suppliers, or purchasers. Instead, the ISO 9001 designation indicates that a supplier has a Quality Management System (QMS) that meets the requirements of the ISO 9000 quality assurance system. In other words, the ISO 9000 regulations focus on the manufacturer's QMS, rather than on the item produced. The purchaser of the product retains the burden of specifying requirements to the manufacturer, as well as the burden of monitoring whether the manufacturer produces items according to the expected standards.

During his visit to China in 2006, one U.S. fasteners manufacturing representative visited 11 fasteners factories, all of which possessed ISO 9001 certificates of conformity. The representative strongly suspected some of those certificates were obtained fraudulently.[52] The same fasteners manufacturing representative is also aware of distributors who switch zinc-coated fasteners for cadmium-coated (which contain carcinogens).[53]

The problem of lack of traceability, an inherent risk of COTS procurement, and the lack of transparency in offshore supply chains complicates the difficulties associated with verifying and enforcing standards. Traceability can be defined as "the concept that a buyer can trace the history of a given lot of fasteners back through any number of distributors or vendors to the original manufacturer(s)."[54] Effective traceability requirements help prevent lot commingling, which risks mixing standard with substandard fasteners in the same lot.[55] The burden of establishing a traceability system currently lies on the customers in the supply chain. Lack of traceability has important implications for national security because it increases the possibility of introducing substandard defense-unique or defense-critical items into the supply chain. Part of the difficulty is that manufacturers have no way to track where their products go once the distributor buys them—the fasteners can go almost anywhere. The question remains how to enforce industry standards without effective traceability.[56]

Not all in the industry agree on the seriousness of the vulnerability or who should be responsible for mitigation. When it comes to the issue of counterfeit fasteners (which has received attention from U.S. government investigators of the counterfeit parts problem[57]) IFI is on record stating that current quality assurance standards are sufficient and effective in largely precluding counterfeit fasteners. These measures include the ISO 9000 and AS 9100 family of quality management standards, as well as the implementation of Qualified Suppliers Lists (QSL), which audit listed manufacturers and distributors.

"Overall, we do not believe there are widespread instances of counterfeit fasteners being supplied to the U.S.

government…IFI believes that existing standards, processes, and procedures in place in the fastener supply chain, including those mandated by OEM customers in the U.S. government itself, are adequate to protect against the introduction of counterfeit fasteners into the government supply chain."[58]

Lacking a clear view of the scope of the problem, it seems premature to determine that existing anti-counterfeiting measures are sufficient. Interestingly, the IFI statement above implicitly acknowledges the existence of counterfeit fasteners in the U.S. government supply chain, albeit minimizing its seriousness, and asserts that existing standards adequately address the potential introduction of counterfeit fasteners into the supply chain.[59] Encouragingly, IFI maintains that the burden of mitigating the risk is shared between the fasteners industry (which holds itself to quality assurance system standards) and OEMs/government procurement officials. The difference of opinion between some in the fasteners industry and IFI on whether vulnerability exists raises the question of whether OEMs and DoD procurement officials should implement additional measures to determine the scope of the problem.

MITIGATING THE RISKS

The health of the domestic fasteners industry is a national security issue: the compromise of defense-unique fastener products threatens operational readiness, and the fasteners industry plays an important role in the American economy. Clearly, there is a necessary balance between maintaining the health of the fasteners industry and implementing quality assurance measures to safeguard the supply chain from introduction of counterfeit and substandard fasteners. How best do we strike that balance? What amount of risk should we assume for defense-unique and defense-critical portions of the fasteners supply chain? What measures should we take to reduce the likelihood of introducing substandard or counterfeit fasteners into the DoD supply chain?

The following are recommendations to address the problem of substandard and counterfeit fasteners in the defense supply chain:

Determine the scope of the problem.
Systematic gathering of data is essential to determine the initial scope, as well as to guide the crafting of any additional anti-counterfeiting measures. The U.S. fasteners industry and U.S. government procurement officials, who together are the major stakeholders in the health of the fasteners industry, must work together to design data collection systems to document causes of fastener failure in defense systems. Collaboration is essential to determine the scope of the problem and avoid unduly burdensome regulations that could potentially harm the domestic fasteners industry.

Review fasteners quality assurance systems ("leverage existing anti-counterfeiting initiatives and practices").[60] Although it makes good financial sense to buy COTS fasteners and to reduce the cost of contracts whenever possible, the integrity of the FQA, including exemptions as appropriate, is critical to protect the quality of the fasteners used in defense applications. The original FQA of 1990 was designed to protect the domestic fasteners industry as well as the public by penalizing substandard suppliers and eliminating counterfeit and substandard fasteners. However, the documentation of fasteners failures in defense systems cited by the 2010 GAO report indicates the standards created by the amended FQA may leave the potential for introduction of counterfeit or substandard fasteners, and calls for potential additional anti-counterfeiting and quality assurance measures to mitigate the risks.[61]

Furthermore, despite manufacturers' best efforts, distributors can (and do) bend the rules in order to turn more profit.[62] The United States needs a more rigorous certification mechanism for the fasteners supply chain—from mill to manufacturer to distributor to OEM/prime contractor.[63] The current FQA exemption regime should be reviewed to eliminate unwarranted exemptions; for example, there appears to be evidence that the ISO 9000 process has, at times, been corrupted by fraudulently obtained certificates provided by foreign authorities..[64] Additional investigation is needed to document how the counterfeiting process works and to better ensure the integrity of the ISO 9000 registration process.

DoD should adopt a department-wide definition and consistently applied means for detecting, reporting, and disposing of counterfeit parts.[65] Collaboration among government

agencies, industry associations, and commercial-sector companies that produce items similar to those used by DoD is necessary to mitigate the risks of counterfeit parts in their supply chains, and offers DoD the opportunity to leverage ongoing and planned initiatives in this area.[66] While industry must take the lead in implementing measures to minimize the introduction of counterfeit and substandard fasteners into the market, DLA also has an important role in safeguarding the defense fasteners supply chain and in addressing the risks noted in the March 27, 2012, GAO report to Congress.[67] An important step would be to strengthen documentation procedures for incidences of substandard fastener and include assessment of the cost to readiness of fasteners failure in specific defense systems.

Address the issue of traceability. DoD's Sector-by-Sector, Tier-by-Tier (S2T2) effort to construct a defense industry-wide database that addresses the issue of supply chain traceability is a step in the right direction. However, we know enough about the scope of the traceability problem to take other steps that would reinforce fastener industry standards. The lack of data on the supply chain trace from manufacturer to distributor to OEM, combined with the industry's essentially self-policing role, demands a more rigorous process. This process should include traceability from mill-to-manufacturer-to-distributor, with a special focus on offshore portions of the supply chain.

Limit fasteners procured for defense purposes to those "made and melted in the U.S.A." Another measure, admittedly more difficult and expensive to implement, would be to strictly limit fasteners procured for the defense industry to those "made and melted in the U.S.A."[68] Although difficult and potentially costly, implementation of this measure would have the greatest impact on protecting our national security interests. Certifying the mills that produce the fasteners would be a first step, and should be combined with a traceability process from steel mill to manufacturer to distributor. Not only would these measures go a long way toward eliminating the introduction of counterfeit or substandard fasteners, but they would have the added benefit of helping to return jobs to the U.S. fasteners industry.[69]

CONCLUSION

Risks to the fasteners supply chain have national security implications, as fasteners from all three subsectors—aerospace, automotive, and industrial—are found throughout DoD systems and end-items. Effective fasteners quality assurance standards are critical to maintain DoD operational readiness, cost effectiveness, and mission success.

The health of the U.S. fasteners industry is important for both broader U.S. economic and national security interests. Failure to adequately address the issue of fastener quality assurance may result in the decline of a domestic supply chain for defense-unique aerospace and Mil-Spec fasteners.

Mil-Spec and aerospace fasteners often have years of R&D behind them. However, declining defense budgets will likely reduce R&D spending, calling into question the ability of the domestic aerospace fasteners industry to remain robust enough to respond to future defense requirements. As the supply chain continues migrating offshore, the risks of operating under a weakened FQA increase. Furthermore, vulnerabilities that affect the broader fastener industry are magnified in the defense sector; as the supply chain moves offshore, production for the defense segment of the market declines.[70] The development side of R&D is most likely to feel the effects of reduced defense spending. This financial squeeze will have consequences not only for the industry's ability to develop new defense-unique products, but also for the United States' capacity to sustain its competitive commercial edge.

ENDNOTES

a   Government Accountability Office (GAO), DoD Supply Chain: Suspect Counterfeit Electronic Parts Can Be Found on Internet Purchasing Platforms GAO-12-375 (Washington D.C.: GAO, February 21, 2012). http://www.gao.gov/assets/590/588736.pdf.

b   Government Accountability Office (GAO), DoD Supply Chain: Suspect Counterfeit Electronic Parts Can Be Found on Internet Purchasing Platforms GAO-12-375 (Washington D.C.: GAO, February 21, 2012). http://www.gao.gov/assets/590/588736.pdf.

1   Government Accountability Office, *DoD Should Leverage Ongoing Initiatives in Developing Its Program to Mitigate Risk of Counterfeit Parts*, *GAO-10-389*, by Belva Martin (Washington, D.C.: Government Accountability Office, 10-389, March 2010), 1. http://www.gao.gov/assets/310/302313.pdf.

2   Ibid., 2.

3   John Nolan, "Military Counterfeit Parts Fraud Exposed," *Dayton Daily News*, March 27, 2012. http://www.daytondailynews.com/news/dayton-news/military-counterfeit-parts-fraud-exposed-1350954.html.

4   Aerospace Industries Association of America, *Counterfeit Parts: Increasing Awareness and Developing Countermeasures* (Arlington, VA: AIA. March 2011), i. http://www.aia-aerospace.org/assets/counterfeit-web11.pdf.

5   Government Accountability Office, *DoD Should Leverage Ongoing Initiatives in Developing Its Program to Mitigate Risk of Counterfeit Parts*, *GAO-10-389*, by Belva Martin (Washington, D.C.: Government Accountability Office, 10-389, March 2010), 9. http://www.gao.gov/assets/310/302313.pdf.

6   "Fastener Quality and the Industry," Industrial Fasteners Institute quoted in Encyclopedia for American Business Online, 2001. http://www.referenceforbusiness.com/industries/Fabricated-Metal/Bolts-Nuts-Screws-Rivets-Washers.html.

7   Ibid.

8   Ibid.

9   Fastener Quality and the Industry," Industrial Fasteners Institute quoted in Encyclopedia for American Business Online, 2001. http://www.referenceforbusiness.com/industries/Fabricated-Metal/Bolts-Nuts-Screws-Rivets-Washers.html.

10 Ibid.

11 "Fastener Quality and the Industry," Industrial Fasteners Institute quoted in Encyclopedia for American Business Online, 2001. http://www.referenceforbusiness.com/industries/Fabricated-Metal/Bolts-Nuts-Screws-Rivets-Washers.html.

12 Ibid.

13 Industrial Fasteners Institute, *2010 Abbreviated Annual Report* (Independence, OH: IFI, 2010), 9. http://www.indfast.org/assets/pdf/2010-Abbreviated-Annual-Report.pdf.

14 Industrial Fasteners Institute, *2010 Abbreviated Annual Report* (Independence, OH: IFI, 2010), 8. http://www.indfast.org/assets/pdf/2010-Abbreviated-Annual-Report.pdf

15 Industrial Fasteners Institute, *2009 Fastener Industry Economics* (Independence, OH: IFI, 2009), 17. http://www.indfast.org/assets/pdf/2009%20Fastener%20Industry%20Economics.pdf.

16 Jennifer Reid, Industrial Fasteners Institute, interview, December 17, 2011.

17 "Fastener Quality and the Industry," Industrial Fasteners Institute quoted in Encyclopedia for American Business Online, 2001. http://www.referenceforbusiness.com/industries/Fabricated-Metal/Bolts-Nuts-Screws-Rivets-Washers.html.

18 David J. Beck *Qualitative Fastener Standards: Procurement Issues* (Monterey, CA: Naval Postgraduate School, December 1989), 18.

19 Ibid.

20 "Fastener Quality and the Industry," Industrial Fasteners Institute quoted in Encyclopedia for American Business Online, 2001. http://www.referenceforbusiness.com/industries/Fabricated-Metal/Bolts-Nuts-Screws-Rivets-Washers.html.

21 Ibid.

22 David J. Beck, *Qualitative Fastener Standards: Procurement Issues* (Monterey, CA: Naval Postgraduate School, December 1989), 34.

23 Ibid., 57.

24 "Fastener Quality and the Industry," Industrial Fasteners Institute quoted in Encyclopedia for American Business Online, 2001. http://www.referenceforbusiness.com/industries/Fabricated-Metal/Bolts-Nuts-Screws-Rivets-Washers.html.

25 *Public Law 101-592, Fastener Quality Act of 1990* (Washington, DC: Government Printing Office, 1990). http://m-i-n-a.org/fqaregs99_2.htm.

26 Ibid.

27 Richard B. Stump, "Following the Fastener Act: A Government and Industry Odyssey," *Quality Digest*, June 2000. http://www.quality digest.com/june00/html/fastener.html.

28 "Fastener Quality and the Industry," Industrial Fasteners Institute quoted in Encyclopedia for American Business Online, 2001. http://www.referenceforbusiness.com/industries/Fabricated-Metal/Bolts-Nuts-Screws-Rivets-Washers.html.

29 Richard B. Stump, "Following the Fastener Act: A Government and Industry Odyssey," *Quality Digest*, June 2000. http://www.quality digest.com/june00/html/fastener.html.

30 David Sharp quoted in "Fastener Quality and the Industry," Industrial Fasteners Institute quoted in Encyclopedia for American Business Online, 2001. http://www.referenceforbusiness.com/industries/Fabricated-Metal/Bolts-Nuts-Screws-Rivets-Washers.html.

31 Subhas G. Malghan, *Highlights of the Fastener Quality Act* (Gaithersburg, MD: National Institute of Standards and Technology, Department of Commerce, 1996). http://www.medey.com/pdf/Fastener%20Quality%20Act.pdf.

32 "Fastener Quality and the Industry," Industrial Fasteners Institute quoted in Encyclopedia for American Business Online, 2001. http://www.referenceforbusiness.com/industries/Fabricated-Metal/Bolts-Nuts-Screws-Rivets-Washers.html.

33 Richard B. Stump, "Following the Fastener Act: A Government and Industry Odyssey," *Quality Digest*, June 2000. http://www.quality digest.com/june00/html/fastener.html.

34 "Fastener Quality and the Industry," Industrial Fasteners Institute quoted in Encyclopedia for American Business Online, 2001. http://www.referenceforbusiness.com/industries/Fabricated-Metal/Bolts-Nuts-Screws-Rivets-Washers.html.

35 U.S. Department of Energy, Office of Health, Safety, and Security, Office of Corporate Safety Analysis, "Suspect/Counterfeit Items Awareness Training," (June 2007), 2. http://www.hss.doe.gov/sesa/corporatesafety/sci/trainingmanual.html.

36 Government Accountability Office, *DoD Should Leverage Ongoing Initiatives in Developing Its Program to Mitigate Risk of Counterfeit Parts*, GAO-10-389, by Belva Martin (Washington, D.C.: Government Accountability Office, 10-389, March 2010), 9. http://www.gao.gov/assets/310/302313.pdf. 25-26.

37 Tom Miller, Nucor Fasteners, interview, March 6, 2012.

38 Government Accountability Office, *DoD Should Leverage Ongoing Initiatives in Developing Its Program to Mitigate Risk of Counterfeit Parts*, GAO-10-389, by Belva Martin (Washington, D.C.: Government Accountability Office, 10-389, March 2010), 4, 9, 10, 12, 14, 19. http://www.gao.gov/assets/310/302313.pdf.

39 Public Law 106-34, 106th Congress, *Fastener Quality Act Amendments Act of 1999* (Washington, DC: Government Printing Office, June 8, 1999). http://www.gpo.gov/fdsys/pkg/PLAW-106publ34/pdf/PLAW-106publ34.pdf.

40 "ISO 9000 Certification Fraud Exposed," *ISO 9000 News*, 1 (1998). www.iso.org /iso/livelinkgetfile-isocs?nodeId=15053000.

41 Tom Miller, Nucor Fasteners, interview, March 6, 2012.

42 Aerospace Industries Association of America (AIA). *Counterfeit Parts: Increasing Awareness and Developing Countermeasures* (Arlington, VA: AIA, 2011), 30. http://www.aia-aerospace.org/assets/counterfeit-web11.pdf.

43 "Counterfeit Parts," Genuine Ford Motor Company Global Brand Protection. http://www.fordbrandprotection.com/counterfeit.asp.

44 Miller, interview (March 6, 2012).

45 Government Accountability Office, *DoD Should Leverage Ongoing Initiatives in Developing Its Program to Mitigate Risk of Counterfeit Parts*, GAO-10-389, by Belva Martin (Washington, D.C.: Government Accountability Office, 10-389, March 2010), 4. http://www.gao.gov/assets/310/302313.pdf.

46 Ibid.

47 Aerospace Industries Association of America (AIA). *Counterfeit Parts: Increasing Awareness and Developing Countermeasures* (Arlington, VA: AIA, 2011), 10. http://www.aia-aerospace.org/assets/counterfeit-web11.pdf.

48 Martin, *DoD Should Leverage Ongoing Initiatives in Developing Its Program to Mitigate Risk of Counterfeit Parts*, 9.

49 David J. Beck, *Qualitative Fastener Standards: Procurement Issues* (Monterey, CA: Naval Postgraduate School, December 1989), 58.

50 Ibid.

51 International Organization for Standardization, *ISO 9001 – What Does It Mean in the Supply Chain* (2012). http://www.iso.org/iso/iso_catalogue /management_and_leadership_standards/quality_management /more_resources_9000/9001supchain.htm.

52 Tom Miller, Nucor Fasteners, interview, March 6, 2012.

53 Ibid.

54 National Aeronautics and Space Administration, "NASA Fastener Procurement, Receiving Inspection, and Storage Practices for Spaceflight Hardware" (Washington, DC: NASA-STD-6008, July 11, 2008). http://standards.nasa.gov%2Fdocuments2Fviewdoc%2F3315592%2F3315592&ei=v6x8T6u_HoeliAKL2oDnDQ&usg= AFQjCNEM8d_ZMtoowYFwHtCXW4rtTDxWJA&sig2= 5dIDEnRFzThZQpbQZQa5vw.

55 Lear Corporation, *Critical Fastener Supplier Expectations* (July 13, 2006). https://lear.portal.covisint.com/c/document_library/get_file?folderId=122853&name=DLFE-105829.pdf.

56 Tom Miller, Nucor Fasteners, interview, March 6, 2012.

57 Government Accountability Office, *DoD Should Leverage Ongoing Initiatives in Developing Its Program to Mitigate Risk of Counterfeit Parts*, GAO-10-389, by Belva Martin (Washington, D.C.: Government Accountability Office, 10-389, March 2010), 1. http://www.gao.gov/assets/310/302313.pdf.

58 As the Director of the White House Intellectual Property Agency, Espinel convened an interagency panel to examine the problem of counterfeit parts, including fasteners. Letter from Industrial Fasteners Institute to Ms. Victoria Espinel, U.S. Intellectual Property Enforcement Coordinator (September 14, 2011).

59 Letter from Industrial Fasteners Institute to Ms. Victoria Espinel, U.S. Intellectual Property Enforcement Coordinator (September 14, 2011).

60 Government Accountability Office, *DoD Should Leverage Ongoing Initiatives in Developing Its Program to Mitigate Risk of Counterfeit Parts*, GAO-10-389, by Belva Martin (Washington, D.C.: Government Accountability Office, 10-389, March 2010), 19. http://www.gao.gov/assets/310/302313.pdf.

61 Ibid., 25-26.

62 Tom Miller, Nucor Fasteners, interview, March 6, 2012.

63 Ibid.

64 "ISO 9000 Certification Fraud Exposed," *ISO 9000 News*, 1 (1998). www.iso.org /iso/livelinkgetfile-isocs?nodeId=15053000.

65 Government Accountability Office, Office of Public Affairs, *DoD Should Leverage Ongoing Initiatives in Developing Its Program to Mitigate Risk of Counterfeit Parts*, Public Affairs Release, by Belva Martin, April 28, 2010. http://www.gao.gov/products/GAO-10-389.

66 Ibid.

67 John Nolan, "Military Counterfeit Parts Fraud Exposed," *Dayton Daily News*, March 27, 2012. http://www.daytondailynews.com/news/dayton-news/military-counterfeit-parts-fraud-exposed-1350954.html.

68 Tom Miller, Nucor Fasteners, interview, March 6, 2012.

69 Ibid.

70 Tom Miller, Nucor Fasteners, interview, March 6, 2012

CHAPTER 7 • SEMICONDUCTORS

EXECUTIVE SUMMARY

Semiconductors, a vital input for high-tech electronics, have been central to U.S. military and economic strength over the past half-century. Without semiconductors, many of the technologies that contribute to U.S. military dominance would not exist. Maintaining U.S. technological dominance requires maintaining a strong presence in the fast-moving semiconductor industry.

Semiconductor design and fabrication increasingly are conducted separately, with firms focusing on one or the other. A few firms—Intel and IBM, for example—remain vertically integrated and perform both functions, but it is far more common to find a division of labor across the sector, with design firms contracting semiconductor foundries to mass-produce chips, for example. The United States remains the leader of semiconductor research and development (R&D), but it has faced a steady decline in fabrication, which is increasingly located in Asia. U.S. firms either establish a facility abroad (offshoring) to take advantage of special grants and subsidies offered by foreign governments, or outsource microchip assembly to an independent third party abroad.

Semiconductor fabrication supports other high-value, export-oriented industries such as consumer electronics. Fabrication is capital-intensive, with a new foundry costing upwards of $10 billion. Beginning in the 1980s, Taiwanese agencies created special incentives to lure foreign and domestic businesses to invest in foundries. Other Asian countries followed suit, resulting in the establishment of an Asia-Pacific semiconductor and consumer electronics hub. As a result, the U.S. share of semiconductor fabrication has decreased from nearly 50 percent in 1980 to only 15 percent in 2012.

The relocation of semiconductor fabrication overseas raises the risk of vulnerabilities in the U.S. defense industrial base. Advanced weapons systems and other military applications often use commercially available semiconductors, but some chips must be designed and fabricated specifically for defense applications in secure facilities. Secure acquisition becomes far more difficult as fabrication migrates overseas, and would likely prove impossible if R&D were to follow.

Unlike consumer electronics, advanced weapons systems have a long lifespan and may need replacement chips that are no longer produced. The Defense Logistics Agency (DLA) increasingly relies on intermediaries to locate obsolete parts—and dishonest disributors may sell counterfeit chips. Recent

SEMICONDUCTORS
BUILDING BLOCKS OF U.S. NATIONAL SECURITY

MANUFACTURING SECURITY

Semiconductors are essential in sophisticated U.S. military electronics

MISSILE GUIDANCE

SATELLITES

SUPERCOMPUTERS

QUALITY ISSUES

Counterfeit electronic components including semiconductors from abroad have become a problem for the U.S. defense industrial base

COUNTERFEIT CHIPS + **MILITARY PLATFORMS** = **WARFIGHTER RISKS**

VULNERABILITIES

U.S. is falling behind the rest of the world in construction of new capacity

35 OUT OF 40 NEW CHIP FACTORIES ARE LOCATED IN ASIA

Japan is the leading producer of silicon wafers, a popular semiconductor material that is essential to U.S. national security and defense platforms

JAPAN MAKES 60% OF THE WORLD'S SILICON WAFERS

INNOVATION

U.S. semiconductor industry is a rich source of innovation

SEMICONDUCTORS HAVE A DIRECT IMPACT ON U.S. INNOVATION

MITIGATING RISKS

Avoiding uncertainty in U.S. semiconductor capacity

COUNTERFEIT DETECTION

GREATER GOVERNMENT SUPPORT

SECURING THE SUPPLY CHAIN

MILITARY EQUIPMENT CHART
SELECTED DEFENSE USES FOR SEMICONDUCTORS

DEPARTMENT	WEAPON SYSTEMS	PLATFORMS	OTHER SYSTEMS
ARMY	■ Javelin FGM-148 anti-tank missile ■ M142 High Mobility Artillery Rocket System (HIMARS) ■ Modular Advanced Armed Robotic System (MAARS) ■ BGM-71 Tube Launched, Optically Tracked, Wire Command Data Link (TOW) guided missile	■ UH-60 Blackhawk helicopter ■ Humvee ■ M1 Abrams main battle tank	Radars, sensors, telecommunications devices, computers, aerospace components, power supplies, motor controls
MARINE CORPS	■ Javelin FGM-148 anti-tank missile ■ M142 High Mobility Artillery Rocket System (HIMARS) ■ Modular Advanced Armed Robotic System (MAARS) ■ BGM-71 Tube Launched, Optically Tracked, Wire Command Data Link (TOW) guided missile	■ AH-1W SuperCobra helicopter ■ SH-60 Seahawk helicopter ■ V-22 Osprey aircraft ■ AA7A1 assault amphibious vehicle	
NAVY	■ 57mm Mk110 naval gun system	■ SH-60 Seahawk helicopter ■ F-14 Tomcat fighter	
AIR FORCE	■ CBU-97 Sensor Fuzed Weapon ■ GBU-39 Small Diameter Bomb ■ Joint Direct Attack Munition (JDAM) precision guidance kit	■ F-15 Fighter aircraft ■ F-22 Raptor fighter ■ F-35 Joint Strike fighter	

studies have found that many simple microchips, as well as more advanced items, are counterfeit.

The semiconductor supply chain is complex and requires detailed understanding to identify bottlenecks that could limit access to defense-critical components. Greater understanding of semiconductor supply chains could also help to identify and curb the growing threat of counterfeit parts. The current U.S. advantage in semiconductor R&D may decline unless the U.S. government supports investment in semiconductor fabrication.

INTRODUCTION

The semiconductor industry is critically important to U. S. national security. Since 2005, the semiconductor industry has been the United States' number one exporter, and is a major source of high-income employment.[1] Investment in the semiconductor industry helped bring about the economic transformation of a number of states, most notably California, Massachusetts, and North Carolina.[2] Nearly all modern conveniences rely on semiconductors, enabling everything from personal computing and electronic communications to climate control and nuclear reactors. U.S. military superiority, which depends on maintaining a technological edge over adversaries, was made possible in recent decades by U.S. leadership in semiconductor technology.

The U.S. armed services are the most technologically advanced military force ever assembled. Many of the military's technological advantages originate from advanced electronic systems, including advanced computing, surveillance, communications, guidance, and propulsion capabilities. It is difficult to name a modern weapon system or capability that is not in some way dependent on microelectronics to function properly, and therefore not dependent on the semiconductors that enable high-tech electronics. Although the United States is the world leader in semiconductor R&D, manufacturing and fabrication are increasingly conducted overseas, which in turn creates incentives for R&D to relocate abroad. U.S. firms either have moved abroad to take advantage of foreign incentive rules or they have contracted with a foreign firm to assemble microchips.

Key themes discussed in this chapter are:

- The U.S. semiconductor industry, formerly dominant, has witnessed slow attrition as governments in other countries subsidized and encouraged the formation of their own national semiconductor sectors.

- Although U.S. companies lead the world in semiconductor R&D, most of the fabrication takes place in the Asia-Pacific region. Over time, R&D may follow because design and manufacturing processes are interlinked, rather than separate, phases of production.

- Military electronics typically have longer lifespans than consumer electronics. To keep its electronic parts operating, the Department of Defense (DoD) seeks to purchase microchips that are no longer being manufactured. DLA faces a massive logistical challenge keeping track of millions of different electronic parts and of restocking many chips no longer available or widely sold in the commercial sector. More important, the risk of counterfeit parts entering DoD supply chains is much higher if DoD must rely on out-of-date microchips, which may be provided by dishonest manufacturers or unscrupulous suppliers.

THE PERILS OF COUNTERFEIT SEMICONDUCTORS (a notional though realistic scenario)

Sabotage, rather than equipment failure, caused an early warning radar system to fail to notify security force personnel of a rocket launch during an attack on Kandahar Airfield. Repair technicians have determined that this radar contained circuits from commercial-off-the-shelf microprocessors and that the circuits turned off the early warning radar. The malicious circuit's origin has not been confirmed, but defense analysts have long stated that the continuous transfer of semiconductor manufacturing and its associated intellectual property to China poses risks to U.S. national security.

A NOTE ON CRITICALITY

A wide range of defense capabilities rely on specialty chips that have limited or no commercial application but that are designed to fulfill specific defense applications. The current trends of outsourcing and offshoring semiconductor fabrication, paired with the potential of lost R&D and design capability, inhibits production of these specialized and often classified components, which would then need to be manufactured abroad. These changes suggest a *high* vulnerability in the defense industrial base. Given the nearly universal importance of advanced electronics and semiconductors to military capabilities, losing the ability to produce specialty semiconductors for advanced weapons systems in secure, trusted foundries could have *incapacitating* consequences on U.S. military capabilities.

BACKGROUND

Silicon-based integrated circuits were first developed in the 1950s when the U.S. Air Force sought sophisticated electronics capable of providing onboard guidance for rockets. Since then, microelectronics built from increasingly sophisticated semiconductors have become essential components of smart bombs, surveillance technology, advanced logistics, intelligence platforms, wireless communication, advanced navigation electronics, sensors, and unmanned aerial vehicles, to name just a few examples.[3] Semiconductor manufacturing makes use of some of the most precise and sophisticated processes currently involving controlled, repeatable, and virtually error-free fabrication of structures at the atomic scale. (Separate elements of semiconductors are so minute that they cannot be discerned with the naked eye.) Among computer hardware engineers, the term "Moore's Law" has been coined to capture the unprecedented rapid rate of innovation in this sector. According to this law, the number of transistors that can be placed in an integrated circuit has doubled approximately every two years, while manufacturing costs remain constant. This

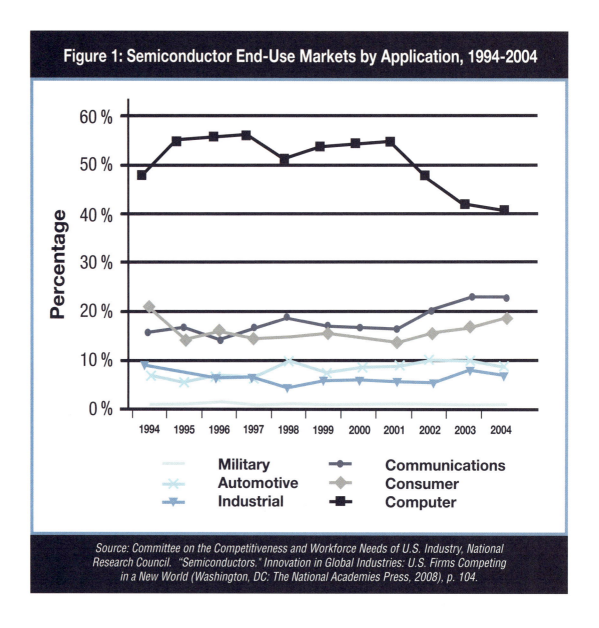

Figure 1: Semiconductor End-Use Markets by Application, 1994-2004

Source: Committee on the Competitiveness and Workforce Needs of U.S. Industry, National Research Council. "Semiconductors." Innovation in Global Industries: U.S. Firms Competing in a New World (Washington, DC: The National Academies Press, 2008), p. 104.

trend was first described in 1965 and has continued to the present day.

Currently, two-and-a-half billion transistors can be placed in an integrated circuit at about the same cost that was required for approximately 2,300 transistors in the early 1970s. It is predicted that in the future, microchips will even be embedded in living organisms, giving rise to a new field of bioelectronics for a wide range of applications, from energy grid systems to transportation to a variety of interactive systems that rely on intelligent sensing.

In 2012, the worldwide semiconductor industry, consisting of manufacturing facilities in more than twenty countries, recorded revenues of $303 billion and supported an electronics industry with sales of over $1.3 trillion.[4] The industry also supports a vast and diverse service sector that spans across energy, health care, aviation, banking, and education, with an estimated annual value of $6 trillion.[5] The U.S. share of the global semiconductor market came to $144 billion in 2011, constituting 48 percent of the global market.[6]

Growth in this sector has been driven by exponential advances in the adaptation of microchips for ever greater levels of capability, reduced power consumption, and higher reliability at lower cost.[7] Maintaining this rapid level of innovation requires that semiconductor companies allocate ever larger sums of money to R&D. According to the U.S. Department of Commerce, semiconductor companies are the leading recipients of corporate patents.[8]

Specialized semiconductors manufactured specifically for military applications represent less than one percent of total semiconductor output on average in the United States (see Figure 1). DoD usually procures commercially-available semiconductors, although occasionally a need arises for specialized chips capable of withstanding extreme conditions or other enhanced defense-related functionality.

A technologically advanced 21st century military relies heavily on microelectronics and computers. There are very few defense-related end-items that do not require microchips or electronics. A short list of semiconductor-enabled capabilities would include all types of navigation systems, aerospace technologies, satellites, and communications systems – in addition to the computing capability needed to operate most of these systems. The U.S. military's network-based approach to warfighting requires extensive systems integration. To establish the needed infrastructure and support systems for advanced systems integration in support of network-based operations, DoD will need even more advanced microprocessors and other semiconductor devices.[9]

Defense Advanced Research Projects Agency (DARPA) plans to create nanochips for monitoring troops' health on the battlefield. The sensors DARPA seeks constitute "a truly disruptive innovation," that could help the U.S. fight healthier and more efficiently than its adversaries.[a]

– Robert Johnson, *Business Insider*

U.S. companies accounted for 48 percent of global semiconductor sales in 2011. U.S. companies still dominate the global market, but they have shifted production abroad and mostly perform R&D in the home market. As a result, U.S. manufacturing capacity in microelectronics and, more recently, semiconductors, is virtually non-existent due to outsourcing and the rise of foreign chip foundries, mostly located in the Asia-Pacific region, a trend that began decades ago. For example, by the mid-1980s, Japan had captured roughly 80 percent of the Dynamic Random Access Memory (DRAM) market, the most important chip market at that time.[10] Korea also gained a significant portion of global market share for semiconductors, as did Taiwan in the mid-1990s.[11]

The attrition of domestic manufacturing has not directly harmed the revenues of the major U.S. semiconductor firms. Instead, to remain competitive and maximize profits, these firms moved fabrication overseas to take advantage of corporate, tax, and labor incentives offered by foreign governments and to be closer to foreign-based electronics industries, the main consumer of microchips, also increasingly concentrated in the Asia-Pacific region (see Figure 2). Firms seek out geographic

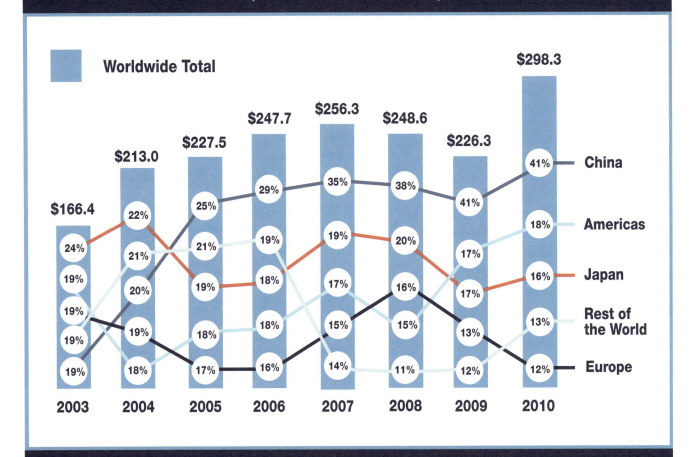

Figure 2: Worldwide Consumption of Semiconductors (2003-2010)
(Billions of U.S. Dollars)

Worldwide Total

	2003	2004	2005	2006	2007	2008	2009	2010
Total	$166.4	$213.0	$227.5	$247.7	$256.3	$248.6	$226.3	$298.3
China		22%	25%	29%	35%	38%	41%	41%
Americas	24%	21%	21%	19%	19%	20%	17%	18%
Japan	19%	20%	19%	18%	17%	16%	17%	16%
Rest of the World	19%	19%	18%	18%	15%	15%	13%	13%
Europe	19%	18%	17%	16%	14%	11%	12%	12%
	19%							

Source: PwC, Continued Growth China's Impact On The Semiconductors Industry
(November 2011). http://www.pwc.com/gx/en/technology/assets/china-semiconductors-report-2011.pdf

clusters that bring consumers and suppliers together because it is advantageous for suppliers and subcontractors to have access to the most favorable conditions and to be in close proximity. Most U.S. firms have now relocated the bulk of their manufacturing to East Asia.

Constructing a new semiconductor production facility requires significant capital investments in order to achieve economies of scale and recoup the initial investment in a reasonable amount of time. A state-of-the-art semiconductor foundry can require close to $10 billion in initial startup costs prior to becoming operational.[12] Recognizing that a vibrant semiconductor industry enables many other export-oriented industries, state agencies in several newly industrializing economies in East Asia generously subsidized the startup costs of new foundries. For example, Taiwan became the global center of semiconductor fabrication with the help of government incentives to encourage investment. Other Asian governments,

Table 1. Preliminary Estimate of World Semiconductor Revenue in 2010 by Company Headquarters Location (Revenue in Millions of U.S. Dollars)

2009 RANK	Company Headquarters	2009 Revenue	2010 Revenue	Percent Change	Percent of Total	Cumulative Percent
1	Americas	$110,936	$147,291	32.8%	48.5%	48.5%
2	Asia-Pacific	$44,598	$65,363	46.6%	21.5%	70.0%
3	Japan	$49,857	$63,766	32.8%	21.0%	90.9%
4	Europe, Middle East, Africa	$24,115	$27,586	32.8%	9.1%	100.0%
	Total	$229,506	$304,006	32.8%	100.0%	

Source: IC INSIGHTS, "Tracking the Top 10 Semiconductors Sales Leaders Over 26 Years," Research Bulletin (December 19, 2011). http://www.icinsights.com/data/articles/documents/359.pdf

impressed by Taiwan's success, emulated its approach by creating incentives for investment in cutting-edge foundries. Subsequently, as shown in Table 1, the Asia-Pacific region (including Taiwan, Korea, Singapore, and China but excluding Japan for comparison) possesses a large number of leading-edge semiconductor companies.

Not only is manufacturing increasingly located overseas, but also large, foreign-owned corporations dominate the industry itself. In 2012, the top 10 semiconductor firms reported sales of $168.4 billion, representing more than 50 percent of global sales. Although the U.S. firm Intel has held the top position since 1993, and seven of the top 15 firms are from the United States, three of those firms—Qualcomm (ranked 4th), Broadcom[13] (ranked 11th), and AMD (13th) are "fabless," meaning that they do not fabricate semiconductors themselves but instead outsource manufacturing to other companies. By comparison, Taiwan Semiconductors Manufacturing Company is ranked third,

and is the world's largest foundry that manufactures chips for other companies, with global sales of $17 billion. The second highest ranked firm is the South Korean company Samsung, with sales of $30.4 billion (about 30 percent less than the 2012 sales figures of Intel at $49 billion).[14]

Offshoring and outsourcing[15] pose genuine risks to the long-term viability of the U.S. defense industrial base. Each could jeopardize U.S. economic competitiveness in Information and Communication Technologies (ICT) if unexpected events caused a sudden or protracted interruption of supply. The following section provides a summary of the potential threats to U.S. national security and economic competitiveness from risks to the semiconductor supply chain.

As various incentives and subsidies offered by overseas producers or governments reduced the comparative cost competitiveness of U.S. products, the incentives for U.S. firms to invest in domestic development of the most advanced technologies

Table 2: Worldwide Semiconductor Sales Leaders (In $ Billions)

RANK	1985		1995		2012 (est.)	
1	NEC	$2.1	Intel	$13.6	Intel	$49.4
2	Texas Instruments	$1.8	NEC	$12.2	Samsung	$30.4
3	Motorola	$1.8	Toshiba	$10.6	TSMC*	$17.0
4	Hitachi	$1.7	Hitachi	$9.8	Qualcomm**	$12.8
5	Toshiba	$1.5	Motorola	$8.6	Texas Instruments	$12.1
6	Fujitsu	$1.1	Samsung	$8.4	Toshiba	$11.1
7	Philips	$1.0	Texas Instruments	$7.9	Renesas	$9.8
8	Intel	$1.0	IEM	$5.7	SK Hynix	$8.8
9	National	$1.0	Mitsubishi	$5.1	Micron	$8.7
10	Matsushita	$0.9	Hyundai	$4.4	STMicroelectronics	$8.4
Top 10 Total ($B)		$13.9		$86.3		$168.4
Semi Market ($B)		$23.3		$154		N/A
Top 10% of Total Semi Market		60%		56%		N/A

*Foundry   **Fabless

Source: IC Insights, Tracking the Top 10 Semiconductor Sales Leaders Over 26 Years, Research Bulletin, December 19, 2011. IC Insights, Top 20 Semiconductor Suppliers' Sales Growth Rates Forecast to Range from Great (+31%) to Terrible (-17%) in 2012, Research Bulletin, November 7, 2012.

declined accordingly. In time, these cost disparities discouraged investment in domestic R&D. In the short term, inefficiencies resulting from obsolescent or outdated equipment can be recuperated by outsourcing to producers with lower operating costs. Over the long run, the focus on maintaining profit margins may come at the expense of investment in higher productivity measures, such as design innovation or incremental improvements in manufacturing processes. Firms make a trade-off between immediate profits in return for slower gains in the future as they may lose the ability to compete against Asian firms.[16]

The semiconductor industry has matured. Currently, the key to retaining competitiveness in semiconductors may be less about achieving rapid innovations than ensuring incremental steps are routinely taken to improve and protect production and manufacturing technologies. As manufacturing activities progressively migrate overseas and become increasingly detached from domestic R&D, it becomes more difficult for scientists and engineers in U.S. research laboratories to fully assess and understand the improvements in production techniques that could be most easily adapted or how to most effectively implement these into the prevailing production systems.[17] The U.S. information technology industry has a tendency to downplay the importance of continuous improvements in manufacturing processes. This attitude poses special risks for the semiconductor industry, which relies on extremely complex manufacturing processes that make it more cost effective to upgrade an existing facility incrementally rather than rebuild from scratch every few years.

Innovation in the development of semiconductors must be adapted to each specific production facility, making it most practical and least costly if R&D and fabrication occur at the same site. It is not clear whether U.S. companies will ever be able to regain this synergy once they stop manufacturing on U.S. soil altogether, even if much of their R&D remains domestic.

While the U.S. government has paid minimal attention to the particular challenges of fabricating at the atomic level, other countries have increased economic and corporate incentives to further the expansion of the semiconductor industry by recognizing the high entry costs of setting up foundries. In Asia, national and regional governments intensely compete to attract semiconductor manufacturing and R&D investments.

As defense electronics increasingly are manufactured and assembled overseas, this trend creates new and unforeseen vulnerabilities at all levels of the defense supply chain. In the last few years, global semiconductor production has been disrupted by unpredictable natural disasters such as flooding, earthquakes, and tsunamis. The industry also has been susceptible to political turmoil in certain countries, causing shortages of key components that are necessary to the integrity of the defense industrial base.

The presence of foreign-supplied counterfeit and defective microchips in both commercial and military products is also a widely acknowledged challenge, and perhaps an inevitable byproduct of offshoring and outsourcing to countries that do not abide by or enforce international standards. Defective chips are surfacing in advanced as well as older technology. Quality control becomes harder as the United States depends on more and more overseas facilities, defense contractors, and subcontractors for vital inputs.

The Pentagon often requires semiconductors that are produced according to Military Specification (Mil-Spec)—highly specialized and custom-produced devices designed specifically for secure computing functions—for which there is no commercial demand. The domestic knowledge base needed to produce these specialty components in a secure setting may eventually disappear if the United States cannot maintain its domestic semiconductor base. This potential loss constitutes a threat to U.S. national security. Deepening reliance on imports may also irreparably damage the viability of smaller, specialized firms that the Pentagon relies on to deliver complex and uniquely capable semiconductor devices.

THE STRUCTURE OF THE SEMICONDUCTOR INDUSTRY

As mentioned earlier, the semiconductor industry enables many other diverse sectors (such as computers, consumer electronics, and telecommunications) playing important roles in the automotive, medical, green technology, and energy industries. The electronics industry is the largest consumer of semiconductors, with Apple, Inc., accounting for 5.7 percent of global sales, Samsung Electronics and Hewlett Packard each accounting for 5.5 percent, and Dell representing 3.2 percent.[18]

Chips designed for semiconductors serve several different functions that can be divided into the following sectors:

Dynamic Random Access Memory chips (DRAMs) are the primary memory for computers. Today's computers may have as many as eight DRAMs on a motherboard to maximize processing power.

Microprocessor Units (MPUs) act as the brains of computers. They include a central processing unit and programmable memory. Microprocessors are also used in other electronic products. Intel Corporation is the leader in microprocessors, though only one-third of its total annual manufacturing remains within the United States. (The other two-thirds are in Europe and the Pacific Rim, with the latter gradually becoming the main production center.)

Application-Specific Integrated Circuits (ASICs) are customized semiconductors designed for very specific functions, most notably the wide range of military and civilian products with touch screen controls. ASICs process and "clean" digitized signals, enabling the microprocessor to interpret them. ASICs are a key element in guided missile systems that need to process large amounts of digitized information. They also are found in automobile air-bag systems and printers.

Digital Signal Processors (DSPs) process signals, including image and sound signals and radar pulses. DSPs convert analog signals (including sound, color, temperature, light, and distance) to digital signals, permitting the high-speed analysis, enhancement, filtration, modulation, and manipulation of those signals. DSPs are widely used in electronic equipment, including cellular phones, scanners, and high-speed modems. Texas Instruments, Inc. is a leader in this industry. DSPs are used in many military applications, including night-vision devices, naval sonar systems, guided missiles, military avionics, tanks, and satellites.

Programmable memory chips (EPROMs, EEPROMs, and Flash) are chips that retain information and programming even when the chip receives no electrical power. They are common in cellular phones, handheld computing devices, memory sticks, and nearly all other products where miniaturization is important.

As technology has progressed, different companies have tended to focus on specific aspects of the development and/or production process. There are three main types of companies that account for the full range of semiconductor manufacturing: design firms; manufacturing firms (called fabricating foundries or "fabs"); and integrated device manufacturers (IDM). The big exceptions are Intel and IBM, two major companies that fall under the category of "generalists" in that they cover all three of these phases and remain vertically integrated, designing both new chips and the technology to manufacture them in-house.

By contrast, Advanced Micro Devices, Inc. (AMD) is a design firm that relies entirely on others to manufacture components according to their specifications. AMD is one of the world's largest fabrication-less ("fabless") firms, generating revenue from royalties from its patents.

The third type of company, known in the business as a "pure play" foundry, manufactures semiconductors for other firms. Pure play foundries are predominantly located in Asia; none are currently located within the United States.

THE SEMICONDUCTOR MANUFACTURING PROCESS

The typical manufacturing process for semiconductors involves more than 300 sequential steps, some of them involving patterning nanometer-length features onto silicon using high-precision and high-volume equipment. The manufacturing process begins with silicon wafers, a natural semiconductor that can either conduct electricity or insulate. Silicon wafers are inexpensive to produce. Silicon is abundant, being the second most available element in the earth's crust (although it must be extracted from compounds such as quartzite and sand in a capital-intensive process). Roughly six tons of inputs are required to produce one ton of silicon. Silicon wafers used in integrated circuits must be refined to 99.999999999 percent purity. High-purity silicon is melted in a crucible and then pulled into a single silicon crystal that solidifies as it is drawn.

Digital Signal Processors can be embedded in special unmanned underwater vehicles to scan for seabed mines. This complex sonar system keeps naval personnel safe and keeps waterways free from explosives.

This crystal is then sliced to produce the individual silicon wafers.

Figure 3 depicts the process for creating individual chips from silicon wafers. Chips are designed in layers, each corresponding to a slice of silicon wafer and subject to three operations: film deposition, which includes Chemical Vapor Deposition CVD), Plasma-Enhanced Chemical Vapor Deposition (PECVD), etc.; lithography (creation of patterns); and etching (Reactive-Ion Etching (RIE) Plasma). A photolithographic process, similar to that of creating a photograph from a negative, transfers the designs for each layer to the silicon wafer. Layers of the chip are "printed" and then etched onto the silicon wafer. This process is repeated for each layer of the chip, generally 20 to 30 times for modern logic devices.

Individual chips are separated from the wafer, tested, and packaged. Testing and packaging are comparatively labor-intensive and often require manual labor. Most often, foundries ship the uncut wafers to a testing and packaging facility, where a machine slices the wafer into single semiconductors that can then be tested and packaged (see Figure 4). The testing and packaging process is usually done in low-labor-cost markets because it requires repetitive, assembly-line labor. The Philippines, Thailand, Vietnam, and Cambodia have emerged as the favorite

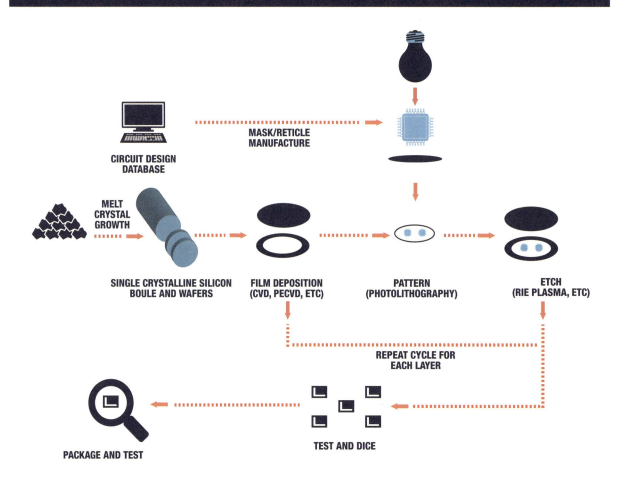

Figure 3: Simplified Overall Process Flow for the Manufacture of Semiconductor Integrated Devices

CIRCUIT DESIGN
DATABASE

MASK/RETICLE
MANUFACTURE

MELT
CRYSTAL
GROWTH

SINGLE CRYSTALLINE SILICON
BOULE AND WAFERS

FILM DEPOSITION
(CVD, PECVD, ETC)

PATTERN
(PHOTOLITHOGRAPHY)

ETCH
(RIE PLASMA, ETC)

REPEAT CYCLE FOR
EACH LAYER

PACKAGE AND TEST

TEST AND DICE

Source: Cliff Henderson, Integrated Circuits: A Brief History.
http://henderson.chbe.gatech.edu/Introductions/microlithography%20intro.htm

destinations for the final phase in the assembly of microchips.

The major challenge of semiconductor innovation arises from the need for miniaturization. For example, in 1960, semiconductor devices were fabricated with components measuring as small as 20 micrometers (μm = 1 millionth of a meter). Today's leading edge semiconductor devices have critical dimensions as small as 18 nanometers (1 billionth of a meter). "Critical dimension" refers to the smallest circuit element in the device or on a particular level of the device. (As a point of comparison, the diameter of a human hair is only about 50 nanometers.)

The manufacturing of semiconductor devices is extremely complex, in part because of the endless scaling of the transistors needed to yield improved

Figure 4: Schematic Presentation of the Entire Process

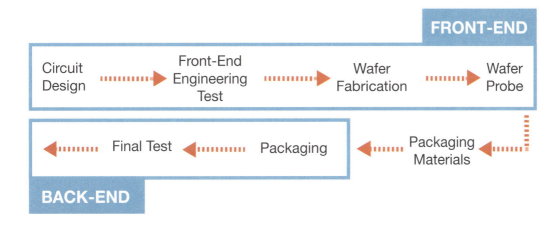

Source: Cliff Henderson, Integrated Circuits: A Brief History.
http://henderson.chbe.gatech.edu/Introductions/microlithography%20intro.htm

performance, reduced power, and lower costs. Currently, a microprocessor may contain more than two billion transistors, each connected by a wiring scheme that is measured in nanometers. Reduction in the size of features allows more devices to be integrated onto a single chip, which in turn results in more functionally powerful products. Reductions in size have also led to dramatic increases in the speed of logic devices such as microprocessors. Advancements in lithographic technologies have been the main force enabling the manufacturing of smaller features.

As more features can be placed on a chip that is either constant or shrinking in size, the overall retail price of semiconductor devices has remained relatively constant. Semiconductor companies face constant pressure to reduce production costs to maintain price stability, which accounts for the interest in relocating to countries where governments provide generous incentive packages.[19]

INVESTMENT IN SEMICONDUCTOR RESEARCH AND DEVELOPMENT

Achieving miniaturization as well as improvements in manufacturing can only succeed if semiconductor companies devote considerable amounts of their capital to R&D. Indeed, in 2008, companies in the semiconductor and other electronic components sectors spent $22 billion in R&D in the U.S., representing 20 percent of domestic sales revenues that year. Most of this R&D was self-financed, with a relatively small fraction coming from outside sources ($734 million).[20] An additional $7.1 billion was spent abroad, but most of the process-technology R&D remains "homebound" in the U.S.[21]

To put this in perspective, the defense industry[22] (aerospace products and parts, navigational, measuring, electromedical, control instruments, and transportation) together spent $40.7 billion on R&D in the United States—nearly double what was spent by the semiconductor industry alone. However, 73 percent of that was paid for by the U.S. federal government/DoD ($29.6 billion).[23] The semiconductor sector remains almost entirely self-financed, receiving only minor direct subsidies or assistance from government agencies.

GLOBALIZATION: OFFSHORING AND OUTSOURCING

In popular discourse, the main explanation for why the U.S. is losing high-tech manufacturing to Asia or why only a small volume of chips are manufactured domestically comes down to labor costs. But labor costs do not account for why so much of the U.S. semiconductor sector has migrated to Asia. Although labor costs are significantly lower in many overseas economies, they are ultimately a minor part of the overall retail price of most consumer electronics. Foundries employ few unskilled workers. Packaging and testing is labor-intensive, and most of that takes place in other countries—not where the fabricators are located (the difference between front-end and back-end assembly).

For example, Apple's iPhone is assembled in China at the Foxconn facility (owned by the Taiwanese Corporation Hon Hai). Through its ownership of the design patents, Apple claims 60 percent of the iPhone's retail price, while China captures only three percent of the retail price of each iPhone produced.[24] The bulk of the components for the iPhone, including the processor and memory, are manufactured in South Korea, Japan, and Germany. A study conducted by the Asian Development Bank Institute summarizing the manufacturing costs of the iPhone reported that, of the $178.96 production cost of each iPhone, nearly 70 percent is attributed to components ordered from various semiconductor firms; 27 percent is "other materials"; and only 3.6 percent of manufacturing cost—$6.50 per unit—is attributed to assembly.[25] The basic assembly of a small device does not generate significant financial benefits for China, because the value rests with the design and popularity of the device, which accounts for the higher retail price.

Rather, there are three distinct trends that account for the extensive off-shoring of semiconductor manufacturing, and indicate the likelihood of this trend continuing into the future.

First are the advancements that have allowed for separate processes for chip design and fabrication. In the early 1980s, Carver Mead and Lynn Conway published a groundbreaking textbook that revolutionized the design of integrated circuits, allowing for the decoupling of chip design and fabrication. Design teams in academia and Silicon Valley used very large-scale integration methods for combining tens of thousands of transistor circuits on a single chip.[26] Although only large, integrated companies such as IBM and AT&T had the technology to execute each step of the vertical chain, from silicon to assembled semiconductors, the decoupling of design from manufacturing allowed for the emergence of smaller and more efficient firms that specialize in chip design, especially for niche and specialty markets.

Second, coming out of World War II, the federal government was a major supporter of the nascent computer industry. Government spending provided a significant portion of U.S. investment on R&D, favoring large, innovative enterprises that developed new and improved products and processes, both in the defense and civilian markets. These major firms possessed enormous financial and human capital resources (e.g. Boeing, General Electric, General Motors, and Ford), including access to capital that could be invested in projects with long-term potential. These large companies were vertically integrated, and different phases of design and production took place in-house.

The recession of the early 1980s was capped by the emergence of newly industrialized countries in Asia (namely Hong Kong, Taiwan, South Korea, and Singapore). Simultaneously, corporate America restructured to strengthen its core competencies, and the federal government withdrew some of its support for basic research. In the wake of major corporate restructuring in which firms emphasized core competencies, vertical and horizontal integration sharply declined in large U.S. firms.[27] Companies began relying on third-party contractors for different services and components that were no longer made in-house. The logical next step was to move production overseas to take advantage of cheaper production costs.

Furthermore, the costs of translating new improvements in lithography into semiconductor fabrication increased from $100 million in 1985 to $10 billion in 2012. Combined with the vertical disintegration of design and manufacturing activities, such start-up costs also exerted pressure to shop around for the most generous incentive packages.

Two types of manufacturing migration have occurred. In one scenario, semiconductor companies moved manufacturing operations to other legal jurisdictions because of tax incentives, subsidies, salaries, and proximity to customers. In another scenario, a portion of operations was contracted out to companies in another country. In each case, substantial investments by local governments in foreign countries persuaded U.S. companies to relocate their fabricating foundries. Eventually, as U.S. government subsidies did not match what foreign authorities were offering, many firms made the logical decision to "follow the money."

Chip-making is extremely complex, but it is also routinized and standardized. With accessible chip production technology, a potential start-up company needs only capital to acquire the technology. Taiwan is a good example of how a country's interest in becoming an offshore producer dovetailed with the new direction of corporate America, and why East Asia as a whole became the primary destination for semiconductor manufacturing. Beginning in the mid 1980s, the Taiwanese government began pursuing compatible development policies—including favorable tax laws, procurement policies, protection for intellectual property, and access to capital sources—that encouraged the rapid growth of semiconductor manufacturing.[28]

Foreign governments recognized that advanced fabrication facilities stimulate the local economy and understood the increasing role of incentives packages in influencing plant location. In the 1990s, Taiwan and other state-interventionist Asian countries began providing incentives packages that sometimes covered up to 25 percent of the approximately $4 billion initial investment. It is not surprising that U.S. corporations, most accountable to shareholders, take advantage of these incentives and relocate the next generation of fabricating foundries to the market with the greatest benefits package. If not, the company runs the risk of losing its comparative advantage vis-à-vis competitors who utilize these incentives.[29]

Asia-Pacific countries also had the advantage of proximity to the consumer electronics industry. As electronics became more advanced and more compact, highly specialized components required increased collaboration between system and component designers. Specialized chips were required for a growing number of applications, increasing the need for direct interactions between chip producers and electronics systems designers located in Southeast Asia.[30]

By comparison, the United States has lost favor among semiconductor companies. Very few new semiconductor fabrication plants are being built domestically. In 2011, 27 large-scale mass-producing semiconductor fabricators were under construction; only one of them was located in the United States (18 were in China and four were in Taiwan). In 2012, construction started on nine new volume foundries worldwide; none of them were in the United States. In 2013, it is expected that six more foundries will be built; none of them will be in the United States. In total, between 2011 and 2013, 42 new fabricators have been built or are slated to begin production soon; only one is in the United States. In 2003, North America ranked in second place globally in semiconductor capacity; in 2013, it is projected to fall behind Japan, Korea, and Taiwan. Although global semiconductor capacity will have risen by close to 500 percent between 2003 and 2013, American capacity will have grown by a comparatively meager 65 percent.[31]

The U.S. semiconductor industry has shifted offshore, together with the supply chain, which now covers the entire region from China, South Korea, Japan, the Philippines, and Thailand. Labor costs no longer matter significantly since so much of semiconductor manufacturing is automated. However, management expertise, customers (e.g, the electronics industry), government-provided tax breaks, and expedited permits can influence the location of a new semiconductor plant. The U.S. government provides minimal incentives, even though a new plant can cost $8 billion to construct.

China has recorded one of the fastest growth rates of semiconductor capacity; 583 percent between 2003 and 2013 (although no advanced semiconductors manufacturer is currently located in China). It is also the largest consumer of semiconductors, accounting for more than 40 percent of global consumption in 2010. All American semiconductor companies have extensive commercial relations with China-based buyers of microchips (see Table 2). For example, between 40 and 50 percent of Intel's total sales were to China, AMD sold between 50 to 70 percent of output to China-based firms, and Freescale more than 70 percent.[32] Even if the largest Chinese semiconductor company sold all of its output within China, it would not rank among the top 40 suppliers to the Chinese semiconductor market in 2010. Rather, studies show that China's strength is in assembly and testing – the back-end of the manufacturing chain.

China lags behind in the area of design because it has not been able to acquire the newest technology. Among its members, the 1996 Wassenaar Arrangement imposes controls on exports of manufacturing technology, including advanced wafer manufacturing technology necessary

Table 3: Semiconductor Suppliers to the Chinese Market (2009-2010)

COMPANY RANK	2009	2010	Revenue in Millions of Dollars				
			2009 IC	% Change	2009 Semi	% Change	Market Share
Intel	1	1	15,570	26.3%	15,570	26.3%	14.9%
Samsung	2	2	5,546	44.5%	5,681	43.9%	6.2%
Hynix	3	3	3,644	52.7%	3,644	52.7%	4.2%
Toshiba	4	4	3,231	28.1%	3,904	27.3%	3.8%
TI	5	5	3,127	35.8%	3,292	36.1%	3.4%
ST	6	6	2,704	22.2%	3,601	23.5%	3.4%
AMD	7	7	3,415	22.2%	3,415	22.2%	3.2%
NXP	8	8	2,281	30.0%	2,891	29.7%	2.8%
Renesas*	9	9	1,272	31.0%	1,716	32.6%	2.6%
NEC*	10	9	739	—	911	—	—
Freescale	11	10	2,057	35.5%	2,430	32.7%	2.4%
Media Tek (MTK)	12	12	2,442	0.5%	2,442	0.5%	1.9%
Total Top 10			44,019	30.6%	46,876	32.1%	46.9%

* Renesas for 2010 is compared with Renesas + NEC for 2009 due to their merger.

Source: PwC, "Continued Growth: China's Impact on the Semiconductors Industry, 2011 update."
http://www.pwc.com/gx/en/technology/assets/china-semiconductor-report-2011.pdf

to manufacture silicon wafer chips with a thickness of 90 nm or less, to certain countries. The United States and Taiwan apply additional restrictions on the export of equipment used to manufacture 65 nm chips. As a result, Chinese chip foundries produce chips that are several generations behind those of the United States, and U.S. companies have continued to maintain careful control over integrated chip design for microprocessors.

VULNERABILITY IN NATIONAL SECURITY AND THE DEFENSE INDUSTRIAL BASE

Defense constitutes only a small portion of the overall demand for semiconductor devices. Consumer electronics, computers, and communication account for 85 percent of demand for chips. Other major consumers are the automobile and commercial aviation sectors. Many defense applications rely on consumer electronics; the specialized defense semiconductor sector represents less than one percent of American semiconductor sales. This segment consists of specialized chips that must be able to withstand extreme conditions or perform extremely sophisticated and specialized tasks. Such specialty chips are commissioned directly by DoD, and enable unique defense capabilities related to remote sensors, radiation-hardened electronics, and missile guidance.

If the current trend continues, semiconductor R&D might soon follow manufacturing overseas, thereby diminishing the United States' ability to design and produce innovative technology, and potentially jeopardizing national security. From the "smart bombs" that allow the U.S. military to minimize collateral damage today to the "super suits" that will protect and enhance the effectiveness of our future warfighters, the important role that micro-chips play in many weapon and communications systems makes maintaining a strong domestic industry a strategic priority.

There are three primary risks, which are described below:

ABILITY TO CONTINUE TO DESIGN AND INNOVATE INFORMATION AND COMMUNICATION TECHNOLOGY

Certain semiconductor devices are unique to defense, with no substitute available in the commercial market. They also are not exported, as they fall under International Traffic in Arms Regulations (ITAR), and are supplied only by trusted original equipment manufacturers (OEMs). This small yet critical segment of the market, featuring production according to Military Specification (Mil-Spec) is controlled by a handful of small firms. By virtue of the specialized nature of these chips that have no commercial market, these firms cannot achieve economies of scale, leading to a higher price per unit when compared to

commercial-off-the-shelf (COTS) components. Because these firms specialize in defense applications and may become dependent on continued DoD business, DoD often provides seed money and other incentives to promote R&D in defense-unique semiconductor devices. Over time, many companies have dropped out of this market due to defense cutbacks, while inadequate support for Mil-Spec semiconductor R&D has deterred companies from entering this sector. Subsequently, DoD may rely on only one vendor for a specific technology, which can lead to inflated prices and little incentive to innovate or develop more efficiency.[33]

Thus, there should be ongoing efforts to nurture and sustain a high-tech base that is capable of undertaking projects critical to ensure the U.S. military's ongoing technological superiority. This specialized military-technology sector is much more likely to survive if the larger commercial technology sector also thrives.[34]

The specialized defense-technology sector is also contending with increased competition for a skilled workforce at the global level. Domestic firms must compete with overseas manufacturing facilities for individuals holding advanced degrees in science and engineering. Although foundations of the microelectronics industry were initially developed in the United States, globalization has diffused this knowledge worldwide, and U.S. firms must compete with foreign counterparts who often benefit from extensive support from their respective governments.

Cross-discipline collaboration and cross-market segment coordination drive innovation. Innovation takes place in geographic clusters where scientists, private companies, start-ups, suppliers, and clients interact and exchange ideas and information. Governments often foster these clusters, recognizing that face-to-face interaction and sharing of ideas and problems have a positive impact on innovation and research. The survival of domestic specialized defense electronics production depends on the health of the general sector, which has been gradually relocating overseas.

DISRUPTIONS IN THE GLOBAL SUPPLY CHAIN

The globalized supply chain has widened the availability and range of affordable chips; however, it has also lengthened the period between design and production. The global semiconductor market is exposed to unintentional and often unforeseen disruptions. Because of to the complexity of the supply chain supporting each primary defense contractor, DoD is largely unaware of who produces which part. Supporting each OEM is a string of dozens of different companies of different sizes and levels of sophistication, often spread out over three continents. Below are several recent examples of how this global supply chain has caused disruptions to the supply of semiconductors.

Heavy rains in Thailand in fall 2011 led to a shortage of external hard drives because a Western Digital plant near Bangkok was flooded. The plant manufactured a component known as a "slider," and accounted for approximately 15 percent of the world's supply of this critical hard drive component. Not only did the extensive damage done to the plant immediately inflate hard drive prices, it will continue to restrict hard drive production for several years until repairs are complete. While the plant was constructed on a flood plain (meaning that occasional flooding is expected), this risk was widely unknown because it was assumed that drives manufactured

> The defense-unique microelectronics sector has shrunk steadily in recent years, creating the risk that there may come a time when Silicon Valley might finally leave the United States, taking with it the firms capable of producing the specialized, defense-specific chips and components needed for the most advanced weapons systems.

by a Japanese company were manufactured in Japan. In reality, many Japanese companies had established production in Thailand due to favorable exchange rates. This relocation was widely unknown to retailers, consumers, and analysts due to the opacity of the supply chain.[35]

In March 2011, a colossal earthquake in Japan and the resulting tsunami took the lives of at least 19,000 people. Additionally, it caused severe disruption to numerous high-tech manufacturing supply chains. Japan produces a large quantity of important chips, such as lightweight flash memory chips used in smartphones and tablet computers. Japan makes about 35 percent of those memory chips, most of which are made by Toshiba. While many companies have back-up facilities, and while more costly electronic components such as flash memory and liquid crystal displays (LCDs) tend to be produced in different factories, many small, specialized parts such as connectors, speakers, microphones, batteries, and sensors are mass produced at specialty plants. The absence of a few small parts can hold up an entire production line, a possibility that becomes more likely for end-products that depend on a large number of suppliers. The earthquake disrupted the supply chain of the

> **Without proper mitigation, the United States and U.S. military security may face a significant crisis in the near future when a part or component is abruptly unavailable because of a natural disaster or political turmoil.**

fourth, fifth, and sixth tier suppliers, many of whom did not have back-up facilities in anticipation of a natural disaster.

Semiconductors consume minute amounts of raw materials, but these special materials are also critical for the semiconductor manufacturing process. Mitsubishi Gas Chemical factory in Fukushima, Japan, produces bismaleimide triazine, a resin used in the packaging for small computer chips in cellphones and other products. These facilities were damaged during the earthquake, leading to questions of whether there would be a shortage of bismaleimide triazine.

Two Japanese companies account for more than 60 percent of the world's supply of silicon wafers needed to fabricate semiconductors. The largest producer, the Shin-Etsu Chemical Corporation, has its main wafer plant in Shirakawa. The earthquake damaged the city and took the factory out of production. Although Shin-Etsu has factories outside Japan that were unaffected, the most advanced manufacturing and silicon-growing processes are done in Japan.[36]

The elongated supply chain has contributed to these new risks, further aggravated by the prevalence of "just-in-time" delivery systems, which refers to inventories that are always kept low to minimize

expenditures.[37] While large established firms may keep inventories on the premises sufficient for several weeks of production, smaller subcontractors often lack the necessary capital to do so and may seek to reduce costs by keeping inventories extremely low. In the event of a major disruption to production, companies at the lower tiers of the global supply chain may fail to deliver parts, paralyzing the production of important semiconductor devices or microelectronics.[38]

DEFECTIVE AND COUNTERFEIT PARTS

Directly related to the health of the global supply chain, various studies have suggested that foreign subcontractors may be tempted to cut corners and supply defective or counterfeit parts. Counterfeiting takes place across the board, involving old and/or discontinued parts as well as state-of-the-art advanced microchips. Sometimes a small firm tries to sell a counterfeit copy of a chip that may function identically to the more expensive authentic version; however, it deprives the original developer of the revenue associated with the intellectual property. In other situations, a substandard or counterfeit chip may be packaged as if it came from a reputable developer/manufacturer.

Clearly problematic in the commercial market, the presence of counterfeit or substandard chips is even more troubling for the defense industrial base. Defense electronics often have a longer lifespan than many consumer electronics. DoD relies on "dated" chips that are no longer manufactured commercially to enable critical weapons systems. Unlike consumer markets, which jettison outdated technology quickly, the military and defense industrial base sticks with older parts and technology, often for decades, as it is too cumbersome and expensive to continuously refit a

Table 4: Top Five Most Counterfeited Semiconductors, 2011

Rank	Commodity Type	Percent of Reported Incidents
1	Analog Integrated Circuit	25.2 %
2	Microprocessor Integrated Circuit	13.4 %
3	Memory Integrated Circuit	13.1 %
4	Programmable Logic Integrated Circuit	8.3 %
5	Transistor	7.6 %

Source: "Combating Counterfeits in the Supply Chain." IHS http://www.ihs.com/info/sc/a/combating-counterfeits/index.aspx.

system to make it compatible with rapidly advancing electronic components. So long as the systems satisfactorily perform their function, there is no need to replace them with newer versions. Thus, DoD sometimes needs replacement parts that are no longer commercially available.

DoD can contact after-market firms specializing in discontinued parts (e.g., Rochester Electronics, which manufactures obsolete parts) or it may tap in-house government production capabilities. But it is often cheaper and faster to acquire the discontinued part from an outside contractor. In such instances, there is the increasing chance that a supplier will provide a cheap, counterfeit chip that was purchased from an untrusted third-party manufacturer.

As more production has moved offshore, more counterfeits have appeared on the market. The Department of Commerce's Bureau of Industry and Security (BIS),[39] the Government Accountability Office (GAO),[40] and the Aerospace Industries Association (AIA)[41] have all examined

the rising prevalence of counterfeiting and have come to the same conclusion: the rising incidence of counterfeit parts is primarily due to the globalized supply chain, which exacerbates bad practices such as weak inventory management, opaque procurement procedures, haphazard inspection and testing protocols, and the absence of communication within and across industry and government organizations. Moreover, even after counterfeit parts have been identified, it is difficult to establish accountability because companies and organizations assume that others in the supply chain are verifying and testing parts. Counterfeit parts not only deprive legitimate businesses of revenue and royalties, which diminishes investment in R&D, but they also result in higher failure rates. These failure rates in turn lead to increased costs associated with maintenance and downtime. More significantly, substandard and defective parts may have catastrophic consequences if they lead to casualties, accidents, and mission failures. When used in military systems, counterfeits (and in this case, counterfeit semiconductors) pose a real and unacceptable risk to the U.S. military and to our national security.

The rise of counterfeit parts is a serious problem, not just a matter of improved reporting. The U.S. Department of Commerce reported a 142 percent increase in counterfeit parts between 2005 and 2008, the majority of which were commercial electronic components widely used across every major technology end-market. In 2011, the five most commonly counterfeited semiconductor types were analog integrated circuits (ICs), microprocessors, memory ICs, programmable logic devices, and transistors. These chips are used widely throughout all major semiconductor applications, including computing, consumer electronics, communications, automotive systems, and defense.[42]

Analog chips, the most commonly reported counterfeit semiconductors category in 2011, convert analog data (including sound, light, distance, and temperature) to digital signals, and are used extensively in consumer and defense electronics. One faulty counterfeit analog integrated circuit can cause a wireless set to malfunction, an aircraft to have a serious accident, or a guided missile to hit the wrong target. Because these chips are so extensively used, the impact of one counterfeit chip can range from barely noticeable to disastrous. Although counterfeit chips themselves are often quite cheap, their use can be incredibly expensive, because they are often unreliable and may cause massive failures of critical systems with catastrophic and potentially deadly consequences.[43]

MITIGATING THE RISKS

Increase our understanding of the semiconductor supply chain, identifying potential chokepoints for disruption and entry points for counterfeit parts. Understanding incredibly complex defense-electronics and semiconductor supply chains is a crucial first step to mitigating risks impacting the availability and reliability of these inputs. This knowledge is critical to address the threat of abrupt disruptions and to combat counterfeiting. The defense-electronics supply chain is probably the most complex of the defense industrial base. The 2011 flooding in Thailand exposed a general lack of understanding of the supply chain among relatively specialized manufacturers producing products that are relatively simple when compared to many military end-items. Still, the downstream supply chain was caught off guard when a natural disaster made inputs unavailable.

At present, DoD and top-tier suppliers have limited understanding of this multi-tiered supply chain, to include the role of silicon extraction and refining, circuit design and fabrication, subsystems design and fabrication, and each subsystem's eventual placement in a more complex weapons system. Without a better understanding of this complex web of manufacturers, designers, and suppliers, it will be nearly impossible to address more specific problems such as counterfeiting. Current DoD efforts, such as the Sector-by-Sector, Tier-by-Tier (S2T2) program, may be helpful as a model for a thorough review of the semiconductor supply chain.

Moreover, a thorough mapping of the semiconductor sector should dramatically reduce exposure to counterfeit parts. Counterfeit parts can be introduced because DoD and defense contractors are unaware of where in the supply chain these parts are coming from, and are therefore unable to hold manufacturers of counterfeit and substandard parts accountable. More thorough mapping of the supply chains for defense end-items will enable identification of the sources of defective or counterfeit chips. In turn, supply-chain mapping will enable identification of other defective components manufactured by those contractors, so that the violators can be punished or removed from defense supply chains.

Supply-chain mapping augments the new anti-counterfeiting regulations included in the FY2012 National Defense Authorization Act (NDAA). The NDAA requires defense suppliers at all levels to enact systems to identify and remove counterfeit components from their supply chains, including the immediate documentation and reporting of suspicion of counterfeit parts. This in turn will support the establishment of a counterfeiting database. Additionally, the FY2012 NDAA criminalizes the deliberate sale of counterfeit components, allowing for substantial fines and extended prison terms for those responsible.

Another potential course of action would be to establish a federal interagency office to monitor and respond to counterfeiting and defective components. The office would also be responsible for monitoring the availability of components and acquiring obsolete parts. (In addition to the difficulty of identifying a secondary supplier for obsolete parts, obsolete parts also present a heightened risk of counterfeiting.) For the sake of efficiency, this single agency or office should oversee the response to counterfeiting as well as the acquisition of obsolete parts most prone to counterfeiting.

Stop the decline in U.S. domestic semiconductor fabrication. The United States must take measures to prevent the future decline of the US. semiconductor industry. Although much R&D is still conducted within the United States, the offshoring and outsourcing trends fostered by the decoupling of semiconductor design and fabrication and encouraged by the generous incentives packages offered by foreign governments and the development of a significant semiconductor community in the Asia-Pacific region provide strong incentives for R&D to follow. Although the United States remains the leader in semiconductor R&D and offers the best education and training in this field, foreign students on temporary visas are obtaining a disproportionate number of advanced degrees in science and engineering, and are then are unable to work in the United States. The growth of the East Asian hub for semiconductors has created a new focal point for these highly trained individuals. This shift may lead to the re-centering of semiconductor R&D away from the United States, which will in turn reduce the technological advantage of the U.S. military.

To prevent the further decay of the U.S. semiconductor industry and maintain the United States' military and technological edge, the U.S. government should support R&D in American universities, research foundations, and corporations, and implement policies that encourage investment in new semiconductor foundries. The combination of subsidies and tax breaks successfully lured semiconductor manufacturing to Taiwan and Korea, and similar policies could prove effective in incentivizing U.S. companies to invest in domestic manufacturing. Because semiconductor manufacturing is capital-intensive, requiring expensive and highly specialized infrastructure, one means of achieving this would be to shorten capital depreciation schedules to allow firms investing in state-of-the-art equipment to write off a larger portion of the investment costs upfront. Presently, corporations may write off investments in new equipment over multiple years. However, in conjunction with Moore's Law, advances in semiconductor fabrication technology outpace the ability of companies to recuperate the expenses of value-added equipment and infrastructure through the current depreciation schedule. Moreover, because startup costs for a new foundry are often in excess of $9 billion, it can be years before new operations are able to turn a profit. Incentives packages to defray the initial costs of foundry development will be necessary to encourage companies to invest in U.S. manufacturing in the high-tech industries.

CONCLUSION

Advances in semiconductors and intensive R&D investments in advanced technologies drove the revolution in consumer and military electronics. American universities, scientists, and ICT companies may still dominate global markets, but it is doubtful whether they can sustain this position if so much fabrication and manufacturing takes place offshore. Experience tells us that eventually R&D will follow, together with scientists and corporate labs. Such a scenario would be detrimental to U.S. military dominance in advanced technologies, as well as for the U.S. economy at large. Much of the U.S.'s military strength, economic prowess, and political leadership still rest on the U.S.'s dominant global position in ICT; however, this dominance is threatened as other countries challenge U.S. technological leadership. Indeed, Asian companies, enabled by state support for their semiconductor industries, have caught up and now rival the U.S. semiconductor sector. Yet it would appear that the U.S. government and DoD do not fully grasp the serious implications of the decline of the domestic semiconductor industry and the ramifications of globalization for the sector. If the current trend continues and the U.S. semiconductor industry continues to wither under the pressure of globalization, not only will U.S. international competitiveness decline, but U.S. military supremacy will be at risk.

ENDNOTES

a. Robert Johnson, "The US Military Wants To 'Microchip' Troops," Business Insider, May 06, 2012. http://www.businessinsider.com/the-us-military-wants-to-microchip-troops-2012-5.

1   Industrial College of the Armed Forces, *Electronics Industry* (Spring 2011), 2. http://www.ndu.edu/icaf/programs/academic/industry/reports/2011/pdf/icaf-is-report-electronics-2011.pdf.

2   Twenty states have significant semiconductors industry direct employment: Ariz., Calif., Colo., Fla., Idaho, Mass., Minn., Mo., N.J., N.M., N.Y., N.C., Ohio, Ore. Pa., Texas, Utah, Va., Wash.

3   Stephanie S. Shipp, *et al*, *Emerging Global Trends in Advanced Manufacturing* (Alexandria, VA: Institute for Defense Analyses, P-4603 March 2012), 23. http://www.wilsoncenter.org/sites/default/files/Emerging_Global_Trends_in_Advanced_Manufacturing.pdf.

4   Brian Bradshaw, "Geeks Informed: The Rise and Fall of the American Chip Industry," May 2009. http://geeksinformed.blogspot.com/2009/05/rise-and-fall-of-american-chip-industry.html.

    Industrial College of the Armed Forces, *Electronics Industry* (Spring 2011), 1. http://www.ndu.edu/icaf/programs/academic/industry/reports/2011/pdf/icaf-is-report-electronics-2011.pdf.

    "IHS Downgrades Semiconductor Industry Market Forecast," IHS press release, December 3, 2012. http://www.isuppli.com/semiconductor-value-chain/news/pages/ihs-downgrades-semiconductor-industry-market-forecast-to-23-percent-decline.aspx.

5   Stephanie S. Shipp, *et al*, *Emerging Global Trends in Advanced Manufacturing* (Alexandria, VA: Institute for Defense Analyses, P-4603 March 2012), 24. http://www.wilsoncenter.org/sites/default/files/Emerging_Global_Trends_in_Advanced_Manufacturing.pdf.

6   Patrick Wilson, "Maintaining U.S. Leadership in Semiconductors," presentation, AAAS Annual Meeting, Washington, D.C., February 18, 2011.http://www.aaas.org/spp/rd/presentations/20110218PatrickWilson.pdf.

7   The implosion of the so-called "new economy" and the Internet bubble in 2000 and 2001 severely affected the semiconductor sector. John Cassidy, *Dot.com: How America Lost Its Mind and Money in the Internet Era* (New York: Harper Perennial, 2003); and Michael Lewis, *The New New Thing: A Silicon Valley Story*, (New York: Penguin, 2001).

8   U.S. Department of Commerce, U.S. Patent and Trademark Office, *Intellectual Property and the U.S. Economy: Industries in Focus* (March 2012). http://www.uspto.gov/news/publications/IP_Report_March_2012.pdf.

9   Peter Dombrowski and Eugene Gholz, *Buying Military Transformation: Technological Innovation and the Defense Industry* (New York: Columbia University Press, 2006).

10 DRAM, Dynamic Random Access Memory, was once considered the most sophisticated component of computer technology. In Japan, government agencies and private companies targeted DRAM as their product of choice and poured money into superior manufacturing technology by relying on lithography technology called "scanners."

11 Texas Instruments created one of the world's first chip foundries in Taiwan in 1989 to manufacture DRAM. The company was called TI-Acer.

12 This assessment is according to Paul Otellini, the CEO of Intel. "A Fab Result," *The Economist*, May 26, 2012, 79.

13 Broadcom is in the wireless and broadband communication business. Qualcomm invented the design of integrated circuits for code division multiple access wireless devices.

14 IC Insights, "Top 20 Semiconductor Suppliers' Sales Growth Rates Forecast to Range from Great (+31%) to Terrible (-17%) in 2012," *Research Bulletin*, November 7, 2012. http://www.icinsights.com /data/articles/documents/466.pdf.

15 Offshoring and outsourcing are two distinct trends. Offshoring refers to establishing a production facility in another country to take advantage of corporate tax incentives or lower labor costs. Outsourcing refers to delegating the task of production, design, or packaging to a third independent party.

16 Dan Breznitz and Peter Cowhey, "America's Two Systems of Innovation: Recommendations for Policy Changes to Support Innovation, Production and Job Creation," *CONNECT Innovation Institute* (February 12, 2012).

    Erica R.H. Fuchs "The Impact of Manufacturing Offshore on Technology Competitiveness: Implications for U.S. Policy," *CONNECT Innovation Institute* (February 12, 2012). http://www.connect.org/programs/innovation-institute/.

17 Gregory Tassey, "Beyond the Business Cycle: The Need for a Technology-Based Growth Strategy," *National Institute of Standards and Technology Economics Staff Paper* (December 2011). http://www.tradereform.org/wp-content/uploads/2011/12/111205-Tassey-article.pdf.

18 "Apple Became the Top Semiconductors Customer in 2011," Gartner Newsroom, January 24, 2012. http://www.gartner.com/it/page.jsp?id=1902414.

19 "Microlithography," Clifford L. Henderson. http://henderson.chbe.gatech.edu/Introductions/microlithography%20intro.htm.

20 National Research Council. "Semiconductors." *Innovation in Global Industries: U.S. Firms Competing in a New World* (Washington, D.C.: The National Academies Press, 2008), table 4-8, 4-22.

21 Ibid., 102.

22 The North American Industry Classification System does not identify a "defense industry," but it is common to group together aerospace, navigation, control instruments, etc. into the defense category.

23 National Science Board, *Science and Engineering Indicators 2012* (Arlington, VA: NSB 12-01, January 2012), 4-21. http://www.nsf.gov/statistics/seind12/pdf/c04.pdf.

24 Jordan Weissmann, "Which Countries Make Money Off the iPad?" *The Atlantic*, January 19, 2012. http://www.theatlantic.com/business/archive/2012/01/which-countries-make-money-off-the-ipad/251654/.

25 Yuqing Xing and Neal Detert, "How the iPhone Widens the United States Trade Deficit with the People's Republic of China," *ADBI Working Paper* No. 257 (2010). http://www.adbi.org/files/2010.12.14.wp257.iphone.widens.us.trade.deficit.prc.pdf

26 Gina Smith, "Unsung innovators: Lynn Conway and Carver Mead," *Computerworld*, December 3, 2007.

27 William Lazonick, "What is New, and Permanent, about the 'New Economy'?" *Sustainable Prosperity in the New Economy?: Business Organization and High-Tech Employment in the United States*. (Kalamazoo, MI: W.E. Upjohn Institute for Employment Research, 2009), 1-38.

28 Stephanie S. Shipp, "Emerging Global Trends in Advanced Manufacturing," *IDA March 2012*. P-4603.

29 Dewey & LeBoeuf, *Maintaining America's Competitive Edge: Government Policies affecting Semiconductors Industry R&D and Manufacturing Activity* (March 2009). http://www.sia-online.org/clientuploads/directory/DocumentSIA/Research%20and%20Technology/Competitiveness_White_Paper.pdf.

30 National Research Council, "Semiconductors." *Innovation in Global Industries: U.S. Firms Competing in a New World* (Washington, D.C.: The National Academies Press, 2008), 105.

31 Richard A. McCormack, "Global Semiconductor Industry Has No Plans To Build Factories In The United States," *Manufacturing & Technology News* 19, September 28, 2012. http://www.manufacturingnews.com/news/semiconductorfabs928121.html.

32 PricewaterhouseCoopers, "Continued Growth: China's Impact on the Semiconductors Industry, 2011 Update," 8. http://www.pwc.com/gx/en/technology/assets/china-semiconductors-report-2011.pdf.

33 Industrial College of the Armed Forces, *Electronics Industry* (Spring 2011). http://www.ndu.edu/icaf/programs/academic/industry/reports/2011/pdf/icaf-is-report-electronics-2011.pdf.

34 Sri Kaza, Rajat Mishra, Nick Santhanam, and Sid Tandon, "The Challenge of China," *McKinsey on Semiconductors* (Fall 2011). http://www.sia-online.org/clientuploads/directory/DocumentSIA/Research%20and%20Technology/, http://www.mckinsey.com/Client_Service/Semiconductors/Latest_thinking/The_challenge_of_China.

35 Thomas Fuller, "Thailand Flooding Cripples Hard-Drive Suppliers," *The New York Times*, November 6, 2011. http://www.nytimes.com/2011/11/07/business/global/07iht-floods07.html?pagewanted=all.

36 Steve Lohr, "Stress Test for the Global Supply Chain," *The New York Times*, March 19, 2011. http://www.nytimes.com/2011/03/20/business/20supply.html?_r=1&pagewanted=all.

37 In 1997, there was a fire at a plant of one of Toyota's main suppliers, Aisin Seiki, which made a brake valve used in all Toyota vehicles. Because of the carmaker's just-in-time system, the company had just two or three days of stock on hand. The fire threatened to halt Toyota's production for weeks.

38 J.T. Macher, D.C. Mowery, and A. Di Minin, "Semiconductors," *Innovation in Global Industries: U.S. Firms Competing in a New World*. eds. J.T. Macher and D.C. Mowery (National Academy Press: Washington, D.C.: 2008): 101-140.

39 U.S. Department of Commerce, Bureau of Industry and Security, *Defense Industrial Base Assessment: Counterfeit Electronics* (Washington, D.C.: BIS, January 2010). http://www.bis.doc.gov/defenseindustrialbaseprograms/osies/defmarketresearchrpts/final_counterfeit_electronics_report.pdf.

40 Government Accountability Office, *Space and Missile Defense Acquisitions: Periodic Assessment Needed to Correct Parts Quality Problems in Major Programs* (June 2011).

41 Aerospace Industries Association, *A Special Report Counterfeit Parts: Increasing Awareness and Developing Countermeasures* (March 2011). http://www.aia-aerospace.org/assets/counterfeit-web11.pdf.

42 Evertiq, "Top 5 Most Counterfeited Parts Represent a $169 Billion Potential Challenge for Global Semiconductors Market" (April 5, 2012). http://evertiq.com/news/21825.

43 Top 5 Most Counterfeited Parts Represent a $169 Billion Potential Challenge for Global Semiconductors Market, IHS-iSuppli, April 4, 2012. http://www.isuppli.com/Semiconductors-Value-Chain/News/Pages/Top-5-Most-Counterfeited-Parts-Represent-a-$169-Billion-Potential-Challenge-for-Global-Semiconductors-Market.aspx

One blatant case: From 2006 and 2010, VisionTech Components knowingly sold counterfeit ICs to approximately 1,101 buyers in the United States and abroad. Headquartered in Clearwater, Florida, it created a website purportedly selling name brand integrated circuits, which were primarily acquired from sources in China bearing counterfeit marks. The circuits were imported into the United States for sale to defense contractors and other manufacturers of critical systems, such as brakes for high-speed trains and instruments for firefighters to detect nuclear radiation. The company sold fake chips for more than 100 to 1,000 times their real value. Caroline Kazmierski, "Counterfeit Semiconductors Importers Pay High Price of Breaking Public Trust, Endangering Americans," Semiconductors Industry Association press release, October 25, 2011. http://www.sia-online.org/news/2011/10/25/news-2011/counterfeit-semiconductors-importers-pay-high-price-of-breaking-public-trust-endangering-americans/.

CHAPTER 8 • COPPER-NICKEL TUBING

EXECUTIVE SUMMARY

Unlike stainless steel and other materials commonly used in commercial ships, copper-nickel (Cu-Ni) tubing is the only material that resists corrosion and biofouling (the accumulation of micro-organisms on wet surfaces) enough to be usable on U.S. Navy vessels. The United States cannot construct modern naval vessels without adequate supplies of Cu-Ni tubing.

For a wide range of U.S. Navy applications, Cu-Ni tubing must have an outside diameter of 4.5 inches or greater. There is currently only one U.S. firm and one foreign firm capable of producing military grade Cu-Ni tubing of this diameter, and there are concerns about the quality of larger diameter Cu-Ni tubing produced overseas.

Just after World War II the United States controlled 85 percent of global Cu-Ni tubing production. Today the sector has declined to the point where only one U.S. firm—Connecticut-based Ansonia Copper & Brass—is capable of supplying the U.S. Navy with large-diameter tubing. The German-Italian conglomerate KME (the only other company that produces large diameter Cu-Ni tubing to U.S. Navy specifications) has been fined by the European Commission for price fixing, dumping, and other "cartel-like" behavior.[1] If KME applied these practices to the U.S. Navy's Cu-Ni tubing supply chain, the unfair competition could place Ansonia's survival as a key U.S. defense industrial base supplier at risk.

If Ansonia were to cease operations, the U.S. Navy would be forced to purchase Cu-Ni tubing essential to construct and maintain its fleet from a single foreign producer.

Reducing the risk of this vulnerability depends on maintaining an equitable playing field so that U.S. firms do not fall victim to predatory and unfair trade practices. One solution is to provide Title III Defense Production Act funding to ensure the long-term viability of domestic Cu-Ni tubing production through a public-private partnership. Additionally, policymakers could strengthen domestic preference legislation, which currently allows for the purchase of foreign-cast tubing that is merely tested in the United States, as a significant portion of the final cost of tubing arises from ultrasonic and hydrostatic testing.

COPPER-NICKEL TUBING
ENABLING U.S. NAVAL DOMINANCE

MANUFACTURING SECURITY

Copper-nickel (Cu-Ni) tubing is an essential component in U.S. naval ship

U.S. SURFACE COMBATANTS

U.S. SUBMARINES

U.S. AIRCRAFT CARRIERS

THE U.S. NAVY DEFENDS **AMERICAN SECURITY** IN KEY REGIONS **ACROSS THE GLOBE**

TUBING QUALITY

Substandard tubing can render U.S. naval forces ineffective

 =

SUBSTANDARD TUBING

U.S. SAILORS AT RISK

U.S. CAPACITY

U.S. production capacity for large-diameter copper-nickel tubing has declined significantly

FROM 85% **GLOBAL** PRODUCTION

TO ONE SINGLE **DOMESTIC** PRODUCER

Cu-Ni tubing prevents biofouling and corrosion

 +

MITIGATING RISKS

Protecting the copper-nickel tubing supply chain

TITLE III DPA SUPPORT

BUY AMERICAN ACT

MAP THE SUPPLY CHAIN

MILITARY EQUIPMENT CHART
SELECTED DEFENSE USES FOR COPPER-NICKEL TUBING

DEPARTMENT	PLATFORMS
NAVY	All U.S. Navy ships, including: ■ Guided missile destroyer ■ Nimitz-class nuclear-powered aircraft carrier ■ SSN-774 Virginia-class nuclear-powered attack submarine ■ Littoral Combat Ship (LCS) ■ LPD-17 amphibious transport ship
COAST GUARD	U.S. Coast Guard ships, including: ■ Legend-class National Security Cutter ■ Polar-class Icebreaker

INTRODUCTION

"It follows then as certain as that night succeeds the day, that without a decisive naval force we can do nothing definitive, and with it, everything honorable and glorious."[2]

– George Washington, Nov. 15, 1781

The United States was not considered a world power until the U.S. Navy became a force to be reckoned with. This change first occurred during the time historians refer to as the "Oceanic Period of U.S. Foreign Policy" (1890-1945).[3] Even today, a critical part of the United States' security lies in its unparalleled access to virtually all areas of the world with its dominant blue-water navy. Maritime power is essential to U.S. national security. China's recent drive to become a major blue-water power, building aircraft carriers and missiles that could potentially be capable of destroying U.S. carriers, helps underscore this point.[4] Accordingly, U.S. policymakers are expanding U.S. naval presence in the Pacific and strengthening partnerships with countries in the region.[5]

Cu-Ni tubing is an essential component of every U.S. Navy ship, and no suitable alternative materials are currently produced. All U.S. Navy vessels incorporate and rely on Cu-Ni tubing. For example, U.S. aircraft carriers incorporate hundreds of yards of Cu-Ni tubing. Additionally, the U.S. Navy's primary attack submarine, the Virginia-class nuclear submarine, uses 500 yards of 1.25 inch-diameter Cu-Ni tubing. Current plans keep the Virginia-class submarine as a critical platform for our naval forces for several decades. Currently nine Virginia-class submarines are in service (the latest, the USS Mississippi, was commissioned on June 2, 2012), and two more submarines will be built per year until 30 total boats are constructed (though cuts due to sequestration could disrupt this schedule). The tubing is used throughout the hulls of these submarines, as well as stocked in shipyards for future construction and refitting of other U.S. Navy ships.

Cu-Ni tubing is manufactured to strict military specifications, including special anti-corrosion and anti-biofouling properties, for use in hydraulic control systems, lubrication systems, fresh water systems, and high-pressure air-injection systems for ballast tanks. To date, no effective alternative has been developed.

Some industry experts believe that two U.S. firms—Connecticut-based Ansonia Copper & Brass and New York-based Lewis Brass—could provide 100 percent of Department of Defense (DoD) requirements if given the opportunity and sufficient U.S. government support. The U.S. Cu-Ni tubing industry is currently in danger of being replaced by foreign-made Cu-Ni products of varying quality and potentially uncertain supply. Only one U.S. company is still capable of producing the large diameter Cu-Ni tubing used throughout U.S. Navy vessels. The second and third tiers of the

U.S. Navy's ship and submarine supply chains are vulnerable to the uncertain quality and potential volatility of foreign suppliers.

Ensuring a reliable source of Cu-Ni tubing, including a domestic production capacity, is in the interest of U.S. national security. In the event of disruptions in supply or decline in product quality, sole reliance on foreign sources for Cu-Ni tubing products could become a critical issue.

This chapter investigates potential vulnerabilities arising from a decline in the country's capability to manufacture Cu-Ni tubing domestically. Globalization fundamentally has changed the characteristics of the defense supply chain. While costs of production have decreased in many instances, deepening the country's dependence on the foreign supply of critical inputs and technologies has introduced inherent vulnerabilities, some of which may prove difficult or impossible to reverse. Current challenges need to be addressed before the U.S. Cu-Ni tubing industry passes a point of no return.

This chapter provides recommendations for bolstering this important industrial capability. These recommended solutions require collaboration and cooperation among defense and commercial industrial sectors and government policymakers.

Key themes discussed in this chapter are:

- The Cu-Ni tubing supply chain has been compromised due to the introduction of foreign-made, sub-standard Cu-Ni that does not meet U.S. government standards.

- With constrained domestic production capability, U.S. access to Cu-Ni tubing supply, especially tubing with a diameter larger than 4.5 inches, could be at risk of disruption. Foreign suppliers—even those from some allied countries—are not fully reliable due to quality control issues, uncertainties regarding future product availability, and cartel-like activities.

- The risks to the Cu-Ni supply chain can be effectively mitigated by increased supply chain visibility, thorough implementation of industry control measures, and stepping up government enforcement policies such as the Buy American Act.

A NOTE ON CRITICALITY

The loss of Cu-Ni tubing production capability and a resulting supply shortage would have a *significant* impact and would reduce the ability to build and repair ships and submarines and damage U.S. Navy readiness. Ships and submarines could neither be built nor repaired without using a less-acceptable substitute. If the United States Cu-Ni tubing were unavailable during a conflict, the U.S. Navy's readiness would be negatively impacted.

Despite the critical role of 4.5 inch or greater Cu-Ni tubing in naval force projection, there is presently only one domestic producer—Ansonia Brass & Copper. If Ansonia were unable to remain in business and KME stopped supplying Cu-Ni tubing, the U.S. Navy would be *highly* vulnerable to a supply disruption. A robust supply of Cu-Ni tubing is critical for the maintenance and functioning of the U.S. Navy; the existence of only one or two suppliers is less than robust.

TUBING CORROSION HINDERS NAVAL OPERATIONS (a notional though realistic scenario)

The USS Independence, one of the US Navy's littoral combat ships, failed to depart homeport as scheduled due to aggressive corrosion of the isolators on its titanium tubing. Navy sources claim that the problem is manageable; however, this is not the first time the USS Independence failed to depart on schedule due to corrosion problems. Titanium can be used a substitute for copper-nickel tubing, but lacks copper-nickel's resistance to corrosion. The USS Independence had been scheduled to join Combined Task Force 150, a multi-national coalition naval force engaged in anti-piracy operations in the Horn of Africa region. The loss of the vessel is expected to hamper coalition efforts to curb illicit activities in this critical region.

BACKGROUND

The special properties of Cu-Ni alloy were recognized by the Chinese as early as 120 B.C.,[6] and Cu-Ni alloys have been the preferred material for applications in seawater since the mid-20th century. Cu-Ni is the alloy of choice for seawater pipework for many of the world's merchant and especially naval vessels, due to its seamless construction, malleability, and inherent and superior anti-corrosion and anti-biofouling properties.[7] Cu-Ni tubing is used for basic piping systems on surface ships and submarines.

Much research and development (R&D) in the copper industry between 1920 and 1960 was devoted to development and enhancement of the stable, protective anti-corrosion film on Cu-Ni alloys.

When wet, certain metals and alloys attract a greater accumulation of micro-organisms at a faster speed than others. This accumulation, called biofouling, speeds up the corrosion process, severely limiting the life of the material in questions. Cu-Ni is the alloy of choice because "high copper alloys—particularly the Cu-Ni alloy with at least 10 percent nickel (UNS C70600)—reveal a unique biofouling resistance,

which can last throughout the service life in marine environments."[8]

Figure 1 demonstrates the anti-biofouling properties of copper, copper-nickel alloy, and titanium. Although copper performs best (indicated by less bacterial film growth over time), the copper-nickel alloy is stronger, and therefore is better suited for use in marine tubing.[9]

Due to Cu-Ni's combination of ductility and high resistance to biofouling and corrosion, naval vessels almost exclusively use Cu-Ni tubing. Industry experts indicate that Cu-Ni tubing is the material of choice for the U.S. Navy due to its seamless construction and manufacturing malleability, biofouling resistant properties, and corrosion resistant properties related to welding and performance in high-temperature seawater applications.[10]

Stainless steel tubing is the main alternative to Cu-Ni tubing for use in ship construction. Even though stainless steel tubing is more susceptible to biofouling and corrosion, commercial vessels (which have a lower risk tolerance than the U.S. Navy) use stainless-steel tubing in order to save costs. Commercial reliance on

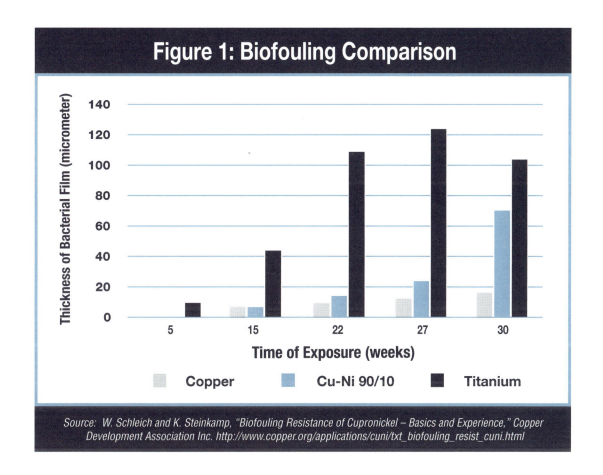

Figure 1: Biofouling Comparison

Thickness of Bacterial Film (micrometer)

Time of Exposure (weeks)

■ **Copper** ■ **Cu-Ni 90/10** ■ **Titanium**

Source: W. Schleich and K. Steinkamp, "Biofouling Resistance of Cupronickel – Basics and Experience," Copper Development Association Inc. http://www.copper.org/applications/cuni/txt_biofouling_resist_cuni.html

stainless steel tubing means there is little international market for Cu-Ni tubing.[11] Beyond supplanting demand due to the potential limited availability of Cu-Ni tubing if domestic production ceased, replacing Cu-Ni tubing with stainless steel tubing would provide a substantial cost-reducing alternative. However, stainless steel tubing historically has been dismissed for naval use due to the fact that stainless steel is so susceptible to corrosion during welding and could form brittle phases in the microstructure when exposed to seawater during high-temperature applications.[12]

Titanium tubing has also been considered as an alternative. The advantages of titanium include an "excellent resistance to corrosion and erosion attack, good yield strength, and relatively low density, permit[ting] use of thinner-walled, reduced diameter piping and heat exchanger tubing (relative to Cu-Ni), thus offer[ing] reduced system weight and volume."[13] However, because titanium has neither the biofouling or corrosion resistance of Cu-Ni nor the cost-advantages of stainless steel, industry experts do not plan to move towards titanium tubing.

Even given titanium's weaknesses in comparison to Cu-Ni tubing, shipbuilders have continued to use titanium tubing when possible. An example is Northrop-Grumman's Landing Platform Dock (LPD)-17 San Antonio-class ship. "The LPD-17 is the first of a new class of 12 684-foot, 25,000-ton amphibious transport dock ships designed to transport and land up to 720 Marines, their equipment, and supplies by means of embarked landing craft or amphibious vehicles augmented by helicopters."[14]

Figure 2: Sample of Plants Closed Since 1980

Year Closed	Plant
1986	Anaconda Brass – Kenosha, WI
1991	Olin Brass (Formerly Bridgeport Brass) – Indianapolis
2007	Wolverine Tube – Montreal and Jackson, TN

However, corrosion problems have become a major issue with these ships. "The LPD-17 encountered a problem with the isolators on titanium piping. The isolators are used to separate different types of metals to keep them from corroding. The problem was discovered in 2006, about a year after the launch of the first ship."[15]

While work continues on examining alternatives, the MIL-T-16420K specification of 90-10 and 70-30 Cu-Ni tubing continues to be the premier standard of tubing for use in U.S. Navy vessels. In spite of the fact that there are several U.S. and foreign companies that produce some Cu-Ni tubing, they cannot produce all varieties of Cu-Ni tubing required by the U.S. Navy.

Following World War II the United States controlled 85 percent of the world production of Cu-Ni tubing. U.S. production capability declined precipitously through the 1960s and 1970s as the majority of production facilities closed. A myriad of factors led to this decline, including rising foreign competition.[16]

The decline of the U.S. Cu-Ni tubing manufacturing capability is symbolized by the decline in number of industry workers in Waterbury, Connecticut (once nicknamed Brass City), where 25,000 workers were employed in the sector following WWII. As of June 1980 approximately 5,600 workers remained; the number has since declined to fewer than 100.[17] Since 1986, many Cu-Ni tubing manufacturing plants have closed due to increased global competition, substantial increases in copper prices, and potential unfair dumping and pricing practices by foreign competitors.[18]

Many of the closings came after the passage of the North American Free Trade Act (NAFTA), which increased competition from Mexican Cu-Ni tubing manufacturers. Mexico was able to produce copper products and Cu-Ni tubing at a cheaper price due to much lower production costs stemming from low wages. U.S. manufacturers have complained that European manufacturers are dumping their product on the market at prices lower than the cost of production in an attempt to drive American Cu-Ni manufacturers out of business.[19]

Since the 1990s, General Dynamics Electric Boat, a major builder of U.S. submarines and current supplier of the Navy's

LPD-17 During Construction

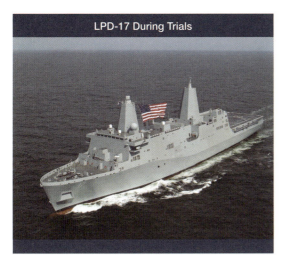
LPD-17 During Trials

Virginia-class submarine, has sourced its Cu-Ni tubing from the American manufacturer, Ansonia Brass & Copper, and the Italian-German conglomerate KME. In May 2010, the U.S. Navy acknowledged that Ansonia Brass & Copper is the only U.S. company capable of producing tubing 4.5 inches or greater in outside diameter that satisfies U.S. military specification: MIL-T-16420K.[20]

According to a 2010 Naval Sea Systems Command (NAVSEA) report, there are four U.S companies capable of producing Cu-Ni tubing to Navy specifications.[21] However, three of these companies could only produce tubing with an outside diameter (OD) range of approximately 0.25 to two inches. Only one domestic company manufactures the large (greater than 4.5 inch-diameter) tubing needed for a wide range of U.S. Navy uses. Moreover, in 2011, the Office of the Under Secretary of Defense for Acquisition, Technology, and Logistics confirmed that only one U.S. company was capable of producing Cu-Ni tubing with an OD greater than two inches. The large-size tubing is used in the construction and repair of almost all U.S. Navy vessels.[22]

While the 2010 NAVSEA report does not cite the name of this company, Ansonia

Brass & Copper can produce tubing over 4.5 inches in diameter, and therefore must be the company to which the report is referring. "Limited research concluded that [Ansonia Brass & Copper] is the only qualified, domestic manufacturer of seamless Cu-Ni tubing, for outside diameter sizes greater than 4.5 inches, that meets the military specification MIL-T-16420K, for naval shipbuilding and submarine applications."[23]

In addition to these domestic companies, the European conglomerate KME and several companies in Mexico also produce Cu-Ni tubing for the U.S. Navy. However, other than Ansonia Brass & Copper, KME is the only company capable of producing this larger diameter tubing according to U.S. military specifications. As a result, the U.S. domestic production capability of Cu-Ni tubing is at risk, potentially leaving the U.S. Navy solely dependent on foreign manufacturers for this important supply chain.

VULNERABILITIES IN THE COPPER-NICKEL TUBING SUPPLY CHAIN

If the United States were to lose its Cu-Ni tubing manufacturing capability, DoD would be dependent on foreign manufacturers. Moreover, as there is little to no commercial demand for Cu-Ni tubing, if the U.S. supplier disappears, the entire supply chain could potentially follow. This loss would complete an unraveling of U.S. Cu-Ni production capabilities. Production of Cu-Ni tubing cannot be restarted overnight, because manufacturing expertise will likely move overseas if the domestic production line closes.[24]

Not only is Ansonia Copper & Brass the sole U.S. producer of large OD seamless

> **Were Ansonia to stop production of copper-nickel tubing, construction of U.S. Navy ships would depend on a sole-source German manufacturer.**

Cu-Ni and alloy tubes, its viability is endangered by stiff competition with foreign competitor KME.[25] Ansonia's future production capability is threatened by KME's ability to undercut its price.

Dumping and price-fixing of products by foreign manufacturers is a major concern for the remaining manufacturers of Cu-Ni tubing in the U.S. In a globalized world, competition is always fierce, but if competing foreign firms in the Cu-Ni industry are actively seeking to drive U.S. firms out of business, the notion of "free and fair competition" becomes little more than cheap talk. The World Trade Organization (WTO) defines dumping as when "a company exports a product at a price lower than the price it normally charges in its own home market."[26] The goal of dumping is to drive competitors out of business in order for the remaining companies to monopolize the market and set prices higher than market prices for the foreseeable future.

There is reason to be wary of major foreign competitor KME's business practices and history of cartel-like behavior. In 2005 the European Commission, under the authority of the European Union, imposed fines totaling 222.3 million euros (approximately $280 million) on major European copper producers (including KME) for operating a 13-year cartel. Indeed, concerns about the KME conglomerate's business practices go back to 1988.[27] More recently, on March 21, 2012, the U.S. International Trade Commission (ITC) extended its antidumping orders on imports of brass sheet and strip from France, Germany, Italy, and Japan, believing that revoking these orders would likely lead to continuation or recurrence of injury to the domestic industry. The ITC votes were 6 to 0 as to Germany, Italy, and Japan and 5 to 1 as to France.[28] Although it concerned brass, a different product category than Cu-Ni, this verdict calls into question the reliability of these producers' business practices.

The disappearance of America's sole manufacturer of large-diameter Cu-Ni tubing could place future construction and refit of the fleet at risk in three ways. From most to least likely, those are:

- The risk of foreign manufacturers significantly raising the cost of production of Cu-Ni tubing;

- The possibility that the foreign manufacturer could fail due to business/credit difficulties;

- A political dispute between the foreign manufacturing country and the United States impeding the sale of Cu-Ni tubing.

In fact, in the foreseeable future the only other producers of large diameter Cu-Ni tubing are located in China—another decidedly problematic scenario.[29, 30]

MITIGATING THE RISKS

DoD needs to understand potential failures that could occur in the second and third tier of this important supply chain. Mapping the supply chain (e.g., DoD's Sector-by-Sector, Tier-by-Tier program) will help identify instances of over-reliance on foreign suppliers, areas of limited competition, and potential "single points of failure."[31] **Address the "single point of failure" problem with the Cu-Ni tubing supply**

chain. This point of failure appears to be Ansonia Brass & Copper, the only U.S. producer of greater than 4.5-inch OD Cu-Ni tubing. Unfortunately at this point, the Assistant Secretary of the Navy holds the position that "the nature of the Cu-Ni tubing industry … does not rise to the level of vital to national defense, and there are foreign sources available, so Title III remedies do not appear applicable in this case."[32]

Fund innovative research to make the supply chain more robust. DoD should invest in developing alternative materials, particularly focusing on overcoming titanium's biofouling and corrosion weaknesses, given that titanium already has several advantages (e.g., weight) over Cu-Ni tubing. A search of studies on titanium tubing indicates it was viewed as a potential alternative for the future during the construction of the LPD ships.[33] However, the corrosion problems experienced during testing appear to have delayed progress.[34]

Leverage existing legislative and policy frameworks to support this important U.S. defense industrial base node. The Buy American Act (BAA), enacted in 1933, requires the U.S. government to buy U.S. made products except in certain instances. However, the law states that a product is considered to be "American" if 51 percent of the cost of the product was incurred in the United States. Cost can be the cost of testing, not necessarily of production. Ultrasonic and hydrostatic testing can represent that majority of cost to a firm that imports the tubing.

This technicality means that much Cu-Ni tubing that is deemed "Made in America" is actually cast overseas. While such practices are consistent with the letter of the law, they clearly violate the spirit in which the law was written. Stronger enforcement of the intent of this law is needed.

CONCLUSION

Strong and smart U.S. policy can ensure the United States' ability to continue to manufacture and use Cu-Ni tubing to strengthen and maintain the U.S. Navy's readiness.

The only other supplier of large-diameter Cu-Ni tubing, Italian-German conglomerate KME, has been convicted of price-fixing, collusion, and being part of a cartel for over 13 years by Europe's highest courts, yet policymakers have ignored allegations of similar practices on the international market. Competition is welcome, but it must be fair. If KME dominates the market, the United States will be at a decided disadvantage.

Another problem resulting from the fact that much "Made in America" Cu-Ni tubing is cast outside of the country is that distributors can circumvent inspection and enforcement of military specifications. Tubing distributors are required to ensure that tubing meets all U.S. government specifications; however, if they are purchasing castings overseas at significantly lower prices than they pay for products in the United Sates, they may be likely to take the word of the foreign firms that it does meet specifications.

The U.S. Cu-Ni tubing industry is important for our navy and for our national security. DoD must ensure that foreign manufacturers and suppliers meet required specifications, laws, and quality standards. In the long run, policymakers need to support the Cu-Ni industrial base in the United States by increasing R&D spending and, in compliance with the BAA, targeting future contracts at existing U.S. firms.

ENDNOTES

1 H. Mische, "Commission Adopts Cartel Decision Imposing Fines on Copper Plumbing Tube Producers," *Competition Policy Newsletter,* Spring (2005), 1. http://ec.europa.eu/competition/publications/cpn/2005_1_67.pdf.

2 *Naval History & Heritage Command.* "Famous Navy Quotes: Who Said Them and When." http://www.history.navy.mil/trivia/trivia02.htm.

3 Michael A. Palmer, "The Navy: The Oceanic Period, 1890-1945," *A History of the U.S. Navy.* http://www.history.navy.mil/history/history3.htm.

4 Bill Gertz, "China Has Carrier-Killer Missile, U.S. Admiral Says," *The Washington Times,* December 27, 2010. http://www.washingtontimes.com/news/2010/dec/27/china-deploying-carrier-sinking-ballistic-missile/.

   "China's First Aircraft Carrier 'Starts Sea Trials'," *BBC News,* August 10, 2011. http://www.bbc.co.uk/news/world-asia-pacific-14470882.

5 David Nakamura, "Obama: U.S. to Send 250 Marines to Australia in 2012," *Washington Post* November 16, 2011. http://www.washingtonpost.com/blogs/44/post/obama-us-to-send-250-marines-to-australia-in-2012/2011/11/16/gIQAO4AQQN_blog.html.

6 Joseph Needham, *Science and Civilisation in China: Spagyrical Discovery and Invention: Magisteries of Gold and Immortality* (Cambridge: Cambridge University Press, 1974), 1-600.

7 Nickel Development Institute, Copper Development Association, and the Copper Development Association Inc., *Copper-Nickel Fabrication* (New York: Nickel Development Institute, 1999). http://www.stainless-steel-world.net/pdf/12014.pdf.

8 W. Schleich and K. Steinkamp, "Biofouling Resistance of Cupronickel – Basics and Experience" (Copper Development Association Inc., March 22, 2012). http://www.copper.org/applications/cuni/txt_biofouling_resist_cuni.html.

9 Nickel Development Institute, Copper Development Association, and the Copper Development Association Inc., *Copper-Nickel Fabrication* (New York: Nickel Development Institute, 1999). http://www.stainless-steel-world.net/pdf/12014.pdf.

10 Department of Defense, Office of the Under Secretary of Defense Acquisition, Office of Manufacturing and Industrial Base Policy, *Annual Industrial Capabilities Report to Congress* (Washington, D.C., September 2011). http://www.acq.osd.mil/mibp/docs/annual_ind_cap_rpt_to_congress-2011.pdf. 40.

11 Department of Defense, Office of Under Secretary of Defense Acquisition, Technology, and Logistics, Office of Manufacturing and Industrial Base Policy, *Annual Industrial Capabilities Report to Congress* (Washington, D.C.: September 2011). http://www.acq.osd.mil/mibp/docs/annual_ind_cap_rpt_to_congress-2011.pdf.

12 Ibid.

13 W.L. Admanson and R.W. Shutz, "Application of Titanium in Shipboard Seawater Cooling Systems," *Naval Engineers Journal* 99, no. 3 (1987): 1.

14 "LPD-17 San Antonio Class," Global Security.org. http://www.globalsecurity.org/military/systems/ship/lpd-17.htm.

15 Ibid.

16 Rachel Guest, "End of An Era: A Myriad of Factors Lead To The Collapse of Brass Production in Waterbury," *The Waterbury Observer,* July 2011, 22-27.

17 Ibid.

18 "U.S. Brass Manufacturers Obtain Favorable Sunset Review in ITC Antidumping Decision," *PR Newswire,* March 21, 2012. http://www.prnewswire.com/news-releases/us-brass-manufacturers-obtain-favorable-sunset-review-in-itc-antidumping-decision-143714606.html.

19 Ibid.

20 Assistant Secretary of the Navy Sean J. Stackley, letter to the CEO of Ansonia Brass & Copper Inc., May 17, 2010.

21 Department of Defense, Office of Under Secretary of Defense Acquisition, Technology and Logistics, Office of Manufacturing and Industrial Base Policy, *Annual Industrial Capabilities Report to Congress* (Washington, D.C.: September 2011). http://www.acq.osd.mil/mibp/docs/annual_ind_cap_rpt_to_congress-2011.pdf.

22 Ibid.

23 Department of Defense, Office of Under Secretary of Defense Acquisition, Technology and Logistics, Office of Manufacturing and Industrial Base Policy, *Annual Industrial Capabilities Report to Congress* (Washington, D.C.: September 2011), 40.

24 J. Yudken, *Manufacturing Insecurity: America's Manufacturing Crisis and the Erosion of the U.S. Defense Industrial Base* (Arlington, VA: High Road Strategies, LLC/AFL-CIO Industrial Union Council September 2010).

25 "Court Backs Fines on Outokumpu, KME, Wieland for Price Fixing," *Metal Bulletin,* May 11, 2009. http://www.metalbulletin.com/Article/2198176/Search/Court-backs-fines-on-Outokumpu-KME-Wieland-for-price.html?PageId=196010&Keywords=KME+fines&OrderType=1.

26 World Trade Organization (WTO), *Understanding the WTO: Anti-Dumping, Subsidies, Safeguards: Contingencies, etc.* http://www.wto.org/english/thewto_e/whatis_e/tif_e/agrm8_e.htm.

27 H. Mische, "Commission Adopts Cartel Decision Imposing Fines on Copper Plumbing Tube Producers," *Competition Policy Newsletter,* Spring (2005), 1. http://ec.europa.eu/competition/publications/cpn/2005_1_67.pdf.

28 "U.S. Brass Manufacturers Obtain Favorable Sunset Review in ITC Antidumping Decision," *PR Newswire,* March 21, 2012. http://www.prnewswire.com/news-releases/us-brass-manufacturers-obtain-favorable-sunset-review-in-itc-antidumping-decision-143714606.html.

29 New Huahong Copper. http://www.jyxhh.net/about.html.

30 W. Wan, K.B. Richburg, and D. Nakamura, "U.S. Challenges China's Curbs on Mineral Exports; China Vows to Push Back," *Washington Post,* March 13, 2012. http://www.washingtonpost.com/world/national-security/us-to-challenge-chinas-curbs-on-mineral-exports/2012/03/12/gIQAV8BX8R_story.html.

31 Brett Lambert, "Presentation to National Defense Industrial Association" (August 2011), 4. http://www.ndia.org/Advocacy/Resources/Documents/NDIA_S2T2_Briefing_AUG11.pdf.

32 Assistant Secretary of the Navy Sean J. Stackley, letter to the CEO of Ansonia Brass & Copper Inc., May 17, 2010.

33 Pat Hoyt, "Titanium Emerges as New Seawater Piping for Navy Vessels," *Welding Journal* (June 2006), 62-65.

34 "LPD-17 San Antonio Class Challenges," Global Security.org. http://www.globalsecurity.org/military/systems/ship/lpd-17-challenges.htm.

CHAPTER 9 • LITHIUM-ION BATTERIES

EXECUTIVE SUMMARY

U.S. ground troops in combat carry an average of 70 pounds of gear, 20 pounds of which can consist solely of batteries to power the range of electronic systems on which modern combat depends. Because the Pentagon has yet to adopt standardized battery specifications, most U.S. military equipment requires customized power sources that typically are not rechargeable–often forcing soldiers to carry multiple units of the same battery.

Shifting to standardized, rechargeable batteries would greatly reduce the physical burden for warfighters while increasing combat efficiency by reducing troop fatigue and freeing space for other portable equipment. Standardization could also provide incentives for mass production and dramatically reduce prices through economies of scale and increased competition.

Rechargeable lithium-ion (Li-ion) batteries weigh one-third as much as conventional batteries and can hold up to three times the charge. The technology for rechargeable batteries was developed in the United States, but today the U.S. battery industry accounts for only two percent of global production. Japan dominates the global market with a 57 percent share, followed by South Korea and China with 17 percent and 13 percent, respectively. Foreign companies now also control many of the patents needed to produce batteries and battery components commercially, creating barriers to U.S. firms interested in entering the commercial market.

The demand for Li-ion batteries has risen in recent years, and their use in electric cars suggests that commercial demand only will increase. Rechargeable batteries are also an integral component of green technology, because they provide storage for the power generated by solar panels or wind turbines. Li-ion battery chemistry is also the preferred method of storing energy generated from non-fossil fuels. However, the United States has been slow in developing and commercializing advanced battery technology. Regardless of how the U.S. government manages this challenge, it faces an uphill struggle to catch up with its Asian competitors.

LITHIUM-ION BATTERIES
POWERING DEFENSE AND NATIONAL SECURITY

MANUFACTURING SECURITY

Lithium-ion batteries are currently integrated in a number of essential military applications and platforms

UNMANNED AERIAL DRONES

BOMB DISPOSAL ROBOTS

PORTABLE WARFIGHTER GEAR

THE BATTERY BURDEN

More non-standardized battery types = Greater burden for the warfighter

AVERAGE **SOLDIER** CARRIES UP TO 20 LBS OF **BATTERIES**

NEW POWER BENEFITS

Lithium-ion batteries offer substantial benefits over old battery technology

   EQUALS 3X THE POWER OF REGULAR **BATTERIES**

EQUALS 1/3 THE WEIGHT OF REGULAR **BATTERIES**

FOREIGN RELIANCE

U.S. security is reliant on foreign lithium sources

CHILE PRODUCES **1/3** OF THE WORLD'S **LITHIUM**

VULNERABILITY

Japan is the leading producer of Lithium-ion batteries

 **JAPAN PRODUCES** 57% OF THE WORLD'S **LITHIUM-ION BATTERIES**

MITIGATING RISKS

Avoiding uncertainties in U.S. Lithium-ion battery supply

RECYCLING BATTERIES

IMPROVED DOD COORDINATION

FUNDING NEW TECHNOLOGY

MILITARY EQUIPMENT CHART
SELECTED DEFENSE USES OF LITHIUM-ION BATTERIES

DEPARTMENT	WEAPON SYSTEMS	PLATFORMS
ARMY	- M1E3 Abrams tank "silent watch" capability	- Long endurance UAVs - Joint Light Tactical Vehicle (JLTV)
MARINE CORPS	- FIM-92 Stinger man-portable, shoulder-fired missile - AIM-120 advanced medium-range air-to-air missile (AMRAAM)	- Integrated trailer environmental control unit generator - Medium Tactical Vehicle Replacement (MTVR) - Man-portable renewable energy system - HMMWV (Humvee) inboard vehicle power system
NAVY	- Electromagnetic rail gun - AIM-120 advanced medium-range air-to-air missile (AMRAAM)	Aircraft batteries: - F-35 Joint Strike fighter - HH-60 Pave Hawk helicopter
AIR FORCE	- AIM-120 advanced medium-range air-to-air missile (AMRAAM) - MIM-104 Patriot anti-ballistic missile	Aircraft batteries: - F-35 Joint Strike fighter - F-18 Hornet fighter - T-45 Trainer aircraft

The United States needs to accelerate funding for research and development (R&D) of new battery technologies in areas where it can compete with existing battery producers. The United States should also explore alternatives to Li-ion batteries in order to hedge against a decline in the global supply of lithium. The Department of Defense (DoD) also should pay more attention on how to integrate power sources into its portable equipment and weapons systems. Because of the lack of interest in battery technology, DoD has allowed suppliers and contractors to determine the power source used with a device. As a result, DoD has thousands of different types and sizes of batteries in use, creating a logistical nightmare.

INTRODUCTION

The United States has retained a decisive global military edge owing a great deal to its exploitation of advanced technologies. Many of these technologies, especially portable devices such as phones, satellites, computers, individual and crew-served weapons, night vision devices, pumps, and heaters rely on advanced power sources. As a result, battery technology, and in particular rechargeable battery technology, is an essential component of the U.S. military's continuing success and dominance.

Li-ion batteries are the smallest and lightest among major rechargeable batteries. Despite being significantly smaller and lighter than alternative rechargeable batteries, Li-ion batteries can store significantly more energy without degrading when recharged, making them especially useful on the battlefield and ideal for portable military devices.[1] Military personnel, especially ground troops in combat, already carry heavy loads, and efficient, rechargeable batteries help to lighten that load.

Developed in the early 1990s, Li-ion batteries currently are used in virtually all consumer electronic devices and many military end-items.[2] Li-ion batteries also are and will continue to be the principal power source for electric vehicles (EVs). Li-ion batteries possess several advantages when compared to alternative types of rechargeable (lead-acid) and non-rechargeable (manganese dioxide-zinc) batteries. Li-ion batteries are more energy efficient and have relatively long lifecycles; they also avoid the gradual decrease of maximum charge capacity over time (known as memory effect) common in other rechargeable batteries.[3] In spite of their growing popularity, the commercial mass production of high-energy lithium systems has been challenging because it requires new innovations in anodes, cathodes, and electrolytes. More importantly, Li-ion batteries are prone to short-circuit and overcharge, resulting in unintentional combustion reactions or explosions.[4] Boeing had to ground its fleet of 787 (Dreamliner) airplanes after two incidents of exploding Li-ion batteries, which are used to start the aircraft's auxiliary power unit. There have been other incidents of batteries catching fire in cars and laptops.[5] Li-ion battery technology is complex, and these incidents indicate that it still has some problems to be resolved.

DoD has been lax about imposing standards and conditions on batteries used by military applications. Ground combat troops, on the average, carry a load of 70 pounds, of which as much as 20 pounds are different types and sizes of batteries.[6] The total load can weigh as much as 130 pounds.[7] A warfighter's combat load could be substantially reduced by imposing uniform battery standards that take advantage of commercial innovations across the military services, especially the U.S. Army and Marine Corps.

Growing energy demands have added to the troops' weight burden from batteries, stressing military logistics. Both the U.S. Marine Corps and Army seek to unburden troops from heavy loads and increase troops' self-sufficiency in energy in the near future.[8] Another benefit of advanced battery research and innovation would be that U.S. niche manufacturers of specialized batteries could benefit from a larger market. The military services (particularly the Army and Marine Corps) must invest in consolidating the array of batteries and in improving the energy efficiency of troops in the field.

In 2007, the Army reported that missions to protect fuel convoys were responsible

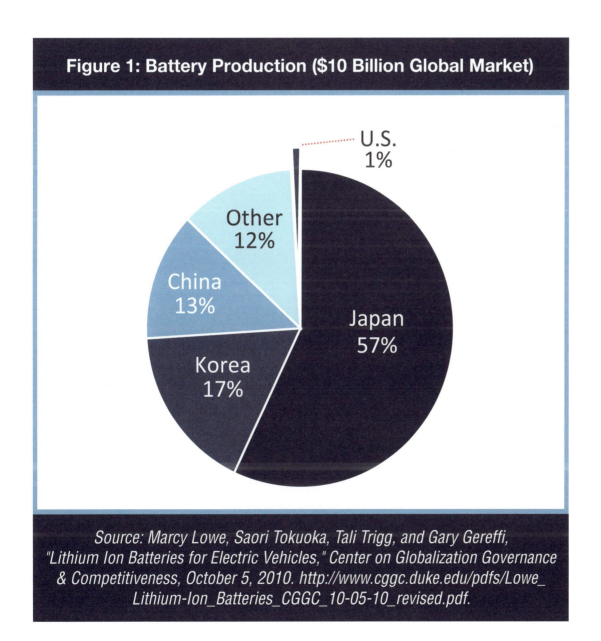

Figure 1: Battery Production ($10 Billion Global Market)

U.S. 1%
Other 12%
China 13%
Korea 17%
Japan 57%

Source: Marcy Lowe, Saori Tokuoka, Tali Trigg, and Gary Gereffi, "Lithium Ion Batteries for Electric Vehicles," Center on Globalization Governance & Competitiveness, October 5, 2010. http://www.cggc.duke.edu/pdfs/Lowe_Lithium-Ion_Batteries_CGGC_10-05-10_revised.pdf.

for one-eighth of all casualties in Iraq. Forward operating bases are most often powered by diesel generators, which require a constant supply of fuel to support critical base operations.[9] Supplying diesel generators is risky because supply convoys are vulnerable to attacks. Although diesel generators and lead-acid batteries have long been the military's main power sources, increasingly there is a search for an efficient way of storing energy to reduce fuel consumption and reduce unnecessary casualties. The most attractive and logical current alternative is to use rechargeable power sources such as Li-ion batteries, which have a high energy density (the amount of energy stored per unit of volume) and thus can last longer.

Although the United States was once the global research leader in basic research on the chemistry, materials, and basic design principles of advanced battery cells, most battery production now takes place in East Asia, primarily in Japan. In 2010, the United States accounted for only one percent of the global production market; in contrast, Japan accounted for 57 percent (see Figure 1). In fact, Kyoto-based GS Yuasa manufactured the batteries used in the Boeing Dreamliner. There is no U.S. manufacturer of these particular types of Li-ion batteries, which reflects the U.S. failure to invest in the manufacturing technology to produce Li-ion batteries and achieve economies of scale.

In addition, Li-ion is expected to be the main source of power for the EV industry. The integration of advanced energy efficient technologies into the U.S. economy and military will improve military readiness and reduce long-term defense costs. Advanced energy efficient technologies means less gear for troops to carry, reduced logistical burdens, and increased battlefield self-sufficiency. While the military is aware

of the need to develop significantly more sophisticated battery storage requirements for its current and future power-generation and energy-management systems,[10] the manufacturing and technological bases for Li-ion batteries are only now being developed (primarily by the automobile industry).[11] However, the military will need a host of different battery chemistries for diverse applications and combat environments. The current focus on vehicle batteries (so-called large-format cells) is a step in the right direction, but it will only address one type of power source.[12]

Key themes discussed in this chapter are:

- The United States still possesses relatively modest battery manufacturing capacity, primarily focused on developing large-format cells to be used in the EV sector. This capacity does not address the U.S. military's reliance on imported Li-ion batteries for many other military applications.

- DoD currently lacks uniform standards for batteries used in similar applications or across the military services. The result is that the different services rely on hundreds of different types of batteries. This lack of consistency creates significant logistical challenges and inadvertently undermines the economic viability of small niche producers of specialized battery applications, none of which can achieve the economies of scale needed to survive in a competitive market.

- The Department of Energy (DoE) has spent billions of dollars to expand battery manufacturing, for EVs in particular, but the future may lie with designing new battery chemistries for smaller-scale applications. There may soon be a global glut of vehicle batteries without a corresponding increase in EV use.

- Virtually all of the advanced batteries currently in use or in development require small amounts of lithium, which is mined mostly in Chile. Rapidly increasing demand for lithium has led to rising concerns about its future availability.

- Recycling of Li-ion batteries continues to be challenging and costly. It will not become economically viable unless the price of raw materials skyrockets and the U.S. government provides funding to develop battery-recycling infrastructure.

A NOTE ON CRITICALITY

Single use, primary batteries remain the principal energy source for most portable military electronics. Often these batteries are not interchangeable, and add significant weight to each warfighter's load. Replacing these batteries with rechargeable, interchangeable batteries will result in

a more efficient and better equipped warfighter. Failure to develop and implement rechargeable and interchangeable battery capabilities will have an *isolated* impact on U.S. defensive capabilities. Current energy requirements are burdensome and require significant space in troop load-bearing equipment. The ability to reduce even partially the weight and space dedicated to single-use batteries permits soldiers to either carry more equipment on an equal weight basis for enhanced capabilities, or less weight reducing fatigue and increasing mobility. Additionally, rechargeable batteries would permit extended missions by enabling mobile recharging through renewable energies such as solar and wind.

The future development and adoption of rechargeable batteries is limited by resources as well as manufacturing capacity, which together constitute a *high* risk. Lithium reserves are geographically concentrated, with the majority of the world's reserves located in South America, China, and Australia. An unforeseen event in any

BATTERY SHORTAGES AFFECT THE MISSION (a real-life scenario)

In April 2003, during the invasion of Iraq, U.S. Marines suffered from a severe shortage of non-rechargeable lithium batteries. These batteries provided a portable power source for nearly 60 critical military communication and electronic systems, including two radio systems, a missile guidance system, and a transmission security device. The shortage was due to the fact that only one supplier supplied these batteries, and the supplier had encountered manufacturing problems and needed months to expand production. However, the Department of Defense was late in notifying the supplier because DoD had not realized that its normal reserves would be inadequate during a period of war. Instead of having the required 30-day supply, the U.S. Marines had less than a two-day supply of certain mission-critical batteries. If they had really ran out of batteries, communications would have been shut down, and their operational capabilities would have been severely degraded. Marines' lives would have been needlessly at risk.

Source: Government Accountability Office, Defense Logistics: Actions Needed to Improve the Availability of Critical Items during Current and Future Operations, GAO-05-275 (Washington, D.C.: April 8, 2005).

of these locations could create a shortage. Moreover, there are concerns surrounding the long-term availability of lithium given fixed supply and growing commercial demand from the automotive industry. A second limiting factor is the lack of domestic manufacturing capacity and the fact that Asian manufacturers are ahead of the design and development curve. These foreign manufacturers continue to lead the race to dominate the next generation of battery technology.[13] The United States possesses only a fraction (about one percent) of global production capacity. As it stands today, it appears likely that the U.S. will depend on foreign battery production based predominantly in Japan, China, and South Korea for the foreseeable future.

BACKGROUND

Despite significant contributions towards the development of Li-ion batteries conducted at U.S. research universities during the 1980s, U.S. firms declined to commercialize the advancements. Instead, Japanese electronics companies that already possessed strong manufacturing bases and heightened demand pursued the technology. Although the U.S. firms might have been able to benefit from a "first mover's" advantage in developing Li-ion battery chemistries, they viewed themselves as unable to compete with well-established, vertically integrated electronics companies in Asia, and declined to invest in research and design.[14]

In the late 1990s, U.S. battery manufacturers Duracell and Energizer began preparations to create domestic Li-ion battery production lines. Energizer expanded its facilities in Florida and acquired the necessary licenses to begin Li-ion production; however, just prior to going operational, the Asian markets collapsed in the 1997-1998 Asian Financial Crisis. The crisis resulted in a significant decrease in the market price of Li-ion cells. Given the established industry in Asia, it was cheaper to acquire Li-ion batteries on the international market than it would have been to manufacture them domestically. This change prompted the U.S. firms to abandon efforts to enter the market.[15]

However, with assistance from the Defense Advanced Research Projects Agency (DARPA) and other government entities, certain smaller companies have continued to research Li-ion battery technology. These smaller firms have had some success in developing batteries for certain medical and military uses, although always on a small production scale. To date, such firms have been unable to translate success in specialty markets into large-scale commercialization. This inability means that production remains costly and market size remains small. Additionally, as many U.S. companies looked offshore to purchase batteries, U.S. venture capital firms have not expressed much interest in financing the manufacturing and commercialization of new battery technology. U.S. battery manufacturers sought to compensate for the high cost of automated production equipment by moving offshore.[16] Interestingly, the United States continues to house leading battery research laboratories such as the Lawrence Berkeley National

While many U.S. companies have looked offshore to purchase batteries, U.S. venture capital has not expressed much interest in financing the manufacturing and commercialization of new battery technology.

Laboratory (Berkeley Lab), Argonne National Laboratory (ANL), Sandia National Laboratory (SNL), National Renewable Energy Laboratory (NREL), Idaho National Laboratory (INL), and Oakridge National Laboratory (ORNL). Thanks to the strong basic research concentration, U.S. based scientists still produce nearly one-fifth of research papers (see Figure 2). Although Japan dominates the filing of new patents, it has only a small lead over the United States in basic research.

While the United States has a strong capacity for research, it lags in manufacturing capacity, skill, and technology. For all practical purposes, the United States lacked meaningful presence in Li-ion manufacturing until the 2009 American Reinvestment and Recovery Act, which awarded $2.4 billion to construct 30 battery plants that will participate in various stages of battery production.[17]

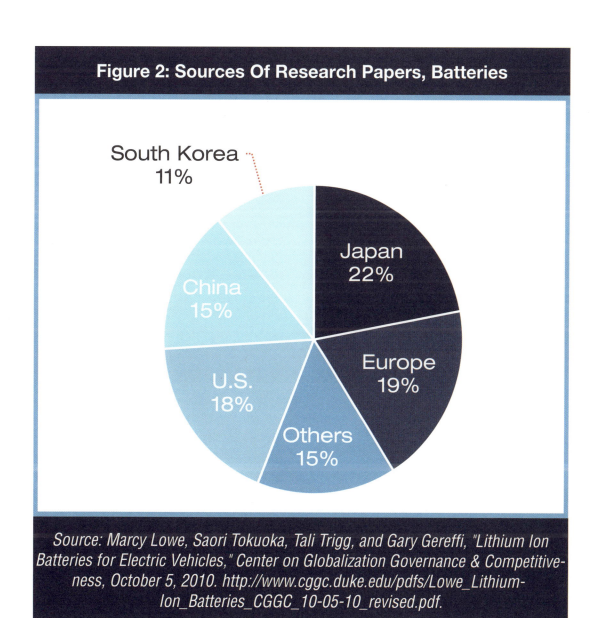

Figure 2: Sources Of Research Papers, Batteries

South Korea 11%

Japan 22%

China 15%

U.S. 18%

Europe 19%

Others 15%

Source: Marcy Lowe, Saori Tokuoka, Tali Trigg, and Gary Gereffi, "Lithium Ion Batteries for Electric Vehicles," Center on Globalization Governance & Competitiveness, October 5, 2010. http://www.cggc.duke.edu/pdfs/Lowe_Lithium-Ion_Batteries_CGGC_10-05-10_revised.pdf.

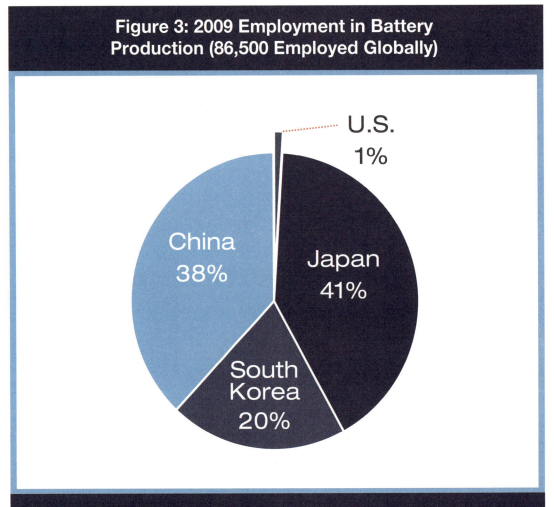

Figure 3: 2009 Employment in Battery Production (86,500 Employed Globally)

U.S. 1%

China 38%

Japan 41%

South Korea 20%

Source: Marcy Lowe, Saori Tokuoka, Tali Trigg, and Gary Gereffi, "Lithium Ion Batteries for Electric Vehicles," Center on Globalization Governance & Competitiveness, October 5, 2010. http://www.cggc.duke.edu/pdfs/Lowe_Lithium-Ion_Batteries_CGGC_10-05-10_revised.pdf.

Until 2009, the main problem had been that the U.S. private sector had struggled to translate basic research into the large-scale production and commercialization of batteries. By contrast, Japanese and South Korean companies invested in the manufacturing technology to commercialize the advanced chemistry of Li-ion batteries. Yet Asian economies have no inherent advantage over the United States in this area. The main minerals used in batteries are lithium, graphite, and cobalt; only graphite is found in the Asia-Pacific region, specifically in China. Neither Japan nor South Korea possesses reserves of the main chemical elements required for the production of advanced batteries. Thus, South Korea and Japan became major producers of Li-ion batteries in spite of the fact that they are fully reliant on the import of raw materials. Rather, Japan's and South Korea's advantage has been their ability to translate basic research into a process for efficiently produced commercialized batteries. In the meantime, U.S. firms have been content to earn royalties from selling their inventions to Asian manufacturers.[18]

The story of A123 Systems illustrates the issues faced by U.S. Li-ion battery producers. A123 was founded by Yet-Ming Chang, a Massachusetts Institute of Technology (MIT) professor who invented a new type battery that is safer and longer-lasting than conventional Li-ion car batteries. Using nanoscale phosphate materials rather than cobalt, Chang was able to situate 600 cells into a space "the size of a carry-on bag." However, when Chang tried to raise money to open a factory in Michigan in 2003 and 2004, he was unable to find investors interested in constructing a new advanced battery factory in the United States. Subsequently, A123 opened five factories in China. U.S. demand for batteries meant steady sales, but the expansion of manufacturing to China also resulted in the loss of intellectual property, as A123 had to teach Chinese producers how to fabricate their advanced batteries. Due to weak enforcement of intellectual property rights, Chinese partners copied the technology and soon competed with A123.[19]

In 2010, A123 received $250 million in aid from the American Recovery and Reinvestment Act as well as tax incentives from the state of Michigan to set up an EV battery factory there. Touted as a success, the company was singled out by the Administration for reviving the "battered" battery manufacturing sector in the Midwest, claiming that A123 would create thousands of jobs.[20] However, after chronic losses and a damaging battery recall, A123 filed for bankruptcy in October 2010. A123's run of bad luck

continued as Fisker Automotive, one of A123's clients, encountered problems and delays with the Karma, its electric vehicle. Because of the problems with the Karma, A123 Systems had to lay off 125 people in January 2012.[21] Then in April 2012, a fire started at a General Motors Tech lab and A123 batteries were suspected to have been the source of the explosion.[22] In June 2012, A123 announced it had developed advanced Nanophosphate(R) lithium iron phosphate batteries and systems, a new Li-ion battery technology capable of operating at extreme temperatures without requiring thermal management. [23]

Finally in January 2013, A123's assets were sold to Wanxiang, a Chinese-owned auto parts maker.[24] The sale excluded A123's business with the U.S. government, which included a DoD contract to develop new battery packs for soldiers. The sensitive nature of the research meant that A123 defense activities were sold to a small Illinois-based energy company, Navitas Systems.

To some extent, A123's experiences mirror the struggles of the nascent U.S. EV battery sector. On the one hand, there is too much Li-ion battery capacity, because demand for electric cars remains weak. On the other, battery factories in Japan and South Korea are extremely competitive, in part because they have had a big head start in the market. A123 also made some costly mistakes. It relied on one customer, Fisker, which was slow in bringing its Karma electric sedan to market. When Fisker finally did place an order for batteries, A123 rushed to fill it and ended up producing defective cells, forcing a costly $55 million recall. In 2013, other U.S. advanced battery producers remain in business (for example, Michigan-based KD ABG) but A123 was the largest company to delve into EV batteries. Instead, because of Wanxiang's acquisition of A123 assets, Chinese manufacturers will now expand their presence in automotive, electric-grid, and commercial businesses in Michigan, Massachusetts, and Missouri.[25]

The rechargeable battery industry's current focus on large, EV batteries has defense applications, but it does not cover the many other sizes and types of batteries used to power smaller military equipment, including communication devices and light sources. Projections show that the market for EVs is still underdeveloped and that consumers are intimidated by the high sticker price of electric vehicles (for example, the Chevrolet Volt costs around $40,000). To prime the EV market, as well as the corresponding EV battery market, the U.S. government may well have to provide further incentives for consumers to buy an EV. The U.S. lack of Li-ion battery production capacity notwithstanding, many industry observers believe that the world could eventually face a glut of EV batteries. The price of electric vehicles (e.g., the Chevrolet Volt, Nissan Leaf, or Ford Fusion BEV) is prohibitive, and many consumers may be reluctant to buy one unless gasoline prices increase sharply in the next few years.[26]

As Figure 4 shows, a standard Li-ion rechargeable battery consists of four separate components: electrolytes, separators, anodes, and cathodes.[27] A viable U.S. advanced battery sector would require robust capacity in all four areas of production.

The United States is already a significant producer of both electrolytes and separators. For example, Ohio-based company Novolyte has a global presence in manufacturing electrolytes, with over 30 years of experience supplying electrolytes for primary cells. Moreover, Celgard, located in North Carolina, is a leading manufacturer of separators, commanding a market share of 20 to 30 percent with plans to double its capacity by 2013.[28] Separators are a key component for Li-ion batteries' performance, safety, and cost. The global separator market is close to $1 billion, and growing at a rate of about 7 percent per year. Until the opening of a brand-new facility in North Carolina, all specialized Li-ion battery separators came from Japan and South Korea.[29]

However, Celgard's separators are unsuitable for the most advanced Li-ion batteries; U.S. firms and DoD must import higher quality separators from Asia. Moreover, because Asian countries dominate the manufacturing of separators, they also lead in separator technology and have extended their capabilities by filing additional patents to protect their lead.[30]

As for the other two key components (cathodes and anodes), the United States has a much smaller role. In contrast, Japan possesses at least a 70 percent market share in all four components,[31] which makes it is relatively easy for Japanese firms to diversify and move into a wider range of Li-ion battery production.

Thus, there are several key areas in which the United States needs to catch up and capture economies of scale to make the commercialization of the battery technology possible. The United States needs more U.S-based cell component and material suppliers in order to capture higher value in the market.

Figure 4: Diagram Of A Standard Li-ion Battery – Four Components

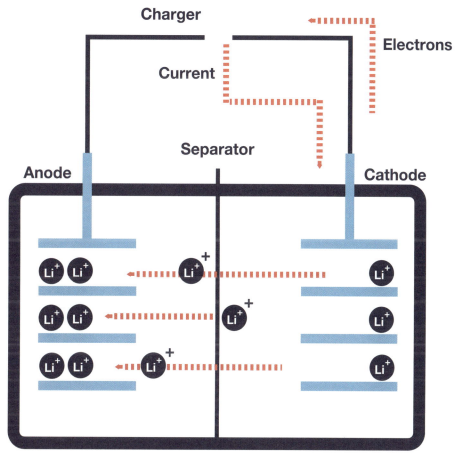

Electrolyte
(Polymer battery: gel polymer electrolyte)

Source: Doris de Guzman, "Chemical Firms Invest in Li-ion," ICIS Green Chemicals, July 18, 2011. http://www.icis.com/blogs/green-chemicals/2011/07/chemical-firms-invest-in-li-io.html.

U.S. DEFENSE APPLICATIONS

The commercial market offers two kinds of batteries: primary (non-rechargeable) and secondary (rechargeable). Primary batteries are single-use batteries that permanently lose their charge through usage, and are the most common type of battery for portable applications such as pocket lights, navigation instruments, smartphones, and small computers. Secondary batteries' charge may be replenished upon depletion, and presently are less common than their single-use counterparts among deployed units. However, the military services are increasingly interested in rechargeable batteries, which are more cost-efficient and have distinct operational advantages such as reducing the weight of troops' equipment.[32] Additionally, rechargeable batteries could reduce the number of supply convoys needed to support a forward operating base, thereby reducing the exposure of troops defending those convoys.

Batteries power many different types of defense equipment, including communications systems, night vision devices, missile guidance systems, munitions, and virtually all other electronic devices needed by the warfighter.[33] Furthermore, it is anticipated that troops will have to carry increasingly heavy loads because they will need to use a growing array of electronic devices. Therefore, there is a keen interest in finding smaller, lighter, and longer-lasting power sources.[34]

Few of the rechargeable batteries used in military equipment are made in the United States. The primary example is the broad category of Li-ion batteries that come from Japan, South Korea, and China. As mentioned earlier, Li-ion batteries consist of distinct components, most of which are not available in the United States. To start, the graphite powder (MesoCarbon MicroBead) often used in Li-ion battery anodes is not manufactured in the United States. Super P Carbon, which serves as a conductive additive to Li-ion batteries, is also imported. Aluminum and copper foils have no equivalent U.S. sources and also come from abroad. Pellon, mercuric oxide, cadmium oxide, and other materials that contribute to high-performance batteries, especially those appropriate for military applications, are only available to the U.S. market through import.[35]

In the long run, U.S. military forces are expected to face increasing demand for rechargeable batteries due to increased reliance on advanced technology and equipment that is becoming smaller, lighter, and more power-dense. Widespread adoption of Li-ion batteries by the U.S. military would considerably reduce warfighters' individual combat loads. Furthermore, Li-ion batteries would reduce the dependence on fuel convoys, both reducing the number of necessary resupplies and reducing casualties resulting from protecting those supplies. Hybrid vehicle technology could greatly improve fuel efficiency, lessening dependence on convoys and reducing mission costs associated with increasingly expensive fuel. Batteries can often be recharged in the field, either using solar or wind sources. Moreover, batteries extend the range of unmanned aerial vehicles (UAVs), increasing their time-on-target.[36]

The U.S. government recognizes the critical role power sources play in reducing operational energy demand, and preserving the flexibility and safety of equipment and soldiers. At the same time, until recently DoD lacked any comprehensive strategic plan for managing operational energy use. The FY2009 National Defense Authorization Act (NDAA) created the Office of the Director of Operational Energy Plans and Programs to advise the Secretary of Defense on matters related to sustaining DoD's forces and weapons platforms during military operations.

Until very recently, DoD did not issue direct stipulations on how an item would be powered or what kind of power sources would be used. Proportionally, batteries represented only a minor fraction of overall mission costs, and received little attention. Equipment manufacturers typically were left to recommend the appropriate means of powering electronic devices, which, in the interest of maintaining steady business, created incentives for the manufacturers to develop proprietary battery units. Thus, contractors often insist that a particular item be powered by a specific type or size of battery.[37]

The result is an array of batteries in use that are difficult to replace because the lack of earlier standardization has produced a plethora of unique proprietary power sources that are now difficult to alter. For example, the U.S. Army and Marines use similar radio systems (AN/PRC-148 Multiband Inter/Intra Team Radio and the AN/PRC-152 Falcon radio respectively) but each system uses a different, proprietary battery. Despite being nearly identical in size and design, "a superficial design characteristic on one battery prohibits the battery from powering the other manufacturer's radio."[38] Likewise, each requires a proprietary charger. Without battery standards, DoD contractors will continue to develop unnecessarily proprietary energy sources and establish single-source dependencies.

To avoid the costs and challenges of unnecessarily retrofitting electronic equipment, standardization is most efficient if done early in the procurement process. However, retrofitting to standard battery units or upgrading to more efficient units may be cost-effective in certain cases. For example, the Government Accountability Office (GAO) reports that replacing an expensive, proprietary battery on the TALON bomb disposal robot resulted in a $7,000 savings per robot, while increasing the battery life of the robot by more than 20 percent.[39]

Recognizing the risks and inefficiencies, DoD (under pressure from Congress) has made progress by establishing the position of Product Director for Batteries. This person will help facilitate central coordination to reduce battery proliferation in the Army, based on its perceived lack of central coordination on battery issues.[40] The Product Director for Batteries' goal is to reduce the number of distinct batteries in use by the Army. Similar, yet more compartmentalized, efforts are currently underway by the Navy and Marine Corps, again aimed at encouraging equipment manufacturers to select a preexisting standard battery unit.

VULNERABILITIES IN THE LI-ION BATTERIES SUPPLY CHAIN

In light of current spending programs and private sector activities, the United States is just beginning to re-launch a domestic advanced battery production sector.

There is another risk, lurking in the future, related to the lithium inside advanced batteries. Inside them, lithium is combined with other chemical elements such as iron, cobalt, manganese, and–potentially, in the future–vanadium. So-called vanadium flow batteries use vanadium ions in different oxidation states to store potential energy.

Lithium is the lightest metal and the least dense solid element in existence, and it is used as an anode material (see Figure 4 for the anode's location in a battery). Because of lithium's high electrochemical potential, it is the preferred material for creating high energy-density rechargeable batteries. There is currently no substitute for this element with similar energy storage capability. Owing to its particular properties, lithium

is used in additional products such as ceramics, pharmaceuticals, and air conditioning (see Table 1).

In light of lithium-based batteries' rising popularity, demand for lithium likely will rise over time. If indeed EVs become the norm for consumers in the United States and elsewhere, demand for lithium will rise steeply (see Figure 5). EVs were first designed to drive with nickel-metal hydride (NiMH) batteries. However, NiMH batteries are heavy, bulky, expensive, and slow to charge. Since 2008, vehicle manufacturers and the battery industry have focused on Li-ion batteries, which seem to

Table 1: World Market Share For Lithium End-Uses, 2007–2012 (Percent Of Global Sales)			
End-Use	2007	2008	2012
Ceramics and Glass	18%	31%	30%
Batteries	25%	23%	22%
Lubricating Grease	12%	10%	11%
Pharmaceuticals and Polymers	7%	7%	5%
Air Conditioning	6%	5%	4%
Primary Aluminum Alloying	4%	3%	1%
Other	28%	21%	23%
Total World Production (Metric Tons)	25,400	25,400	37,000*

*excluding U.S. production

Thomas Goonan, "Lithium Use in Batteries," U.S. Geological Survey Circular 1371 (Reston, Virginia: USGS, 2012). http://pubs.usgs.gov/circ/1371/; U.S. Geological Survey, Mineral Commodity Summaries January 2013. http://minerals.usgs.gov/minerals/pubs/commodity/lithium/mcs-2013-lithi.pdf

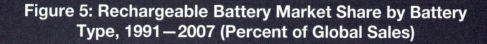

Figure 5: Rechargeable Battery Market Share by Battery Type, 1991—2007 (Percent of Global Sales)

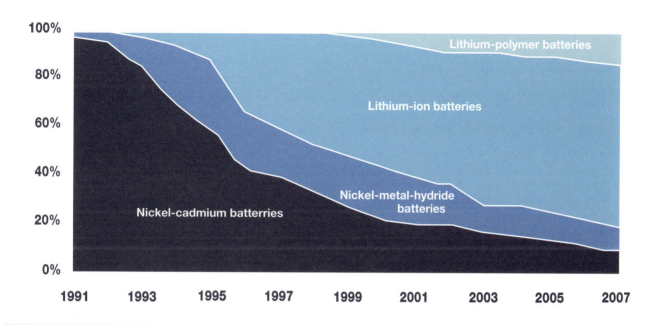

Source: T. Goonan, Lithium Use in Batteries: U.S. Geological Survey Circular 1371 (Reston, VA: USGS, 2012).

be a better option than Ni-MH batteries, although safety and costs continue to be concerns.[41]

Estimates of future demand for lithium vary and are usually derived from its expected use in next-generation EVs.[42] Thus, future demand and possible supply risks are based on rough projections of the EV market in a decade or two. In 2012, identified lithium resources totaled 5.5 million tons in the United States and approximately 34 million tons in other countries. Bolivia and Chile account for 9 million tons each, and China holds 5.4 million tons of proven reserves of lithium. In short, more than 65 percent of the world's lithium reserves are located in the salt lakes of Bolivia, Chile, and China. The United State relies on imports for over 70 percent of its lithium

consumption, and Argentina and Chile together meet close to 100 percent of its import needs. However, China mines sizable deposits of lithium, while production in Argentina has stagnated. [43]

With demand for large batteries for EVs expected to rise steadily, it is possible that global lithium reserves will not be sufficient to meet demand. Expert opinion is divided, but one projection based on future forecast of global production of portable electronic devices and EVs argues that lithium resources will be depleted by 2025. According to some estimates, this scenario could be the case even if 100 percent of all Li-ion batteries were recycled today.[44] Other analysts are even more pessimistic, projecting a global lithium shortage by 2017.[45] These concerns have led many

The United States is currently almost completely dependent on imported vanadium from China, Venezuela, South Africa, and Russia. A small percentage comes from burning coal and heavy oil, a byproduct of uranium mining, and imported pig iron slag.

Asian electronics companies to establish joint ventures and alliances with lithium companies in the hopes of discovering either expanded lithium supply or guaranteed access to existing lithium supplies.[46]

The potential for a lithium shortage should be very disconcerting to the U.S. military. The U.S. Army Tank Automotive Research, Development and Engineering Center (TARDEC) together with the defense research team of A123 Systems, (which was bought by Illinois-based Navitas Systems, as mentioned earlier in reference to EV batteries), developed a new 6T battery that is lighter than the stock battery, but provides twice the reserve time without suffering from memory effects. Additionally, since there is no acid inside this Li-ion battery to cause corrosion, vehicle deterioration and human contamination from potential chemical or gas spills are eliminated.[47] The most unusual element of these new batteries is that they are fireproof even when hit by bullets, and they are claimed to hold up under extreme heat or cold.[48] Currently, the U.S. military requires roughly 800,000 6T batteries, one-third of which were purchased in 2010. Because the older batteries are mostly lead-acid batteries, converting to Li-ion batteries will only expedite the depletion of known global reserves.[49]

Asia-Pacific electronic companies and battery producers are concerned about possible restrictions on lithium supplies. Among themselves, these companies have pursued corporate strategic alliances and joint ventures with private lithium exploration companies worldwide to ensure a reliable, diversified supply of lithium for Asia's battery suppliers and vehicle manufacturers. With lithium carbonate being one of the lowest cost components of a Li-ion battery, the issue is not cost difference or production efficiency but rather supply security attained by acquiring lithium from diversified sources.[50] The United States should be equally concerned and assign top priority to guaranteeing access and supply of the chemical elements needed to produce advanced batteries.

While the world may be exhausting lithium reserves, the next generation of super-dense batteries relies on vanadium pentoxide, which improves mass energy storage. The state-of-the-art vanadium redox (flow) battery essentially is a large liquid-filled tank to store energy from solar panels or wind turbines for the power grid. Currently, China, Russia, and South Africa account for nearly all vanadium production. Vanadium is also used to strengthen steel (see this report's chapter on titanium). Already, there are concerns that there will not be enough supply to meet that demand if vanadium redox batteries significantly increase demand.[51]

Vanadium redox batteries represent the next wave in advanced batteries. The current design is based on vanadium-charged lithium batteries, but the vanadium redox battery is even more promising because it is ideal for the storage of energy from renewable energy sources. (Wind and solar power do not provide a consistent flow of electricity, necessitating the storage of surplus energy.) Vanadium is a

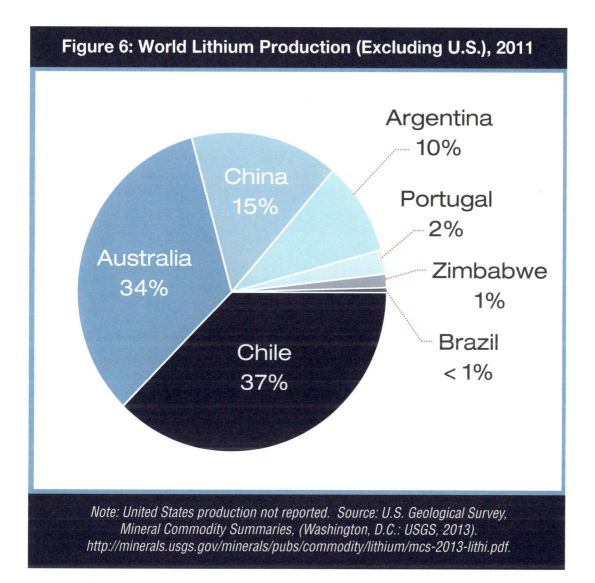

Figure 6: World Lithium Production (Excluding U.S.), 2011

Argentina 10%

Portugal 2%

Zimbabwe 1%

Brazil < 1%

China 15%

Australia 34%

Chile 37%

Note: United States production not reported. Source: U.S. Geological Survey, Mineral Commodity Summaries, (Washington, D.C.: USGS, 2013). http://minerals.usgs.gov/minerals/pubs/commodity/lithium/mcs-2013-lithi.pdf.

transition metal (see this report's chapter on Specialty Metals), which can support four energy states. Vanadium is used in the electrolyte state, which is the element that stores the energy and that flows through or past an electrode. This movement of electrons or protons across a membrane allows heat to be taken out of the battery and separates the energy storage from the power. It can be scaled up without having to increase the power output, and it lasts for decades.[52] Thus, as with lithium, vanadium provides special advantages that are

unique to its chemistry and that are highly suited for defense applications.

The United States is currently almost completely dependent on imported vanadium from China, Venezuela, South Africa and Russia, while a small percentage comes from burning coal and heavy oil, a byproduct of uranium mining, and imported pig iron slag.

Figure 7: Identified Resources of Lithium in 2012 (40 Million Tons)

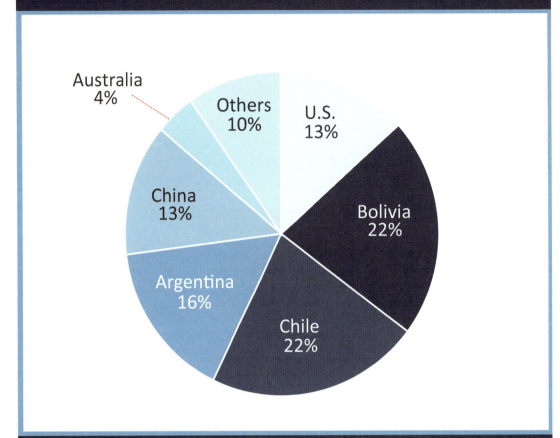

Australia 4%

Others 10%

U.S. 13%

China 13%

Bolivia 22%

Argentina 16%

Chile 22%

Source: U.S. Geological Survey, Mineral Commodity Summaries, (Washington, D.C.: USGS, 2013). http://minerals.usgs.gov/minerals/pubs/commodity/lithium/mcs-2013-lithi.pdf.

Although the United States imports over 80 percent of its vanadium from several different countries and regions, there is no shortage of vanadium globally. However, the vanadium used in batteries is expensive to produce because it requires additional processing. The vanadium must be converted into an electrolyte, and all contaminants must be removed. In the future, the United States may become reliant on a handful of countries that mine and extract vanadium to supply domestic advanced battery producers.

MITIGATING THE RISKS

How can the U.S. government work with U.S. defense industrial base firms to respond to this emerging reliance on a few mineral resources? A first step would be to improve and standardize recycling.

To date, recycling of lithium content has been insignificant, but discussion of the potential of recycling has increased. Since 1992, one U.S. company has recycled lithium metal and Li-ion batteries at its facility in British Columbia, Canada. In 2009, DoE awarded the company $9.5 million to construct the first U.S. recycling facility for Li-ion batteries in an existing recycling plant in Ohio. So far, the plant still only focuses on nickel and lead batteries. The Belgian company Unicore is the only company in the world that has refined and scaled-up the recycling of used cobalt-based lithium batteries.

Recycling faces several immense technical and financial hurdles. First, battery developers have focused intensely on bringing down the costs of raw materials and producing standardized common consumer batteries. Lower raw materials costs make recycling less economically feasible. More expensive and advanced batteries, such as those based on cobalt and lithium, would be more likely candidates for recycling. However, recycling Li-ion batteries is both a complex and expensive undertaking, and requires stopping the internal chemical process, separating the various metals, and then refining the recoverable metal content. To make matters worse, lithium itself is still relatively inexpensive, and contributes very little to the overall cost of the battery.[53]

If the batteries contain cobalt, an expensive raw material, it may be possible to set up an economically viable recycling system. However, for batteries that do not contain cobalt, it makes no economic sense to invest in recycling. In fact, most battery recycling programs focus almost exclusively on recovering nickel and cobalt.[54]

The low level of Li-ion battery recycling is somewhat counterintuitive because the most commonly recycled product in the world is the lead-acid battery. However, the process for recycling lead-acid batteries is simple and well established, making

it profitable. On the other hand, cobalt-based lithium batteries require expensive recycling technology, which may only be profitable if the price of cobalt remains high. Lithium-based batteries without cobalt (including certain Li-ion batteries) may never be candidates for recycling unless the price of lithium surges.

Because of these economic uncertainties, the U.S. private sector likely requires additional incentives to take on recycling. Incentives could include stronger battery collection laws, such as those that exist in many other advanced industrial countries.

The United States should focus on improving the network for recycling EV batteries. An existing vehicle battery recycling system (based on the lead-acid model), could be adapted to process Li-ion batteries. Second, large EV batteries contain more lithium, which could mean that battery recycling can be more profitable in spite of the currently low price of lithium.

Still, it will be expensive to recycle batteries, especially small consumer batteries. The U.S. government will have to provide substantial support to finance comprehensive battery recycling programs.[55] In this effort, the U.S. government has targeted loans and grants of about $11 billion for research and development (R&D) and for production and recycling facilities.[56]

Second, together with DoE, DoD should encourage firms to specialize in a particular range of battery products and focus on niche markets including military, aerospace, and medical. DoE has already begun to incentivize Asia-based U.S. battery companies to return to the United States, along with their manufacturing expertise.[57]

Third, reducing the number of types and sizes of batteries used by troops would boost domestic production of specialized batteries. DoD must engage more in specifying which power source it desires and supply sufficiently large economies of scale to stimulate domestic production and R&D. As the GAO documents, DoD currently lacks awareness of its total investment in power sources.[58] Power and batteries programs fall into three separate categories: science and technology, logistics support, and acquisition programs. There is no program that aggregates all the data; indeed, DoD has limited understanding of its own acquisition programs and requirements. A relatively easy solution would be to establish a supervisory authority to ensure basic accountability, anticipate future funding, and measure performance.[59]

Moreover, changes in the acquisition, procurement, and standardization of batteries would have three immediate benefits. First, standardization and uniform adoption of high-tech batteries such as Li-ion batteries would greatly reduce item unit costs; would reduce logistical, supply, and organizational challenges; and would substantially reduce the reliance on sole-source suppliers for mission-critical items.

Second, standardization and movement towards more capable batteries would offer manufacturers greater production volumes and may result in a healthier industrial base. This progress also would provide incentives for the Li-ion battery industry to pursue greater economies of scale to bring down the costs of manufacturing high tech batteries, especially in the area of non-EV batteries. Many smaller firms would prefer propriety power sources. However, in the absence of this option, most firms would like to compete to produce standard power sources, in order to stabilize their production volumes and revenue. This would secure the health of the battery sector by encouraging economies of scale in the production of these standard batteries.[60]

Third, standardization would improve operational readiness, reducing the potential for unforeseen shortages.[61] Standardization also would guarantee the long-term viability of military battery technology, which is also used in the commercial sector, and it would free the warfighter from having to carry large loads of different sorts and sizes of batteries.

Finally, U.S. government and industry need to collaborate "beyond lithium-ion" and focus on the next generation of rechargeable, high-density batteries. The U.S. has numerous high-powered laboratories already working on future batteries that are environmentally benign, low-cost, less dependent on rare chemical elements, and have longer life cycles. Research money and continuous support in translating new laboratory ideas into commercial ventures may yield more benefits than seeking to catch up with established Japanese and South Korean Li-ion battery manufacturers.

CONCLUSION

Years ago, U.S. battery companies made a conscious decision to forego research and investments in commercializing advanced battery design and technology. That decision has come to haunt U.S. military strategists, the automobile sector, and the green technology industry. South Korean, Japanese, and Taiwanese companies took the lead to perfect the commercial manufacture of Li-ion batteries, and their batteries are now largely used in laptops, smartphones, and other electronics. DoD also ignored the importance of ensuring that it relies on standardized and uniform battery types and sizes, with the result that warfighters must carry dozens of different batteries to power electronics, weapons, and communication devices.

The U.S. government has invested heavily in starting new production lines for Li-ion batteries, as it has identified EVs as the wave of the future. Not all of these efforts, as we have seen in the case of A123 Systems, have been successful. It may be more useful to target specific niche areas in which the nascent U.S. Li-ion battery industry can compete, such as battery power for defense use and specifically for the warfighter on the battlefield. Since some countries and foreign firms are well ahead of U.S. firms in the advanced battery market, the U.S. government and private sector should be more selective and strategic about where they wish to invest time and resources.

In the short term, the current global manufacturing capacity for battery manufacturing far exceeds the demand for EVs. There is little doubt that many battery manufacturers, both in the U.S. and abroad, will not survive. But at least the debate has focused attention on an important vulnerability, relevant not only to defense supply chains, but also to the health of the U.S. economy. In the long run, the United States cannot abandon battery innovation and technology and continue to rely on imports from Asia-Pacific countries (whether allies like Japan and South Korea, or competitors like China). As equipped today, U.S. armed forces cannot effectively fight without access to power sources in the form of batteries. Thus, the fresh attention to the advanced battery sector is a positive development, because it will stir interest in battery innovation and technology. In the short run, the United States can build up expertise in more esoteric areas of advanced battery research and slowly build up its own EV sector.

For the U.S. military, power sources are a mission-critical technology. DoD has been slow to recognize the rising importance of rechargeable batteries, with the result that many battery-powered devices do not use the lighter, denser kind of batteries. The current debate on green technology, as well as reports published by the GAO, have created a new awareness and a new commitment to remedy the state of affairs in the supply of battery power to military vehicles, aircraft, and electronic devices. As a matter of national security, we need concerted government and industry efforts to bolster the U.S. advanced battery industry.

ENDNOTES

1 Doug Moorehead, "The Merits of Lithium Ion Energy Storage on the Battlefield," *National Defense Magazine*, May 2011. http://www.nationaldefensemagazine.org/archive/2011/May/Pages/TheMeritsofLithiumIonEnergyStorageOntheBattlefield.aspx.

2 Masaki Yoshio, Ralph J. Brodd, Akiya Kozawa (eds), *Lithium-ion Batteries: Science and Technologies* (New York: Springer, 2009).

3 Memory effect in NiCd batteries refers to a decrease in energy capacity after the battery has been disconnected before reaching a full charge. The battery remembers the smaller capacity and thereafter can no longer fully charge. Li-ion batteries do not have this memory effect, so the battery can always be recharged even before its stored energy has been depleted.

4 In 2008, Li-ion batteries powering the U.S. Navy's Advanced SEAL Delivery System mini-submarine ignited, prompting concerns over the safety of the batteries, especially in a military setting. Similar problems were reported in various laptop computer models, and resulted in several recalls between 2006 and 2008.

5 Doug Moorehead, "The Merits of Lithium Ion Energy Storage on the Battlefield," *National Defense Magazine*, May 2011. http://www.nationaldefensemagazine.org/archive/2011/May/ Pages/TheMeritsofLithiumIonEnergyStorageOntheBattlefield.aspx.

Andrew Parker, "US Regulators to Review Boeing Dreamliner," *Financial Times,* January 11, 2013. http://www.ft.com/intl/cms/s/0/0e0a35c0-5be9-11e2-bf31-00144feab49a.html#axzz2Myxm7gt7.

6 Eric Beidel, "Soldier Energy Needs Outpacing Technology Policy," *National Defense Magazine*, March 2012. http://www.nationaldefensemagazine.org/archive/2012/March/Pages/SoldierEnergyNeedsOutpacingTechnology,Policy.aspx.

7 Sandra I. Erwin, "Army, Marines Face Uphill Battle to Lighten Troops' Battery Load," *National Defense Magazine*, May 2011. http://www.nationaldefensemagazine.org/archive/2011/May/Pages/ArmyMarinesFaceUphillBattleToLightenTroops%E2%80%99BatteryLoad.aspx.

8 For example, the average multiband "manpack" radio requires 216 AA batteries every two days for continuous operation. Ibid.

9 David McShan, "Advanced Battery Technology Shrinks Military Energy Costs," *Journal of Military Electronics and Computing* (November 2011). http://www.cotsjournalonline.com/articles/view/102156.

10 According to the U.S. Government Accountability Office (GAO), DoD spent at least $2.1 billion on power sources between 2006 and 2010. However, that investment so far has not resulted in significantly more efficient and lighter batteries for warfighters. Sandra I. Erwin, "Army, Marines Face Uphill Battle to Lighten Troops' Battery Load," *National Defense Magazine*, May 2011. http://www.nationaldefensemagazine.org/archive/2011/May/Pages/ArmyMarinesFaceUphillBattleToLightenTroops%E2%80%99BatteryLoad.aspx.

11 "The U.S. Department of Energy Takes Credit for Reviving U.S. Battery Industry," *Manufacturing & Technology News* (August 30, 2010), vol. 17, No. 13, p. 6(1). http://www.manufacturingnews.com/cgi-bin/backissues/backissues.cgi?flag=show_toc&id_issue=295&id_title=1.

12 Large format batteries offer up to 70 times the capacity of traditional Size D lithium batteries.

13 David Sedgwick, "Asian Suppliers Lead Battery Race," *Automotive News*, March 19, 2012. http://www.autonews.com/article/20120319/OEM01/303199984.

PRNewswire, "Advanced Batteries Market to 2020 - Demand for Electric Vehicles to Drive Growth, Asia Pacific to Remain the Major Producer," June 6, 2012.

http://www.statejournal.com/story/18715994/advanced-batteries-market-to-2020-demand-for-electric-vehicles-to-drive-growth-asia-pacific-to-remain-the-major-producer.

14 Marcy Lowe, Saori Tokuoka, Tali Trigg and Gary Gereffi, *Lithium Ion Batteries for Electric Vehicles: The U.S. Value Chain* (Durham, NC: Duke University Center on Globalization Governance & Competitiveness, October 2010). http://www.cggc.duke.edu/pdfs/Lowe_Lithium-Ion_Batteries_CGGC_10-05-10_revised.pdf.

15 Energizer sold its Gainesville, Florida, facility to Moltech Corporation in 1999 after the plant sat idle for two years. In 2002, Moltech sold the facility to U.S. Lithium Energetics, which is producing select Li-ion batteries for the consumer and military markets. Ralph J. Brodd, "Factors Affecting U.S. Production Decisions: Why are There No Volume Lithium-Ion Battery Manufacturers in the United States?" *ATP Working Paper Series* 5, n. 01 (June 2005). http://www.atp.nist.gov/eao/wp05-01/contents.htm.

16 Ralph J. Brodd, "Factors Affecting U.S. Production Decisions: Why are There No Volume Lithium-Ion Battery Manufacturers in the United States?" *ATP Working Paper Series* 05, n. 01 (June 2005). http://www.atp.nist.gov/eao/wp05-01/contents.htm.

17 DoE offered $2.4 billion of funding to battery-related manufacturers, including auto manufacturers, battery material suppliers, and battery recycling companies. These funds will help establish 30 U.S. manufacturing plants, all playing key roles across the value chain, including materials, components, and production of cells and battery packs. The funding also supports several of the world's first demonstration projects for EVs. An additional $2.6 billion has been provided in Advanced Technologies Vehicle Manufacturing (ATVM) loans to Nissan, Tesla, and Fisker to establish EV manufacturing facilities in Tennessee, California, and Delaware, respectively. DoE also has offered $25 billion in low-interest loans to battery companies. To help consumers pay the higher purchase price for EVs, the government offers a $7,500 tax incentive. Marcy Lowe, Saori Tokuoka, Tali Trigg and Gary Gereffi, *Lithium Ion Batteries for Electric Vehicles* (Durham, NC: Duke University Center on Globalization Governance & Competitiveness, October 2010). http://www.cggc.duke.edu/pdfs/Lowe_Lithium-Ion_Batteries_CGGC_10-05-10_revised.pdf.

18 Larry Thomas, "Advanced Batteries for Energy Security" (presentation at delivered at the Defense Energy Security Caucus, Ithaca, NY, November 3, 2011). http://primetprecision.com/remarks-from-primets-cto-larry-thomas-before-the-defense-energy-security-caucus-desc/.

19 Don Lee, "Fighting for 'Made in the USA,'" *Los Angeles Times*, May 08, 2010. http://articles.latimes.com/2010/may/08/business/la-fi-green-manufacturing-20100509.

20 Michael Grabell, *Money Well Spent? The Truth Behind the Trillion-Dollar Stimulus, the Biggest Economic Recovery Plan in History* (New York: Public Affairs, 2012).

21 Mark Rechtin, "Fisker Battles Problems as the Karma Lurches Toward launch," *Automotive News*, January 23, 2012. http://www.autonews.com/article/20120123/OEM01/301239965#ixzz1wE1F7nxq.

22 Sebastian Blanco, "Five People Hurt in GM Lab Explosion, A123," *Auto News*, April 12, 2012. http://www.autoblog.com/2012/04/12/five-people-hurt-in-gm-lab-explosion-a123-battery-reportedly-re/.

23 "A123 Systems Introduces Breakthrough Lithium Ion Battery Technology That Optimizes Performance in Extreme Temperatures," *The Wall Street Journal*, June 12, 2012. http://www.marketwatch.com/story/a123-systems-introduces-breakthrough-lithium-ion-battery-technology-that-optimizes-performance-in-extreme-temperatures-2012-06-12.

24 Wanxiang won the secret bid, but several other companies were attempting to buy A123's assets, including an U.S. company, Johnson Controls of Wisconsin, and the electronics manufacturers NEC Corporation of Japan and Siemens AG of Germany. The Committee on Foreign Investment in the United States (CFIUS) approved the sale in January 2013.

25 Bill Vlasic, "Chinese Firm Wins Bid for Auto Battery Maker," *The New York Times*, December 9, 2012. http://www.nytimes.com/2012/12/10/business/global/auction-for-a123-systems-won-by-wanxiang-group-of-china.html?_r=0.

Brad Plumer, "A123 Systems Files for Bankruptcy: Here's What You Need to Know," *Washington Post Wonkblog*, October 16, 2012. http://www.washingtonpost.com/blogs/wonkblog/wp/2012/10/16/a123-systems-files-for-bankruptcy-heres-what-you-need-to-know/.

26 Rod Lache, Dan Galves, and Patrick Nolan, "Vehicle Electrification: More Rapid Growth; Steeper Price Declines for Batteries," *Deutsche Bank Industry Update* (March 7, 2010). http://gm-volt.com/files/DB_EV_Growth.pdf.

27 Lithium-ion batteries create electricity as Lithium ions move between the anode and cathode. In the discharge cycle, lithium in the anode (carbon material) is ionized and emitted to the electrolyte. Lithium ions move through a porous plastic separator and insert into atomic-sized holes in the cathode (lithium metal oxide). At the same time, electrons are released from the anode. This process creates an electric current that travels to an outside electric circuit. Since this is a reversible chemical reaction, the battery can be recharged, causing lithium ions to go back from the cathode to the anode through the separator. . Akira Yoshino, "The Birth of the Lithium-Ion Battery," *Angewandte Chemie International Edition* 51, no. 24 (2012): 5798-5800.

28 "Celgard to Expand Battery Separator Capacity for EDVs," *Global Carolina Business Journal*, March 15, 2011. http://gcbusinessjournal.com/index.php?option=com_content&view=article&id=1042:celgard-to-expand-battery-separator-capacity-for-edvs&catid=61:automotive&Itemid=89.

29 Ibid.

30 "America's First New Lithium Ion Battery Separator Manufacturing Company in Decades Set to Open with Breakthrough Technology," *BusinessWire*, June 14, 2010. http://www.businesswire.com/news/home/20100614005260/en/Americas-Lithium-Ion-Battery-Separator-Manufacturing-Company.

Bill Canis, *Battery Manufacturing for Hybrid and Electric Vehicles: Policy Issues* (Washington, D.C.: Congressional Research Service, March 22, 2011).

31 Marcy Lowe, Saori Tokuoka, Tali Trigg, and Gary Gereffi, *Lithium Ion Batteries for Electric Vehicles: The U.S. Value Chain,* (Durham, NC: Duke University Center on Globalization Governance & Competitiveness, October 2010), 20. http://www.cggc.duke.edu/pdfs/Lowe_Lithium-Ion_Batteries_CGGC_10-05-10_revised.pdf

32 Government Accountability Office (GAO), *Defense Acquisitions: Opportunities Exist to Improve DoD's Oversight of Power Source Investments GAO-11-113*, by Michael J. Sullivan (Washington, D.C.: GAO, December 2010). http://www.gao.gov/new.items/d11113.pdf.

33 Ibid.

34 U.S. Army, Capabilities Integration Center, Research, Development, and Engineering Command, Deputy Chief of Staff, G-4, *Power and Energy Strategy White Paper* (Washington, D.C.: U.S. Army, April 1, 2010).

35 Jim Gucinski, Chair, Committee on Military Power Sources Committee and Reserve Battery, Minutes from the Ad-hoc Committee, "Military Power Sources" held May 11, 2010. http://www.ndia.org/Divisions/Divisions/Manufacturing/Pages/Militarypowersources.aspx.

36 Larry Thomas, *Defense Energy Security Caucus Advanced Batteries for Energy Security* (November 3 2011). http://primetprecision.com/remarks-from-primets-cto-larry-thomas-before-the-defense-energy-security-caucus-desc/.

37 Government Accountability Office (GAO), *Defense Management: Overarching Organizational Framework Needed to Guide and Oversee Energy Reduction Efforts for Military Operations, GAO-08-426*, by Zina Dache Merritt (Washington, D.C.: GAO, March 13, 2008). http://www.gao.gov/products/GAO-08-426.

Government Accountability Office (GAO), *Best Practices: Stronger Practices Needed to Improve DoD Technology Transition Processes, GAO-06-883*, by Michael J. Sullivan (Washington, D.C.: GAO, September 14, 2006). http://www.gao.gov/products/GAO-06-883.

38 Government Accountability Office, *Defense Acquisitions: Opportunities Exist to Improve DoD's Oversight of Power Source Investments GAO-11-113*, by Michael J. Sullivan (Washington, D.C.: Government Accountability Office, December 2010), 22. http://www.gao.gov/new.items/d11113.pdf.

39 Ibid.

40 Sandra I. Erwin "Army's Energy Battle Plan: Attack Fuel Demand," *National Defense Magazine,* May 2011. http://www.nationaldefensemagazine.org/archive/2011/May/Pages/Army%E2%80%99sEnergyBattlePlanAttackFuelDemand.aspx.

41 Thomas Goonan, "Lithium Use in Batteries," *U.S. Geological Survey Circular 1371* (2012). http://pubs.usgs.gov/circ/1371/.

42 William Tahil, *The Trouble with Lithium: Implications of Future PHEV Production for Lithium Demand* (Martainville, France: Meridian International Research, December 2006). http://www.evworld.com/library/lithium_shortage.pdf.

William Tahil, *The trouble with Lithium 2—Under the Microscope* (Martainville, France: Meridian International Research, May 29, 2008). http://www.meridian-int-res.com/Projects/Lithium_Microscope.pdf.

43 Australia has become a major producer of lithium, and directly supplies the Asian market to meet the needs of battery suppliers and vehicle manufacturers. U.S. Geological Survey (USGS), *Mineral Commodity Summaries 2013* (Reston, VA: USGS, 2013). http://minerals.usgs.gov/minerals/pubs/mcs/2013/mcs2013.pdf.

44 Thomas Cherico Wanger, "The Lithium future—Resources, Recycling, and The Environment," *Conservation Letters* 4 (2011): 202–206.

45 John Petersen, "Grid-Scale Energy Storage: Lux Predicts $113.5 Billion in Global Demand by 2017," *Alt Energy Stocks* (April 2012). http://www.altenergystocks.com/archives/2012/04/gridscale_energy_storage_lux_predicts_1135_billion_in_global_demand_by_2017.html.

46 U.S. Geological Survey, *Mineral Commodity Summaries 2010* (Washington, D.C.: USGS, 2011). http://minerals.er.usgs.gov/minerals/pubs/commodity/lithium/mcs-2010-lithi.pdf

47 Y. Ding, S. Zanardelli, D. Skalny, D., and L. Toomey, "Technical Challenges for Vehicle 14V/28V Lithium Ion Battery Replacement," *SAE Technical Paper* (2011) 01-1375. http://papers.sae.org/2011-01-1375.

48 Julie Wernau, "Navitas Key to Sale of A123 to Chinese Firm," *Chicago Tribune*, January 30, 2013. http://articles.chicagotribune.com/2013-01-30/business/ct-biz-0130-navitas-20130130_1_navitas-systems-wanxiang-group-microsun-technologies.

49 Dave Brown, "Military Applications Could Increase Lithium Demand," *Lithium Investing News*, March 1, 2012. http://lithiuminvestingnews.com/5358/military-applications-could-increase-lithium-demand-a123-battery.

50 Thomas Goonan, *Lithium Use in Batteries*, (Reston, VA: USGS Circular 1371, 2012). http://pubs.usgs.gov/circ/1371/.

51 U.S. Geological Survey, *Mineral Commodities Summary 2011* (Reston, VA: USGS, 2012).

"Flow Batteries Seen Increasing Vanadium Demand By 10% To 15% in Coming Years," *The Street*, June 13, 2012. http://www.thestreet.com/story/11581983/1/flow-batteries-seen-increasing-vanadium-demand-by-10-to-15-in-coming-years.html.

Richard Mowat, "Redox Batteries May Cause Sharp Increase in Vanadium Demand," Vanadium.com, November 18, 2011. http://www.vanadiumsite.com/vanadium-redox-battery-demand/.

52 Damon van der Linde, "Vanadium Batteries for Sustainable Energy," *Vanadium Investing News*, July 18, 2011. http://vanadiuminvestingnews.com/1811/vanadium-batteries-for-sustainable-energy/.

Damon van der Linde, "US Could Face Vanadium Shortage as Demand Increases in China," *Vanadium Investing News*, August 15, 2011. http://resourceinvestingnews.com/20805-us-could-face-vanadium-shortage-as-demand-increases-in-china.html.

53 "Battery Recycling as a Business," *Battery University*, March 2012. http://batteryuniversity.com/learn/article/battery_recycling_as_a_business/.

John Peterson, "Why Advanced Lithium ion batteries won't be recycled" AltEnergyStocks.com, May 16 2011. http://www.altenergystocks.com/archives/2011/05/why_advanced_lithium_ion_batteries_wont_be_recycled.html.

54 Thomas Goonan, "Lithium Use in Batteries," in *U.S. Geological Survey Circular* 1371 (Reston, VA: U.S. Geological Survey, January 2012). http://pubs.usgs.gov/circ/1371/.

55 Isidor Buchmann, "Recycling Batteries," *Battery University* (2009). http://www.batteryuniversity.com/print-partone-9.htm.

56 Thomas Goonan, "Lithium Use in Batteries," in *U.S. Geological Survey Circular 1371* (Reston, VA: U.S. Geological Survey, January 2012). http://pubs.usgs.gov/circ/1371/.

57 Marcy Lowe, Saori Tokuoka, Tali Trigg and Gary Gereffi, *Lithium Ion Batteries for Electric Vehicles: The U.S. Value Chain* (Durham, NC: Duke University Center on Globalization Governance & Competitiveness, October 2010). http://www.cggc.duke.edu/pdfs/Lowe_Lithium-Ion_Batteries_CGGC_10-05-10_revised.pdf.

58 Government Accountability Office (GAO), *Defense Acquisitions: Opportunities Exist to Improve DoD's Oversight of Power Source Investments GAO-11-113*, by Michael J. Sullivan (Washington, D.C.: GAO, December 2010). http://www.gao.gov/new.items/d11113.pdf.

59 Ibid.

60 "Army Battery Standardization: Rechargeable Batteries Power the Future Force," *Case Study*, Defense Standardization Program Office (Fort Belvoir, VA: DLA, 2002). https://assist.daps.dla.mil/docimages/A/0000/0021/6353/000000451055_000000151687_QZCTGRDGRE.PDF?CFID=2270626&CFTOKEN=45823119&jsessionid=84307f228d8282049a53153340e48772c556

61 Government Accountability Office (GAO), *Defense Acquisitions: Opportunities Exist to Improve DoD's Oversight of Power Source Investments GAO-11-113*, by Michael J. Sullivan (Washington, D.C.: GAO, December 2010). http://www.gao.gov/new.items/d11113.pdf.

CHAPTER 10
HELLFIRE MISSILE PROPELLANT

EXECUTIVE SUMMARY

The AGM-114 HELLFIRE air-to-ground missile (AGM) is one of the most widely used and effective weapons in the U.S. arsenal. This guided missile carries a payload capable of defeating any modern armored vehicle and is accurate enough for use in urban environments where it is important to limit collateral damage and civilian casualties. The missiles rely on a solid rocket fuel, Butanetriol trinitrate (BTTN), manufactured in Pennsylvania. BTTN, in turn, requires the chemical Butanetriol (BT), which is currently only manufactured by a single Chinese company. The last U.S. producer of BT, Cytec Industries, discontinued production in 2004.

Currently there are no viable alternatives to BTTN, which is more stable than nitroglycerin (NG). NG can be more robust in low-temperature environments, but it is also significantly more expensive. Given the high costs associated with producing BT that meets military standards and the lack of commercial applications for solid rocket fuel, Department of Defense (DoD) demand has been insufficient to sustain U.S. domestic production of BT.

The FY2006 National Defense Authorization Act (NDAA) prohibits acquiring munitions from Chinese military companies. In 2008, a one-time waiver allowed the purchase of BT from China to avoid an imminent shortage of HELLFIRE propellant. However, it is unclear how the U.S. will acquire additional BT once current supplies are depleted, or whether a second waiver permitting acquisition from China will be issued. Dependence on Chinese BT gives the Chinese government potential leverage over the United States and could lead to situations in which China could withhold or threaten to withhold BT exports to the United States, thereby restricting U.S. access to this essential component of an important weapons system. For example, should the United States mount military operations that China does not support, trade restrictions could deny access to HELLFIRE missile propellant subcomponents, potentially limiting U.S. military capabilities.

A long-term solution to this vulnerability will require reestablishing a domestic source of BT, which may only be possible with U.S. government support.

HELLFIRE MISSILE PROPELLANT
ENABLING A CRITICAL CAPABILITY

MANUFACTURING SECURITY

HELLFIRE missile propellant is essential to the primary U.S. air-to-ground missile

MULTIPLE DEFENSE PLATFORMS USE HELLFIRE MISSILES

PROTECTING SOLDIERS

HELLFIRE missiles can effectively defeat any enemy tank

HELLFIRE PROPELLANT + **HELLFIRE MISSILES** = **DEFEATED ENEMY TANKS**

UNDERSTANDING SUPPLY

The U.S has one domestic manufacturer for HELLFIRE missile propellant

1 MANUFACTURER FOR HELLFIRE PROPELLANT **PRODUCTION IN THE U.S.**

Planned U.S. solid rocket motor programs are projected to purchase more than 1,000,000 pounds of HELLFIRE missile propellant per year

1 MILLION POUNDS ANNUALLY

DOMESTIC SUPPLY

The U.S. maintains a national defense stockpile equal to an 18 month supply of HELLFIRE missile propellant

18 MONTH SUPPLY

VULNERABILITY

The United States produces zero butanetriol

NO DOMESTIC PRODUCTION **FOR BUTANETRIOL**

One Chinese firm manufacturers Butanetriol, a key chemical for HELLFIRE missile propellant

ONE CHINESE FIRM MANUFACTURERS **100% of U.S.** BUTANETRIOL SUPPLY

MITIGATING RISKS

Assuring U.S. HELLFIRE missile propellant supply

DEVELOP U.S. CAPACITY   **INVEST IN RESEARCH**   **SECURE THE SUPPLY CHAIN**

MILITARY EQUIPMENT CHART
SELECTED DEFENSE USES FOR HELLFIRE PROPELLANT

DEPARTMENT	PLATFORMS
ARMY	**PLATFORMS THAT FIRE THE HELLFIRE MISSILE:** ■ AH-64 Apache helicopter ■ OH-58D Kiowa Warrior helicopter ■ RAH-66 Comanche helicopter ■ AH-6 Little Bird helicopter ■ UH-60 Blackhawk helicopter ■ MQ-1C Grey Eagle drone
MARINE CORPS	**PLATFORMS THAT FIRE THE HELLFIRE MISSILE:** ■ AH-W1 Super Cobra helicopter ■ AH-1Z Viper helicopter ■ KC-130 aerial refueling tanker
NAVY	**PLATFORMS THAT FIRE THE HELLFIRE MISSILE:** ■ SH-60 / MH-60R / MH-60S Seahawk helicopter ■ MQ-9 Reaper drone
AIR FORCE	**PLATFORMS THAT FIRE THE HELLFIRE MISSILE:** ■ AC-208 Combat caravan ■ MQ-1B Predator drone ■ MQ-9 Reaper drone

INTRODUCTION

Since they were put into U.S. military service in 1985, thousands of HELLFIRE missiles have been used to defeat adversaries of the United States. The AGM-114 HELLFIRE missile, manufactured by Lockheed-Martin, is one of the most effective and widely used weapons in the U.S. arsenal and is a key contributor to U.S. national security. Given the HELLFIRE's success in combat and the degree to which it is considered an important U.S. military weapons system, it is surprising that there is no domestic supplier of the chemical Butanetriol (BT), one of the essential components of the missile's propellant. Even more troubling is the fact that the only known manufacturer of BT is in China.

The AGM-114 HELLFIRE missile is the U.S. military's primary air-to-ground, anti-armor, precision-guided missile system. The HELLFIRE can also be used as an air-to-air missile. Moreover, the HELLFIRE provides precision strike capability against tanks, reinforced structures, and bunkers, and has proven capable against any currently deployed tank or armored vehicle. It can be guided to its targets by aircraft-mounted remote control or by lasers mounted outside aircraft.[1] The HELLFIRE missile has no short-term replacement.

The HELLFIRE missile is propelled by a Thiokol TX-657 solid-fuel rocket motor (or solid rocket motor [SRM]), using BTTN as a key component of its solid-state rocket fuel. BT, an essential energetic plasticizer in certain propellant formulations, is a chemical precursor needed to produce the BTTN required to fuel the HELLFIRE missile. BTTN and BT are used as replacements for NG because they are more stable and thus easier to handle.

In a 2011 report, then-Under Secretary of Defense for Acquisition, Technology, and Logistics (USD[AT&L]) Ashton Carter argued that the U.S. military's need for SRMs only will increase in the future.[2] BT is a key ingredient for most small SRMs, including the HELLFIRE missile. Current military programs using SRMs are already coming dangerously close to outstripping all potential supply of BT. "Specifically, planned small SRM programs will purchase more than *one million pounds of propellant per year* [emphasis added]."[3] If the United States is to sustain the military capability provided by the HELLFIRE missile, access to BT is essential.

Since it was introduced into service in 1985, the HELLFIRE missile (and, by extension, its propellant) has been used in almost every U.S. conflict. Two versions of the HELLFIRE missile, Version I and the updated Version II, are now being used on unmanned aerial vehicles (UAVs) and are key weapons in U.S. counter-terrorism operations.

HELLFIRE missiles have been successfully fired from Apache, Seahawk, and Cobra attack helicopters; Kiowa scout helicopters; Predator and Reaper UAVs; ground-based tripod mounts; ground vehicles; and even boats.

The HELLFIRE II air-to-ground missile system (AGMS) gives attack helicopters heavy anti-armor capability. The HELLFIRE II is capable of striking multiple targets with great lethality and precision.[4] Initially designed "to defeat tanks and other individual targets, while minimizing the exposure of the launch vehicle to enemy fire," the HELLFIRE also is effective in urban areas because its relatively small warhead can reduce the risk of civilian casualties. Used in conjunction with laser guidance, a skilled operator can strike targets with precision, including putting a missile through the window of a building.[5] Additionally, the AGM-114's solid rocket fuel engine allows the HELLFIRE to have an operational range of six miles, with a top speed of Mach 1.3.[6]

This chapter examines the vulnerabilities associated with the U.S. ability to acquire BT, a key subcomponent of the HELLFIRE missile propellant. This chapter provides recommendations for preserving this critical capability in an age of evolving threats and constrained resources, and proposes solutions, which requires collaboration between the commercial and defense sectors as well as government policy-makers.

Key themes discussed in this chapter are:

- The HELLFIRE missile propellant supply chain is at risk of disruption due to reliance on a sole, foreign source of the key subcomponent BT.

- DoD is inadequately aware of the second and third tiers of its defense industrial supply chain, which can quickly lead to critical shortages of BT and HELLFIRE systems at inopportune times.

- DoD should focus on creating a robust domestic production capability of BT and BTTN to mitigate the risks of supply chain disruption for the HELLFIRE missile.

A NOTE ON CRITICALITY

When Cytec Industries discontinued production of BT, the United States military became dependent on a single Chinese producer of this important commodity. The Chinese monopoly over this commodity creates an *extreme* risk of future unavailability; China has already demonstrated its willingness to restrict access to resources to obtain political and economic concessions. For example, in 2010 China ceased exports of rare earth metals to Japan over a maritime dispute (further discussed in this report's chapter on high-tech magnets). Chinese control over BT means that as U.S. supplies diminish, China could inflate the price of BT, demand policy concessions from the United States in exchange for access, or strategically diminish the U.S. military's ability to use many of its advanced rocket systems. This latter possibility becomes even more likely in the event that Chinese and U.S. foreign policy interests are opposed, which is increasingly the case as the U.S. focuses foreign policy attention towards the Asia-Pacific region.

The inability to acquire BT would have *significant* consequences for U.S. defense capabilities, as HELLFIRE missile systems and other weapons systems requiring solid rocket propellant would become unavailable. The unavailability of these weapons systems would diminish U.S. military capability by limiting military commanders' force projection and support options.

HELLFIRE MISSILES USED BY U.S. MILITARY IN THE MOST SENSITIVE OPERATIONS (a notional though realistic scenario)

Pentagon sources announced that two Predator drones carrying HELLFIRE AGM-114C missiles attacked a convoy carrying senior leadership cadre of the East African terrorist group Harakat al-Shabaab al-Mujahideen, the Somalia-based cell of the militant Islamist group al-Qaeda. According to a senior U.S. official, the operation was carried out by Joint Special Operations Command, under the direction of the CIA. The Predators hovered above the Al-Shabaab convoy as it left an urban area, and the drones fired the HELLFIRE missiles that killed the terrorist leader. Among the casualties of this attack was one of the faction's key leaders, Sheikh Moktar Ali Zubeyr, also known as Muktar Abdirahman "Godane." The United States has designated Godane, who received training and fought in Afghanistan, as a terrorist. The HELLFIRE missiles destroyed the target, killing the terrorists but causing no damage other than to the target itself.

BACKGROUND

The name HELLFIRE comes from the initial designation as a "Helicopter Launched, Fire and Forget Missile." The HELLFIRE missile system is a short-range, laser- or radar- guided air-to-ground missile system (AGMS) designed to defeat armored targets. It was designed and developed in the 1970s, with advanced development continuing through 1976 when the U.S. Army awarded an engineering contract to Rockwell International.[7]

The first guided launch took place from an AH-1G Cobra helicopter in 1978, and later in 1979 from an AH-64 Apache. In 1982, Rockwell received the contract to produce HELLFIRE launchers and missiles; the HELLFIRE missile system entered service in 1985. The first three generations were laser guided, while the fourth generation "Longbow" HELLFIRE uses a radar frequency seeker.[8]

The current supplier of BTTN to DoD is Copperhead Chemical Company, located in Tamaqua, Pennsylvania. Copperhead does not manufacture the chemical precursor BT, but purchases it from another supplier. Until 2004, Cytec Industries (headquartered in Woodland Park, New Jersey) produced BT and provided it to Copperhead. When Cytec discontinued production of BT, Copperhead purchased its remaining stockpile.[9]

Due to continued high demand for BTTN, DoD joined with Copperhead in 2007 to locate another supplier of BT. The only source they identified that could produce the quantity and quality needed was located in China. This realization prompted the U.S. Army to look for domestic BT sources. The U.S. companies ATK-Radford Army Ammunition Plant, AFID Therapeutics, and BAE-Holston Army Ammunition Plant are being considered as potential suppliers of BT to Copperhead.[10]

The HELLFIRE missile and by extension BT are important for current and future U.S. combat operations. The U.S. military spent up to $1 billion in contracts for the upgraded HELLFIRE II missiles from 2008 to 2011. It is considered to be the U.S.'s standard anti-armor missile.[11]

The HELLFIRE II missile has six different variants. Each variant has a specific purpose, and is effective at achieving its mission goals, which explains why they have become the U.S. military's standard AGM and a part of so many different launching platforms. Given its versatility, however, the R type has been designated as the mainstay of the future HELLFIRE fleet, as explained by a U.S. Army representative:

"One of the most noticeable operational enhancements in the AGM-114R [HELLFIRE] missile is that the pilot can now select the [blast type] while on the move and without having to have a pre-set mission load prior to departure…. This is a big deal in insurgency warfare, as witnessed in Afghanistan where the Taliban are fighting in the open and simultaneously planning their next attacks amongst the local populace using fixed structure facilities to screen their presence."[12]

On March 28, 2011, Lockheed Martin announced that the sixth and final proof-of-concept test for the new AGM-114R HELLFIRE II missile was concluded successfully at Eglin Air Force Base in Florida. The test used ground launch lock-on after launch mode at a standoff distance of 2.5 kilometers. "The AGM-114R baseline design is now defined and allows us to go into system qualification…The R model remains on cost and on schedule, and meets all performance objectives,"[13] said U.S. Army Lieutenant Colonel Mike Brown, HELLFIRE Systems product manager at the Army's Joint Attack Munition Systems project office.

There have been discussions and preliminary bids for a new weapon system entitled the Joint Air-Ground Missile (JAGM) program. However, contractors are still in the early testing phases and the program is years away from deployment, assuming it is successfully developed.[14] Even if JAGM did replace the HELLFIRE missile system, it would still most likely use a similar SRM, requiring BTTN (which, again, requires BT). In sum, the HELLFIRE missile system is one of the most important and widely used weapons in the U.S. arsenal. Its production is unlikely to be discontinued anytime soon, and any future replacement will also likely use BT as a key component of its propellant.

STANDARDS AND ALTERNATIVES

A 1982 report by the Naval Surface Weapons Center in Dahlgren, Virginia, outlined the benefits of BTTN over NG when used as an SRM propellant. Key among BTTN's advantages are: lower freezing point, meaning the propellant is less likely to crack; lower volatility (it is six times less volatile than NG); and greater thermal stability. While BTTN is not the only NG substitute with these properties, BTTN is substantially more energetic, and is therefore preferred as a propellant.[15]

However, BT was found to be considerably more expensive to produce than NG, at $25 a pound, compared to NG's $1.50 a pound. Because of this prohibitive production cost, U.S. commercial producers are not interested in manufacturing BT at levels less than one million pounds per year due to economies of scale, as this was the amount of production where they would begin to see a profit. The report also discovered that purity levels being produced were about 95 percent, and its contaminants varied in type and level from lot to lot. A consistent and high quality of BT is essential for safe nitration because the process can often be affected drastically by even trace contaminants.[16]

The 1982 report concluded:

"The cause of the BT supply problem can be attributed in part to the choice of manufacturing process which facilitates shifting production from one to another product with a minimum of equipment modification and down-time. Since the military demand for BT has been relatively low by industrial standards, it has not been cost-effective to optimize reaction conditions, evaluate alternate methods of synthesis, or improve the product purity. As a result, very little effort has been made in this direction."[17]

In 2011, the Under Secretary of Defense for Acquisition, Technology, and Logistics claimed that up to a million pounds of propellant would be needed each year, meaning that the required amount of BT would approach the amount that, 30 years ago, industry claimed necessary to produce to be commercially viable.

The one advantage of NG is its lower cost. Beyond NG, there appears to be no viable chemical substitute for BT. Research is underway that could lead to a breakthrough for BT production; its success would allow for a plentiful stockpile. This research comes in two variants: advanced chemistry and cutting-edge biochemistry.[18] The biochemical method, funded by the Office of Naval Research, and taking place at Michigan State University, uses microbial bacteria to produce the BT.[19] Although promising, successful production of BT from the biochemical process lies in the future, as a patent for the bio-chemical

process was only filed in 2011. The standard chemical manufacture of BT is still the easiest method of production.

VULNERABILITIES IN THE HELLFIRE MISSILE PROPELLANT SUPPLY CHAIN

As identified earlier in the chapter, the sole known U.S. producer of BT, Cytec Industries, stopped BT production in 2004. Copperhead Chemical Company (which uses BT to create BTTN) and the Army were able to locate just one other supplier.

"Only one source was identified that could produce at the quantities and quality required, Shanghai Fuda Fine Chemicals[20] located in China. Section 1211 of the National Defense Authorization Act of 2006 has a prohibition on U.S. companies buying items listed on the U.S. Munitions List from 'Communist Chinese military companies.' Because Shanghai Fuda Fine is part of the defense industrial base of the People's Republic of China, it is a prohibited source."[21]

While the level of the current U.S. stockpile of BT is unclear, at one point in 2008 it was estimated there was only an 18-month supply left.[22] In November 2008, the Secretary of the Army approved a one-time waiver to allow the Army to buy BT from China. A later estimate took into account a stockpile of BT that the Naval Surface Warfare Center (NSWC) Indian Head Division had procured but not used; their stockpile indicated that Copperhead would have enough BT to produce BTTN through March 2010.[23]

After this supply of BT is exhausted, it is unclear where Copperhead will acquire more; another waiver may be necessary. However, a waiver to acquire more BT from Shanghai Fuda Fine would only be a stopgap measure—and one that highlights a potentially disastrous situation: DoD is currently dependent on a Chinese company for an important defense material.

As of 2008 this view did not appear to be shared by top DoD officials, as evidenced by a Department of the Army official's statement that the BT/China issue is "very minor."[24] The cause of this disconnect appears to lie in the Pentagon's present level of understanding of the lower levels of the supply chain. DoD has not systematically tracked defense industrial base supply chains below the third tier. Additionally, below the first tier, DoD and many defense contractors often do not realize the source of subcomponents.[25] Perhaps most disconcertingly, it appears that the U.S. Army and DoD failed to examine the BT and BTTN supply chain situation until prompted by Copperhead.

MITIGATING THE RISKS

Develop a greater understanding of DoD's lower tier supply chain links to mitigate "unexpected" critical shortages. DoD's Sector-by-Sector, Tier-by-Tier (S2T2) effort to map supply chains will help to identify instances of over-reliance on foreign suppliers and "single points of failure."[26] However, supply chain mapping will not be enough. The DoD must also be determined to address problems in the lower tiers of the supply chain and must prioritize efforts to address supply chain risks. Effective collaboration between industry and DoD is essential to solving the risks to the HELLFIRE missile propellant supply chain.

Continue public funding to develop a supply chain for BT production. DoD should invest more research dollars in developing better production methods of BT, and continue to fund ATK-Radford Army Ammunition Plant, AFID Therapeutics, and BAE-Holston Army Ammunition Plant programs to create a domestic supply of BT. While it may not be the cheapest short-term option, it is important to build up a robust domestic supply chain to reduce the risk of supply chain disruption. In this vein, even though Copperhead Chemical Company does not appear to be in danger of closing, the DoD should research alternative manufacturers of BTTN, as well as take steps to ensure the viability of Copperhead, so they do not abruptly find themselves in short supply.

Continue public funding of innovative research into more cost-effective domestic BT production methods. In general, the U.S. Army, U.S. Navy, and DoD appear to understand the dilemma arising from the lack of a domestic BT supplier. DoD is offering funding in the form of a Small Business Innovative Research (SBIR) program to fund microbial research. In 2008, the Army solicited bids (through the Federal Business Opportunities program) from domestic manufacturers capable of making BT to the desired quality and quantity.[27] These efforts must be bolstered in order to spur development and innovation.

CONCLUSION

Once U.S. manufacturing capability is eroded, it is difficult to ramp up again. This challenge is illustrated by how long it has taken for a domestic BT supplier to become viable again after Cytec ceased production. It took DoD three years to realize that disruption to the HELLFIRE missile supply chain might be a problem; after nearly six years, the United States still lacks a viable domestic BT producer. The United States must prioritize the domestic production of this important defense materiel.

ENDNOTES

1   "AGM-114B/K/M HELLFIRE Missile," *United States Navy Fact File*. http://www.navy.mil/navydata/fact_display. asp?cid=2200&tid=400&ct=2.

2   Under Secretary of Defense for Acquisition, Technology, and Logistics, Office of Industrial Policy, "Report to Congress on the Solid Rocket Motor Industrial Base Sustainment and Implementation Plan – Redacted Version" (Washington, D.C., May 2011). http://www.acq. osd.mil/mibp/docs/Final_Redacted_SRM_Sustainment_ Plan_6-6-11.pdf.

3   Ibid., 4.

4   "AGM-114 HELLFIRE II Missile, United States of America," *Army-Technology.com*. http://www.army-technology.com/projects/hellfire-ii-missile/.

5   "AGM-114 HELLFIRE Missile," Boeing. http://www.boeing.com/history/bna/HELLFIRE.htm.

6   "AGM-114 HELLFIRE," History, Wars, Weapons (November 14, 2010). http://historywarsweapons.com/ agm-114-HELLFIRE/.

7   "AGM-114 HELLFIRE Missile," Boeing, http://www.boeing.com/history/bna/HELLFIRE.htm.

8   Ibid.

9   Office of Under Secretary of Defense Acquisition, Technology, and Logistics, Office of Industrial Policy, "SRM Industrial Capabilities Report to Congress – Redacted Version" (Washington, D.C., June 2009). http://www.acq.osd.mil/mibp/docs/srm_ind_cap_report-redacted_6-12-09.pdf.

10 Ibid.

11 "Up to $1B+ for HELLFIRE II Missiles," *Defense Industry Daily*, March 15, 2011. http://www.defenseindustrydaily. com/3567M-for-HELLFIRE-II-Missiles-05043/.

12 "Up to $1B+ for HELLFIRE II Missiles," *Defense Industry Daily*, March 15, 2011. http://www.defenseindustrydaily. com/3567M-for-HELLFIRE-II-Missiles-05043/.

13 Ibid.

14 "JAGM: Joint Common Missile Program Fired – But Not Forgotten," *Defense Industry Daily*, June 6, 2011. http:// www.defenseindustrydaily.com/joint-common-missile-program-fired-but-not-forgotten-0229/#jagm-milestones-army-contract-awards.

15 Frank J. Piscane, "1,2,4-Butanetriol: Analysis And Synthesis," Naval Surface Weapons Center-Dahlgren (NSWC TR 82-380) (July 20, 1983).

16 Ibid.

17 Ibid., 6.

18 B.S. Kwak et al., "Continuous Production Method of 1,2, 4-BUTANETRIOL," *WIPO Patent Application WO/2005/061424* (PCT/KR2004/003359), July 07, 2005. http://www.wipo.int/patentscope/search/en/ WO2005061424.

19 W. Niu et al., "Microbial Synthesis of the Energetic Material Precursor 1,2,4-Butanetriol," *Journal of the American Chemical Society* 125 (October 29, 2003), 12998-9. http://www.ncbi.nlm.nih.gov/ pubmed/14570452?dopt=Abstract.

K. Ruder, "Military Microbes: Scientists Recruit Bacteria for Missile Research," *Genome News Network,* January 23, 2004, http://www.genomenewsnetwork.org/ articles/2004/01/23/missile_microbe.php.

20 Shanghai Fuda Fine Chemicals has been renamed Shanhai Fuda Fine Materials.

21 Office of Under Secretary of Defense Acquisition, Technology, and Logistics, Office of Industrial Policy, "SRM Industrial Capabilities Report to Congress – Redacted Version" (Washington, D.C., June 2009), 44. http://www.acq.osd.mil/mibp/docs/srm_ind_cap_report-redacted_6-12-09.pdf.

22 J. Buchanan, "Mission Impossible," *Forward Online – Global Perspective From MSCI* (January/February 2008). http://forward.msci.org/articles/?id=175.

23 Office of Under Secretary of Defense Acquisition, Technology, and Logistics, Office of Industrial Policy, "SRM Industrial Capabilities Report to Congress – Redacted Version" (Washington, D.C., June 2009). http:// www.acq.osd.mil/mibp/docs/srm_ind_cap_report-redacted_6-12-09.pdf.

24 J. Buchanan, "Mission Impossible," *Forward Online – Global Perspective From MSCI* (January/February 2008). http://forward.msci.org/articles/?id=175.

25 J. Buchanan, "Mission Impossible," *Forward Online – Global Perspective From MSCI* (January/February 2008). http://forward.msci.org/articles/?id=175.

26 Brett Lambert, "The Manufacturing and Industrial Base Policy" (Presentation to National Defense Industrial Association, Washington, D.C., August 15, 2011), Slide 4. http://www.ndia.org/Advocacy/Resources/ Documents/NDIA_S2T2_Briefing_AUG11.pdf.

27 Small Business Innovation Research, Interactive Technical Information System Archive, "Manufacture of Energetic Materials From Renewable Feedstocks Using Engineered Microbes" (SBIR: N06-078). http:// www.dodsbir.net/sitis/ archives_display_topic. asp?Bookmark=28847.

CHAPTER 11 • ADVANCED FABRICS

EXECUTIVE SUMMARY

Advanced fabrics are engineered to protect U.S. troops against the hazards of combat. They include flame-resistant fabric that shields troops against fire and explosive threats. Lightweight body armor made from para-aramid fibers can stop handgun rounds, with additional protection against fragmentation and larger ballistic projectiles offered by adding supplemental armor inserts. In short, advanced fabrics save lives and enhance battlefield effectiveness.

Advanced fabrics supply chains are at risk. Flame-resistant U.S. Army uniforms use flame-resistant (FR) rayon fibers manufactured by a single company in Austria. Additionally, in 2003, an unforeseen para-aramid fiber shortage, combined with unprecedented surge in demand for ballistic protection for vests and ground combat vehicles, slowed the deployment of Interceptor Body Armor vests. This delay forced more than 40,000 U.S. soldiers to operate with protection that used previous generation technology. DuPont is the sole U.S. producer of the para-aramid fiber Kevlar used to manufacture the Interceptor vest. A second Kevlar producer, Japan-based Teijin, operates production facilities in the Netherlands.[1]

The 1941 Berry Amendment requires the Department of Defense (DoD) to purchase textiles from domestic sources when available, but permanent waivers exist for both FR rayon and para-aramid fibers. There is no domestic FR rayon production, necessitating imports of foreign-produced FR rayon or the use of a domestic alternative. Although there is domestic production of para-aramid fibers, concerns over reliance on a single source and limited production capacity resulted in a waiver for those textiles to the Berry Amendment's domestic source restrictions.

The permanent Berry Amendment waivers permit continued dependence on foreign providers despite the existence of domestic alternatives and discourages the development of new fabrics and investment in domestic production capacity. DoD has identified a domestic alternative to the imported FR rayon fibers used in U.S. flame-resistant combat uniforms, but continues to purchase fabrics containing non-domestic FR rayon due in part to concerns relating to appearance. The permanent para-aramid waiver discourages U.S.

ADVANCED FABRICS
PROTECTING U.S. FORCES

MANUFACTURING SECURITY

Advanced fabrics protect U.S. warfighters from explosive and ballistic threats

UNDERSTANDING SUPPLY

U.S. production of flame-resistant rayon ended in 2005

0 DOMESTIC PRODUCTION FACILITIES

The U.S. is 100% dependent on one foreign source for FR rayon

 **AUSTRIA** SUPPLIES 100% OF **FR RAYON FIBER**

PRODUCTION SHORTFALLS

In December 2003 production shortfalls meant that 40,000 out of 130,000 U.S. soldiers in Iraq were without necessary armor

40,000 OUT OF 130,000 SOLDIERS **WITHOUT ARMOR**

DOMESTIC PRODUCTION

Waivers to import para-aramid fibers from foreign sources continue to negatively impact U.S. production

WAIVERS FOR PARA-ARAMIDS IMPACT **U.S. PRODUCTION**

PERFORMANCE ASPECTS

Synthetic fibers can be combined to create fabrics that are comfortable, lightweight, and protective

 ALLOWS AIR

 ALLOWS MOISTURE

 REPELS FIRE

MITIGATING RISKS

Mitigating risks to U.S. advanced fabrics supply

DEVELOP U.S. INDUSTRIAL CAPACITY   **REASSESSING BERRY AMDT WAIVERS**   **MAINTAIN EQUIPMENT RESERVE**

MILITARY EQUIPMENT CHART
SELECTED DEFENSE USES FOR ADVANCED FABRICS

DEPARTMENT	DEFENSE PRODUCTS
ARMY	■ Advanced Combat Helmet (ACH) ■ Tactical Communications Helmet (TCH) ■ DH-132 AS helmet ■ Special Operations Headset Adaptable Helmet (SOHAH) ■ PM High Altitude-Low Opening (HALO) and High Altitude-High Opening (HAHO) helmets ■ Interceptor Body Armor (IBA)
MARINE CORPS	■ Lightweight Helmet (LWH) ■ Interceptor Body Armor (IBA)
NAVY	■ DH-132 AS/RHIB helmet ■ MK-7 Navy Battle helmet ■ Tactical Communications Helmet (TCH) ■ Special Operations Headset Adaptable Helmet (SOHAH) ■ Phonetalker Type III helmet
AIR FORCE	■ Aircraft ballistic seat
DOD	■ Tent liners ■ Vehicle seats ■ Vehicle interior blast shielding systems ■ Spill liners ■ Blast curtains ■ Bomb disposal blankets ■ Bomb disposal protective gear

firms from investing in the expansion of domestic production capabilities. DuPont has invested over $500 million to increase Kevlar production since 2003, but did so only after witnessing several years of high and stable demand. The success of the Mine-Resistant Ambush-Protected (MRAP) vehicle program depended in part on preferential treatment of "rated" DoD Kevlar orders.

The best and most immediate way to strengthen the advanced fabrics sector in the United States is to reevaluate the wisdom of permanent waivers for key inputs. While temporary waivers occasionally may be necessary to ensure that U.S. troops have timely access to the best equipment, permanent waivers can institutionalize undue dependency on foreign sources and risk potential disruption of U.S. supply chains.

INTRODUCTION

A range of technologies make U.S. troops effective on the modern battlefield. Every piece of equipment worn by a U.S. service member is engineered to increase protection and maximize effectiveness. Advanced fabrics, a product of the U.S. textile industry, are an important but often overlooked part of the U.S. defense industrial base, and they provide critical protection against the hazards U.S. troops must confront.

At the onset of World War II, Japan was the world's largest exporter of silk, controlling roughly 80 percent of global production by the early 1930s. Silk is a strong, lightweight fiber and, although best known for its use in luxury clothing, served a vital role in the U.S. war effort as the primary material for parachutes. The onset

of hostilities with Japan demonstrated the dangers of single-source reliance for resources critical to the defense industrial base. U.S. access to Japanese silk, which had been severely limited during the 1930s, was finally cut off.

In 1939, the American firm DuPont introduced the synthetic Fiber 66, more commonly known as nylon. An alternative to silk, Fiber 66 originally was marketed as an affordable substitute for silk stockings, but by 1942 it was declared a defense critical product. Domestic nylon production was redirected towards the production of parachutes and rope, both of which were in short supply due to the unavailability of Japanese silk.[2] Without the successful substitution of silk with nylon, airborne troops who spearheaded the D-Day invasion of Normandy would not have been able to land by parachute, jeopardizing this pivotal operation in World War II. Not only was the development of nylon important to the U.S. war effort during World War II, but it also demonstrated the important role that synthetic fibers could serve in reducing U.S. dependence on foreign-produced defense products.

Today, synthetic fibers have replaced naturally occurring fibers (such as silk and cotton) in virtually all military textiles. Synthetic fibers can be engineered to possess superior strength, durability, and flame and heat resistance at lower weight when compared to the natural fibers they are designed to replace. Nylon remains the standard material used in parachutes[3] and is the most common material used for the mooring and towing of naval vessels.[4] More significant is the use of advanced fabrics in uniforms issued to U.S. troops to add resistances to fire, abrasions, hazardous materials, and even bullets and other ballistic threats, while maintaining the comfort and functionality of the natural fibers they replace.

Over the last decade there has been renewed attention to military fabrics due to U.S. troops' increased exposure to threats posed by long-term combat deployments. Insurgent use of improvised explosive devices (IEDs) has resulted in hundreds of deaths and injuries to U.S. and coalition forces.[5] In 2010, the deadliest year in Afghanistan, IEDs were responsible for nearly 60 percent of all U.S. fatalities.[6] IEDs are increasingly being used in other conflict zones around the world.[7]

Responding to the growing IED threat, in 2007 the U.S. military introduced uniforms made from Defender M,[8] a flame-resistant fabric composed primarily of FR rayon, to troops serving in the field. When exposed to flame, Defender M is designed to self-extinguish, significantly reducing burns. Use of these uniforms reduced U.S. casualties and injuries in Iraq and Afghanistan.[9]

The operations in Afghanistan and Iraq also increased the need to adopt more effective anti-ballistic protection for U.S. troops on the frontline. By 2005, the Army and Marine Corps transitioned from the Personnel Armor System for Ground Troops (PASGT) flak vest to Interceptor Body Armor. A significant upgrade over its predecessor, the Interceptor System uses a multi-layer Kevlar vest capable of stopping small arms fire. The Interceptor System also offers additional protection and greatly reduces casualties, through the incorporation of Small Arms Protective Insert (SAPI) boron-carbide ceramic plates that are capable of stopping 7.62mm assault rifle bullets. According to a 2005 Marine Corps statement, following adoption of the Interceptor system by the Marine Corps, only five percent of casualties resulted from torso wounds, demonstrating the dramatic

improvement the Interceptor system made over PASGT.[10]

This chapter focuses on advanced fabrics' important defense applications, with a specific focus on FR rayon (standard in ground combat uniforms) and ultra-strong protective fabrics made with Kevlar, Twaron,[11] Spectra,[12] and Dyneema.[13] Textiles for defense applications, including synthetic fibers, have long been protected by domestic preference legislation (the 1941 Berry Amendment). Despite this protection, advanced fabrics supply chains still face significant risks primarily related to limited production capacity by a small number of producers, including a single foreign company being the sole producer of FR rayon used in flame-resistant uniforms.

Figure 1: Desired Performance Characteristics for Flame-Resistant Uniforms

AIR — ALLOWED

WATER — ALLOWED

FIRE — REPELLED

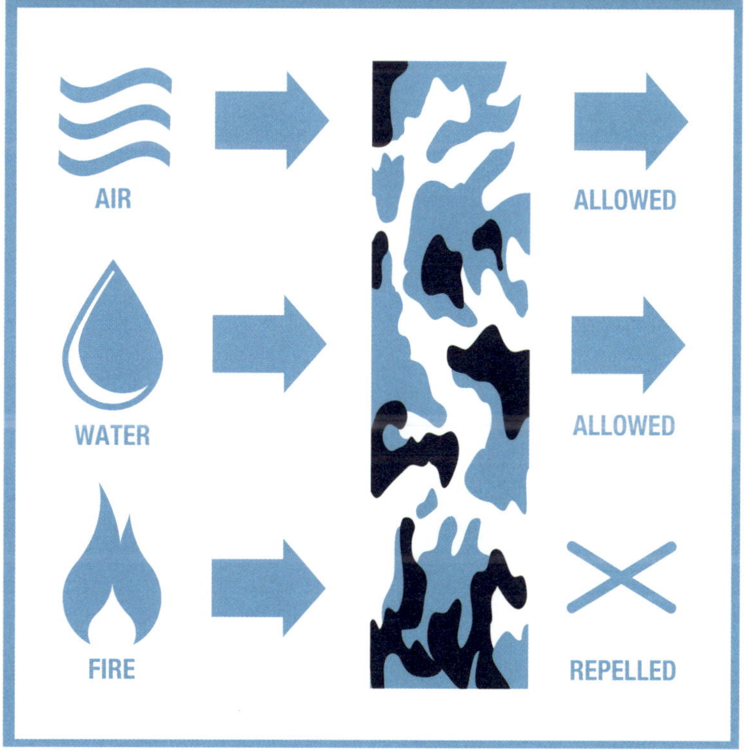

Key themes discussed in this chapter are:

- Advanced fabrics used in military uniforms are an essential form of protection for U.S. soldiers, and greatly reduce casualties from battlefield hazards.

- Advanced fabrics supply chains are at risk and may be interrupted or delayed due to foreign source reliance, changes in market conditions, or insufficient production capacity.

- The fabric used in the Flame-Resistant Army Combat Uniform (FR-ACU) is imported under a permanent waiver to the Berry Amendment. Entirely domestic alternatives exist, but have not been not adopted because the waiver permits the continued acquisition of foreign-produced FR rayon.

- A Berry Amendment waiver for the para-aramid fibers used to make ballistic fabrics was established in 1999, and is now permanent. Since then, domestic production capabilities have expanded significantly, and may now be adequate to meet full DoD demand. However, because the waiver is permanent, there is no requirement to reassess domestic production capability or revisit the need for a waiver. This fact puts domestic para-aramids at a disadvantage and reduces the incentive for further investment in new production capacity.

A NOTE ON CRITICALITY

U.S. soldiers operate in hazardous environments. Although the 2003 shortage of the para-aramid fiber needed to manufacture body armor did not prevent U.S. soldiers from carrying out their missions, it did expose them to increased risk of injury or death. In light of these and other considerations, discussed in more depth below, we assess the impact of an advanced fabric/para-aramid shortage as *isolated*, reflecting the role of military textiles in enhancing the effectiveness of the U.S. warfighter by

WHAT IF TROOPS LACK PROTECTION AGAINST UNANTICIPATED THREATS? (a notional but realistic scenario)

Early in the morning, grenades explode at Camp Lemonnier in the capital of the East African country of Djibouti, killing two and wounding 12 other U.S. service members. Incidents of this nature are unusual given the close relationship between the United States and Djibouti. Many believe that the al-Qaeda-affiliated group al-Shebaab is responsible for this attack.

The attack occurs within the sleeping area for transient troops. People familiar with the camp later express concern that those killed and wounded were in tents lined with anti-blast curtains that apparently failed. The U.S. Department of Defense Inspector General already had been pursuing an investigation into allegations that unapproved textile mills in Southeast Asia had been subcontracted to meet supply demands.

increasing their resistance to many battle-field threats.

The likelihood of an additional protective fabric shortage rates is *high.* Currently, there are several global producers of para-aramid fibers, the principle fiber used in anti-ballistic body armor, with only one producer (DuPont) located in the United States. Furthermore, over the past decade military and commercial demand for para-aramid fibers has increased dramatically, leading to uncertainty over the global production's ability to fulfill that increasing demand. In response to these supply concerns, DuPont recently constructed a new production facility, ultimately capable of expanding production capacity of para-aramid fabric by up to 40 percent. This expanded production may be adequate to downgrade the risk to *low.*

FLAME-RESISTANT FABRICS FOR UNIFORMS

In response to the growing IED threat that exposes military personnel to injury from the initial explosion as well as secondary burns, the U.S. military redesigned its combat uniforms to make use of flame-resistant fabrics. These fabrics have physical properties that prevent, terminate, and/or inhibit ignition. The fabric's production requires fibers that do not melt, and that retain form when exposed to extreme heat. Additionally, to resist ignition, fibers should have a limiting oxygen index (LOI) (the percentage of oxygen needed in surrounding gasses to fuel combustion) of at least 25. Air naturally contains approximately 21 percent oxygen, and an LOI above 25 is considered flame-resistant. [14] In addition, flame- resistant uniforms must meet a range of performance characteristics

including: comfort, breathability, the ability to be dyed, similarity in appearance and texture to non-flame resistant uniforms already in use, durability, and the ability to retain flame resistance throughout the life of the fabric.[15]

In 2006, the Army evaluated 18 potential fabrics before selecting Defender M fabric for its FR-ACU. Defender M fabric is a blend of 65 percent FR rayon, 25 percent para-aramid (Twaron) to add strength, and 10 percent nylon— Defender M is manufactured in the United States by the U.S. subsidiary of TenCate, a multinational textile company headquartered in the Netherlands. The U.S. Marine Corps also selected Defender M. The Army and Air Force jointly attempted to identify alternatives to FR rayon-based Defender M fabric in 2007. Milliken's Abrams V[16] fabric (made from flame resistant Nomex fibers blended with chemically treated, flame retardant cotton) is made from 100 percent domestic U.S. fibers, and was recommended as a potential alternative to Defender M. The Army continued to prefer Defender M-based uniforms. Although the Air Force agreed to use the Abrams V fabric instead of Defender M, this decision resulted from concerns that the FR rayon supply would not be able to meet surge demand for uniforms made with Defender M.[17] Defender M is also used by defense forces in countries such as Australia, Italy, and Norway.[18]

Although the Defender M fabric is woven in the United States, two of the three fibers used in the Defender M blend (FR rayon and the para-aramid Twaron) are produced overseas. Waivers circumventing the Berry Amendment's domestic source requirement currently exist for both fibers.

In 1999, the Under Secretary of Defense for Acquisitions, Technology, and Logistics issued a Domestic Non-Availability

Determination (DNAD) in response to concerns over the supply of domestically produced para-aramid fiber Kevlar. The DNAD allowed DoD acquisition of products containing para-aramid fibers produced in the Netherlands (see section on para-aramid fibers and yarns).[19] In contrast, although rayon initially was developed by DuPont, in 1924,[20] domestic production of rayon was gradually phased out. In 1989, Avtex Fibers, at one point the world's largest rayon producer, ceased production due to environmental concerns.[21] In 2000, the North American Rayon Corporation's Elizabethton, Tennessee, plant burned down in a fire that took over a week to extinguish.[22] According to the United States International Trade Commission, the last remaining U.S. rayon producer closed its doors in 2005.[23] A DNAD was issued in 2001 for the procurement of rayon yarn for use in military clothing. A waiver included in the FY2008 National Defense Authorization Act (NDAA) allowed for the purchase of fabrics containing FR rayon fibers produced in Austria due to the heightened military demand for flame resistant uniforms and the U.S. Army's selection of Defender M fabric for its flame-resistant combat uniform. Lenzing, an Austrian company that is the sole producer of weavable FR rayon fibers, expanded its domestic production facility in response to the waiver.[24] The FY2012 NDAA made the FR rayon fiber waiver permanent.

PRODUCTION OF FLAME-RESISTANT UNIFORMS

FR rayon is the primary raw material used to manufacture the fabric for the flame-resistant combat uniforms most commonly issued to deployed troops. Rayon is a semi-synthetic fiber manufactured from a natural cellulose polymer derived from wood pulp. The cellulose undergoes a series of chemical treatments that convert it into a soluble compound that then is forced through a spinneret to create filaments. These wet filaments then are stretched to align the polymer chains along the fiber axis, and are chemically reconstituted to create rayon fibers.[25] The chemical alterations the cellulose undergoes during production render it permanently resistant to fire, because they incorporate a flame-resistant substance into the cross-section of the fiber.[26] These fibers are in many ways similar to unprocessed cotton, and can be spun into yarn and later woven into fabrics.

There is no domestic production of flame-resistant rayon fibers. Domestic firms do not seem interested in restarting domestic production, citing the costs of maintaining an environmentally compliant plant, the uncertainty of DoD demand, the lack of significant commercial demand, and the general economic feasibility of recuperating startup costs.

Lenzing produces all FR rayon fibers used in U.S. military uniforms in its facility in Austria. Although several other countries (including Finland, China, and Japan) produce FR rayon fibers, these fibers are not suitable for woven products such as fabric and are instead used to add flame-resistance to mattresses and other products.[27] There is no U.S. domestic production of FR rayon fibers, nor do U.S. firms seem interested in restarting domestic production. Firms cite the costs of maintaining an environmentally compliant plant, uncertainty over DoD demand, the lack of significant commercial demand, and the general economic feasibility of recuperating startup costs. However, much of the machinery used to produce rayon fibers is not specific to rayon, and may be used to manufacture other fibers.[28] Additionally, similar environmental regulations exist in Austria, where Lenzing nevertheless has found it profitable to expand production of FR rayon in response to increasing DoD demand.[29] Because Lenzing appears able to meet global demand for FR rayon fibers, a low-price commodity with low commercial demand, domestic producers currently do not have an economic incentive to begin domestic production, especially in light of the permanent Berry Amendment waiver.

TenCate, the manufacturer of the Defender M fabric used in flame-resistant uniforms, imports the raw FR rayon fibers in bulk. To create the Defender M fabric the FR rayon's fibers are then blended with para-aramid and nylon fibers and spun into yarn. The addition of para-aramid and nylon fibers greatly improves the durability of the fabric. The fabric is then dyed and printed (a requirement for military fabric) and delivered to the garment manufacturer who fabricates the final uniforms ordered by the Defense Logistics Agency (DLA), the agency within DoD responsible for acquiring and distributing military uniforms.

Apart from the initial production of the fibers, the remainder of the production chain is located in the United States. According to correspondences addressed to members of the Senate and House Armed Services Committees from the Textile Manufacturers for Rayon and bearing signatures of 24 U.S. textile companies, 90 percent of the value-added content of flame-resistant military garments originates domestically and FR rayon

Both the U.S. Army's and U.S. Marine Corps' flame-resistant uniforms are made with TenCate's Defender M fabric. Although the fabric is manufactured within the United States, it relies on imported fibers for two of its three main components.

imports support approximately 10,000 U.S. jobs.[30]

ALTERNATIVES TO FR RAYON

Testing potential fabrics for flame-resistant military uniforms is a complex and time-consuming process. At the Natick Soldier Research Development and Engineering Center, mannequins outfitted with different uniforms are exposed to four-second blowtorch blasts to simulate an IED explosion, while sensors record where and to what extent the soldier would experience burns. Using the FR-ACU made from the Defender M fabric, such tests showed that a soldier would sustain burns on 29 percent of his or her body, a marked improvement over a non-flame-resistant combat uniform, yet obviously with room for improvement.[31]

Since 2006, numerous efforts to test and evaluate flame-resistant materials for use in uniforms have been conducted independently and jointly by the U.S. Army, Marine Corps, and Air Force. While many fabrics (including some that are produced entirely from domestically sourced materials) meet or exceed military flame-resistance requirements, military fabrics must also fulfill other requirements, including durability, suitability to a variety of environments, comfort, ability to be dyed, and cost. In nearly all cases, submitted fabrics are blends of various fibers such as FR rayon, para-aramid (Kevlar and Twaron), chemically treated cotton, Nomex, and polybenzimidazole (PBI). According to a 2011 Government Accountability Office (GAO) report, over half of the flame-resistant fabrics tested by the U.S. Army at the time contained FR rayon.[32]

Blending fibers allows desirable traits to be combined. For example, FR rayon is readily dyed and absorbs moisture better than cotton, but on its own lacks the durability necessary for military use. On the other hand, para-aramids are incredibly strong and fire-resistant, but are much more difficult to color. Fabrics created by blending these fibers assume the strength of the para-aramid coupled with the absorbency of the rayon, overcoming the limitations of the individual fibers.

Nomex is a synthetic meta-aramid fiber manufactured by DuPont and most commonly known for its use in firefighter equipment. Nomex also has been used for military flight suits and combat vehicle uniforms. The fiber possesses an LOI of 26-38, meaning that it will not ignite until the atmospheric oxygen reaches 28 percent, well above the 21 percent oxygen content in air.[33] The Abrams V fabric approved for use in flame-resistant uniforms by the Air Force is comprised of Nomex blended with chemically treated, flame-resistant cotton. With the recognition of the long lead-time associated with Nomex production, DLA's "Warstopper" program invested in a strategic buffer stock of Nomex fibers in 2009.[34] In addition to long lead-times previously associated with Nomex, its price has been prohibitive for widespread military use as the sole fiber in a fabric, but it provides a strong backbone for flame resistance and durability in multi-component blends. The Abrams V fabric blends Nomex fibers with flame-resistant cotton, which possesses an LOI between 28 and 30.[35] To make the fiber inherently flame-resistant, cotton can either be topically treated with a flame-resistant agent,[36] or bonded with a flame-resistant polymer.[37] Despite obtaining similar levels of flame-resistance, FR cotton may shrink when laundered, react to chlorine bleach, and will fade over time. If the FR cotton is topically treated with chemicals to add flame-resistant qualities, these qualities will diminish with use and laundering.

Para-aramid fibers such as Dupont's Kevlar or Teijin's Twaron (produced in the Netherlands) possess similar flame-resistance to Nomex (an LOI between 28 and 29); however, they are up to seven times stronger than Nomex by weight. Kevlar and Twaron are more difficult to dye. PBI fiber neither ignites nor melts (its LOI is 41); however, it is a relatively weak fiber that is often blended with aramid fibers to increase its strength.[38] PBI fibers are produced in the United States and incorporated into military applications such as headgear, in addition to civilian usage as firefighting equipment (a 60/40 Kevlar/PBI blend). PBI is gold in color, and difficult to dye, limiting its potential use for some military uniforms.

The sheer number of flame-resistant fabrics submitted for testing since the rise of the IED threat indicates that a variety of fiber blends are available, including those that contain 100 percent domestic U.S. fibers. However, because the Berry Amendment waiver for FR rayon is now permanent, the military has no obligation, and little incentive, to adopt a domestic-sourced fabric, even if a suitable domestic substitute fabric already has

> The domestic non-availability of flame-resistant rayon is not the result of U.S. inability to produce rayon; rather, it is the result of minimal economic incentives to invest in the infrastructure necessary to reestablish domestic production capability.

been identified. Because the military may continue to purchase fabrics containing foreign fibers indefinitely, the U.S. defense industrial base has minimal incentive to develop new, domestically sourced fabrics for flame-resistant uniforms.

BALLISTIC FABRICS AND BODY ARMOR

The Interceptor Outer Tactical Vest (OTV) is a multi-layered Kevlar containing vest designed to replace the PASGT flak jacket. Unlike the PASGT, which was designed to stop shrapnel and other fragmentation projectiles, the OTV system is capable of stopping 9mm rounds, and 7.62 and 5.56 mm rifle rounds when supplemental ceramic plate inserts are used.[39] Weighing only 16.4 pounds, the OTV is also significantly lighter than the 25.1-pound PASGT, dramatically reducing the load troops must carry on the battlefield.

In 2003, when only part of a unit had received modern OTVs, U.S. troops often adopted a swap and share approach, rotating who wore the OTV and who wore outdated PASGT Flak vests. Other troops reported taping SAPI plates to their outdated PASGT flak jackets, while some scavenged for plates Iraqi soldiers had discarded, which were of inferior quality, often damaged, and did not properly fit U.S. issued vests.[40] In some cases, family and friends of deployed troops pooled money to purchase Interceptor OTVs as well as SAPI plates, costing upwards of $1,600.[41] The Congressional Budget Office estimated that up to 10,000 vests and corresponding ceramic inserts had been purchased privately for soldiers not adequately outfitted by the military services.[42] To partially rectify this shortcoming, the FY2005 NDAA authorized the reimbursement of up to $1,100 for protective armor bought by or for any deployed warfighter who was not issued adequate equipment. By January 2004, Interceptor body armor had been issued to all troops deployed in Iraq.[43]

The transition from the PASGT flak vest to the Interceptor armor system has greatly increased the survival rate of U.S. troops from small arms fire. The combination of the OTV with SAPI plate inserts has reduced the casualty rate associated with torso wounds to only five percent during Operations Iraqi Freedom and Enduring Freedom, compared to fatality rates of 33 percent for abdominal wounds and 70 percent for chest wounds during the Vietnam War.[44] In one case, a soldier sustained minimal injury despite being shot four

> In December 2003, the Pentagon confirmed that 40,000 of 130,000 soldiers deployed in Iraq were lacking either the Interceptor Outer Tactical Vest or the ceramic Small Arms Protective Inserts plates, which added supplemental protection from rifle rounds.

times in the abdomen by rounds from an AK-47. In another case, an Army Specialist was shot in the stomach by rifle fire, detonating three ammunition clips and a smoke grenade on his person, but his vest stopped both the bullet and subsequent explosion.[45]

In 2005, GAO conducted an investigation into the 2003 shortage of Interceptor tactical vests and SAPI plates. The GAO concluded that the shortage was caused by a lack of adequate production capacity in conjunction with the fact that the OTVs and SAPI plates would be given to all select military personnel to all military and civilian personnel, not just select military personnel. The expanded (quarterly) demand for Interceptor OTVs rose from 8,593 vests in December 2002 to 77,052 vests in March 2003. Quarterly demand rose to 210,783 vests in December 2003, at which point GAO reports that monthly production was only 23,900 vests. This lack of supply resulted in significant backorders. The U.S. military experienced similar shortages in SAPI plates, with demand increasing by a factor of 10 during first quarter 2003, peaking at 478,541 in December 2003, and representing a 50-fold increase in one year's time. Production was able to meet only a small portion of this demand, with monthly output reported at 40,495 SAPI plates in December 2003.[46]

According to the GAO report, insufficient supplies of the fabrics Kevlar and Spectra Shield[47] limited production of both the vest and plates. Manufactured by DuPont, Kevlar is a para-aramid fiber that is roughly five times stronger than steel fiber on an equal weight basis. SpectraShield is made from Spectra, a polyethelene fiber also many times stronger than steel by weight. SpectraShield creates a mesh backing for the SAPI ceramic tiles capable

of absorbing the force of the projectile.[48] While synthetic fibers, including Kevlar and Spectra, must be purchased domestically under the Berry Amendment (10 U.S.C. 2533a), the FY1999 NDAA included a waiver for para-aramid fibers. The waiver was issued out of concern that reliance on a sole supplier (DuPont) would result in conditions not in the best interest of the United States. Although the legislation enabled the acquisition of products containing the para-aramid Twaron, manufactured in the Netherlands, the waiver was still insufficient to avoid the para-aramid shortage in 2003 that greatly reduced delivery of life-saving Interceptor body armor to personnel serving in Iraq and Afghanistan.

BALLISTIC FIBERS

There are several types of fibers that are stronger than steel and suitable for military use. These fibers include para-aramids (Kevlar and Twaron), ultra-high-molecular-weight polyethylene (UHMWPE, including Spectra and Dyneema), poly para-phenylene benzobisoxazole (PBO, also known as Zylon),[49] and polypropylene (PP, including Innegra[50]). These fibers are synthetic polymers, which are large molecular chains that consist of one or more repeated chemical structures (monomers) held together by covalent bonds. The order and configuration of the monomers influence the polymer's properties, including strength, melting and boiling points, durability, and chemical resistance. Polymers may be engineered at the molecular level to possess performance characteristics exceeding those found in naturally occurring substances. Each of these fibers possesses high tensile strength (tension the material can withstand while maintaining its

PARA-ARAMID

Aramid fibers, possessing excellent fire-resistant and strength-to-weight qualities, are widely used for ballistic protection. Para-aramid fibers, a type of aramid fiber with increased strength-to-weight characteristics compared to the aramid fibers from which they are derived, are synthetic fibers that possess high tensile strength and low elasticity despite being relatively easy to weave. Para-aramids are highly stable, and maintain their physical properties when exposed to flame, extreme heat, and many chemicals. Para-aramid fibers are about five times stronger and 2.75 to 3.85 times more stretch-resistant than steel on an equal weight basis.[51] Developed in the 1960s by DuPont and released commercially in 1971, Kevlar is the best-known para-aramid fiber. Originally, Kevlar was developed as a replacement for steel in tires.[52] Twaron, a para-aramid fiber almost identical to Kevlar, was released in the 1980s by the AKZO Company in the Netherlands. AKZO sold its industrial fibers division to the Japanese firm Teijin in 2000. Para-aramids are inherently flame-resistant and are blended with FR rayon and nylon to create the FR-ACU. Since the 1991 Gulf War, all U.S. military personnel have been issued a helmet containing para-aramid fibers.[53] The Interceptor OTV is fabricated from multiple layers of high strength para-aramid fabric that stops handgun bullets by the second or third fabric layer, diffusing the force of the impact among the remaining layers of the vest.[54] Para-aramids also are used in a variety of civilian applications: they can be wrapped around support structures to increase weight-bearing capacity; they are common in sports equipment, including helmets and hockey sticks; and they are also used in protective coatings for fiber-optic cables.[55] Production of para-aramid fibers is capital intensive. For example, DuPont estimates that it had invested well over $1 billion in research and development (R&D) and infrastructure prior to the wars in Afghanistan and Iraq,[56] and at least another $550 million in the past decade.[57] Para-aramid fibers are created by dissolving chemical monomers in a solvent that fuses the monomers, creating tangled and suspended polymer chains. The excellent strength-to-weight ratio of the para-aramid fiber comes from the polymers' parallel reorientation, a result of a process that stretches the fiber's polymer chains and causes a crystalline structure to form within the fiber.[58]

ULTRA-HIGH-MOLECULAR-WEIGHT POLYETHYLENE (UHMWPE)

UHMWPE fibers are more commonly known by the brand names Spectra and Dyneema. They are stronger, more rigid, and lighter than para-aramid fibers, and possess better resistance to acids. However, they are not nearly as heat resistant. Unlike para-aramid fibers, which are thermally stable above temperatures of 500°C, polyethylene-based fibers are flammable in air (an LOI below 20) and will melt at about 155°C.[59] Polyethylene-based fibers also are difficult to dye. Due to their vulnerability to heat and flame, UHMWPE fibers are not suitable for use as primary materials in military apparel. They are generally reserved for applications that do not require a high level of flame protection.

Until 2004, Honeywell's Spectra was the only UHMWPE manufactured in the United States, and its main military use was as a fabric mesh backing for ceramic armor, including the SAPI inserts used in the Interceptor armor system. (The ceramic plate is lined with a layer of high-strength fabric that absorbs the force of the projectile after the ceramic plate erodes the projectile.[60]) Due to its advantageous strength-to-weight ratio, UHMWPE is preferred over para-aramid-based fabrics for use with the Interceptor SAPI inserts. When the supply of SpectraShield limited the production of SAPI tiles, Kevlar and Twaron were temporarily used to allow continued fabrication of the tiles. However, inserts using para-aramid weighed about half a pound more than those made from UHMWPE fibers.[61] In 2004, DSM Dyneema (Netherlands) opened a U.S. plant in Greenville, North Carolina, to supply increasing U.S. demand (especially from the U.S. military) for UHMWPE fibers.[62]

HIGH MODULUS POLYPROPYLENE (HMPP)

High modulus polypropylene fibers are the lightest available structural fiber, and are increasingly blended with para-aramid fibers to reduce the weight and costs associated with fabric.[63] The HMPP fiber Innegra S was developed by Innegra Technologies (formerly Innegrity), a South Carolina-based advanced materials company, and has only recently become commercially available.[64] Innegra HMPP fiber was designed to be blended with other fibers. Using para-aramid fiber as a baseline for comparison, the Innegra S fiber is 60 percent the weight, but only about one-quarter the strength and one-tenth the rigidity. Because HMPP fibers are thermoplastic, they will melt and are much less heat-resistant than para-aramid fibers, and are roughly equivalent to UHMWPE fibers. [65] The advantage of the HMPP fiber is that it is less brittle than aramid fibers, absorbing about twice the energy prior to snapping.[66] While not independently suitable as ballistic fibers, a 50/50 Kevlar/Innegra S blend performed at 97 percent

the effectiveness of a 100 percent aramid fabric at similar weight,[67] but costs about 35 percent less.[68] While not a replacement for para-aramid, HMPP fibers may significantly reduce the costs of ballistic fabrics while also partially mitigating the risk of another para-aramid shortage.

POLY PARA-PHENYLENE BENZOBISOXAZOLE (PBO)

PBO fibers are manufactured under the trade name Zylon, by the Japanese company Toyobo. These fibers are at least 1.5 times stronger than aramid fibers, but only slightly heavier.[69] PBO fibers remain stable in extremely high temperatures, and like para-aramids are resistant to chemical exposure and abrasion.[70] The higher strength-to-weight ratio means that ballistic vests made from Zylon have equal stopping power with about half the thickness of para-aramid vests. However, vests made from PBO fibers are also several times more costly than those made from para-aramid.[71]

In 2003, a police officer was critically wounded when his Zylon vest failed to halt a bullet it was designed to stop. This incident raised concerns over the long-term durability of PBO fibers and caused a recall of ballistic vests containing Zylon. Subsequent investigation revealed that PBO fibers can lose tensile strength after prolonged exposure to humidity, temperature, and ultraviolet light.[72] According to the manufacturer, PBO fibers may experience 15 percent degradation in strength when exposed to temperatures of about 100°F at 80 percent relative humidity for just 150 days.[73] In 2005, after the vest recall and in light of the vest's rapid degradation, Zylon manufacturer Toyobo reached a $29 million settlement to replace ballistic vests containing the fibers.[74] Despite concerns over the shelf life of PBO-based vests, they may be an appropriate for short-term use, especially when the thickness of the vest is a concern.

VULNERABILITIES TO THE BALLISTIC FABRIC SUPPLY CHAIN

The underlying cause of the 2003 shortage in Interceptor OTVs was the lack of sufficient quantities of ballistic fabrics needed to manufacture the vest and SAPI ceramic plates. Surging demand, combined with an immediate need for ballistic protection for ground combat vehicles against proliferating IED encounters, caused the fabric shortage. Concerns over the supply of para-aramid fibers were not new; the FY1999 NDAA contained a waiver allowing the import of foreign para-aramid fibers. The waiver passed for fear of a single domestic source of para-aramid fibers and because of concerns that the lack of competition for governmental contracts would result in elevated prices and unfavorable delivery schedules.[75] The waiver did not prevent the significant supply shortage that occurred in 2003.

Under the 1950 Defense Production Act (DPA), domestic producers can be required to meet DoD demand prior to filling other commercial orders. DoD reports that much of DuPont's domestic production of Kevlar has been allocated to defense applications in recent years, placing the company in a difficult position, as imported Twaron began to satisfy a significant portion of commercial demand. In 2008, the para-aramid waiver was expanded to allow importation from any Memorandum of Understanding-qualifying country, in part to allow DuPont to use its Northern Ireland plant to fulfill DoD contracts more quickly.[76]

The demand surge for para-aramid fibers dramatically exceeded production capacity so much that even with imported fibers, supply lagged considerably behind demand. This disparity was due in part to the initial decision to provide Interceptor body armor only to the most exposed troops. Nevertheless, even with advance warning, it does not appear that para-aramid manufacturers would have been able to satisfy the 2003 surge in demand. Well before the 2003 shortage, the two para-aramid producers reportedly were having difficulty meeting global demand. A 2001 U.S. International Trade Commission report notes that despite disputes over patents and claims of economic dumping, DuPont and Teijin had been referring customers to one another and have declined to bid on many contracts due to low profit margins. While the commission cites extensive documentation of an "acute shortage in aramid fiber," specific figures on the extent of the shortage are redacted.[77]

Since the 2003 shortage, demand for Kevlar has expanded by at least 10 percent annually, with more than half of that increase stemming from defense-related orders.[78] As a result, DuPont has invested in several expansions of its Kevlar-producing facilities, including a $500 million facility in South Carolina that became operational in 2011. This new plant is expected to increase immediate production of para-aramid fabric by 25 percent, and up to 40 percent over the next two years as the plant becomes fully operational.[79]

Similarly, in response to the shortage of SpectraShield used as backing for SAPI ceramic plates, and with hopes to gain access to expanding DoD demand, DSM Dyneema opened a U.S. plant in 2004. (Their product, Dyneema, is a close substitute for SpectraShield; the two fibers are the only two to meet both the strength and weight specifications for the SAPI inserts used with the Interceptor vest.[80]) The opening of a domestic plant to make Defense Federal Acquisition Regulation Supplement (DFARS)-compliant Dyneema fiber created 300 jobs as well as a second domestic UHMWPE producer that will prioritize military orders.

The expansion of production for both para-aramids and UHMWPE fibers were reactions to insufficient defense industrial base capacity. Although recent expansions will increase potential output, it is unclear whether expanded production capacity is adequate to meet a future surge in demand. Moreover, there was a four-year lag between the decision to develop a new Kevlar facility and that plant actually becoming operational. The decision to construct a new facility lagged an additional four years behind the 2003 shortage that placed U.S. soldiers at risk in Iraq and Afghanistan.

Because production is capital intensive, producers of ballistic fibers appear hesitant to invest in additional production capacity until sustained increased demand has been well established. Moreover, patents protect the significant investments necessary to develop these advanced fibers, significantly reducing the likely emergence of alternative producers of para-aramid and UHMWPE fibers. Together, these factors suggest that production will increase only when there has already been an increase in demand adequate to maintain prices. The risk remains that if demand were to surge again, production capacity would lag.

MITIGATING THE RISKS TO THE ADVANCED FABRIC SUPPLY CHAIN

Recognizing the advanced fabrics' important role in protecting U.S. service members, there are multiple courses of action that would minimize the risk that a supply disruption would create shortages of either flame-resistant uniforms or body armor. In addition to production capacity, long lead times have been an issue with the flame-resistant and ballistic fibers needed to manufacture critical military apparel. In the past these delays postponed adequate outfitting military personnel deployed in Iraq and Afghanistan. The following steps would mitigate the risks of future disruptions to the availability of critical protective apparel.

DoD must establish a better understanding of its supply chains. Although the principal causes of the 2003 Interceptor body armor shortage were limits on the production capacity of para-aramid fibers and a simultaneous and unprecedented surge in demand, the military's expectation that production could be ramped up readily to meet increasing military demand proved to be faulty. While the ceiling on production capacity was well known to the manufacturers, that information was not shared with the military and was not apparent until the production bottlenecks had emerged. A better understanding of how production capacity is limited could have avoided the 2003 shortages. In addition, had DoD been familiar with the relevant supply chain issues, it could have found other ways to obtain the necessary quantities of protective apparel prior to the engagements in Iraq. In the future, DoD's Sector-by-Sector-Tier-by-Tier (S2T2) program, which maps defense supply chains, could help with this problem. However, better communication and coordination between the military and this sector of the defense industrial base are essential to reduce any remaining uncertainty surrounding production capacity, lead times, performance requirements, and other barriers.

In the short term, maintaining a reserve of vests and inserts can reduce the risk of a future shortage of body armor. Providing soldiers with the highest level of protection requires a continuous supply of body armor and inserts. Due to significant time needed to produce the Interceptor body armor, on-demand production is not possible. Instead, DLA should acquire body armor in advance so they can distribute it immediately to troops as needed.

Long term risk mitigation will require revoking or "sunsetting" waivers that circumvent the Berry Amendment's domestic source requirement. Waivers to the Berry Amendment allow for the importation of foreign-produced FR rayon and para-aramid fibers for use in defense applications. In both cases those waivers are permanent, allowing for the ongoing importation of foreign para-aramid and FR rayon fibers. While these waivers reduce manufacturers' uncertainty surrounding the availability of inputs in the short term, the effect of making the waivers permanent–rather than requiring periodic reassessment of their necessity–is to discourage investment in domestic production capacity and innovation.

On the one hand, the U.S. military services have actively tested and evaluated different fiber blends for use in flame-resistant uniforms. On the other hand, the permanent Berry Amendment waiver discourages domestic manufacturers from designing and producing flame-resistant fabrics

containing fibers produced in the United States.

Additionally, the permanent waiver means that the U.S. military will not necessarily seek out fabrics that contain 100 percent domestic fibers. According to the GAO, the Army's continued preference for Defender M over the Berry-compliant Abrams V fabric results primarily from the pattern of the weave rather than the performance characteristics of the fabric. Because uniforms made with the Defender M fabric were already in use, switching to the domestically produced Abrams V fabric would, GAO reported, "creat[e] variations in appearance that Army leadership found unacceptable."[81]

Unless the permanent waiver for FR rayon is revoked and subject to reevaluation, domestic manufacturers can be reasonably sure that an alternative fabric will not be selected. This knowledge reduces their incentive to innovate and develop new, potentially superior products that use 100 percent domestic inputs.

Similarly, the permanent waiver permitting the acquisition of products containing foreign-produced para-aramid fibers originated from concerns that a single-source for para-aramid fiber was not in the best interests of the United States and that an additional source of para-aramid would reduce prices and ensure adequate supply. However, this approach does not appear to have been effective, because the existence of a second producer of para-aramid fibers did little to prevent the 2003 shortage of Interceptor vests. Moreover, the waiver may have influenced DuPont's decision not to expand Kevlar production prior to the 2003 shortage, given that the company expected foreign manufacturers would service expanded DoD demand.

Although the Berry Amendment waiver allows defense contractors to purchase para-aramid fibers from Teijin, as a foreign producer the company is under no obligation to prioritize DoD orders ahead of other customers. Under the DPA, domestic producers such as DuPont must prioritize defense contracts over commercial contracts. "Rated," or prioritized, orders such as the Mine-Resistant Ambush-Protected (MRAP) vehicle have taken up a large proportion of DuPont's production capacity. According to former Under Secretary of Defense for Acquisition, Technology, and Logistics John Young, DuPont lost commercial business to Teijin by servicing DoD orders, especially with the expanded need for para-aramid fibers during the wars in Iraq and Afghanistan.[82] If military demand were to abruptly decrease, DuPont might struggle to reacquire commercial business that it has lost to Teijin.

Despite this uncertainty, DuPont has invested $500 million in a new Kevlar plant in South Carolina, and made numerous other upgrades at its existing facilities. The company has expanded Kevlar production by 25 percent in 2011 and up to a projected 40 percent by 2013.[83] Once these facilities are operating at full capacity, DoD should reevaluate the appropriateness of a permanent waiver for para-aramid fiber. With DuPont's expanded production capacity, defense-related para-aramid demand could be fully met with domestically produced Kevlar.

Together, these waivers to domestic preference frameworks serve to perpetuate a status quo that provides insufficient incentive for innovation, R&D, or investment in additional domestic production.

CONCLUSION

Advanced fabrics' contribution to the safety and effectiveness of the U.S. military can often be overlooked. Advanced fabrics enable U.S. troops to routinely survive situations that would have been fatal only a few years ago. The cost of providing this protection to U.S. service members was affordable, but would have proved impossible without a well-established textile industry capable of rapidly identifying and manufacturing a solution.

American warfighters bravely place themselves in harm's way, and deserve the best available equipment to minimize their vulnerability to the spectrum of battlefield hazards. Inattention to the defense industrial base can result in an inadequate level of preparedness, especially if U.S. forces are deployed rapidly in response to an unforeseen crisis. Although U.S. troops proved versatile during the 2003 body armor shortage, improvising jury-rigged temporary protection and voluntarily sharing the risk by rotating who wore the limited number of available Interceptor vests, they should have been given the proper equipment from the start.

The shortage of Interceptor body armor resulted in part from inattention to U.S. defense industrial base capabilities and an incomplete understanding of the factors limiting the production of ballistic fabrics. While the rapid innovation and deployment of flame-resistant uniforms is a clear success of the textile industry, the body armor shortage demonstrates that better coordination and communication between the military and manufacturers could result in a greater degree of protection to our fighting men and women. In both cases, the U.S. military was not able to react quickly enough to protect U.S. troops, because it lacked awareness of how long it would take the defense industrial base to surge vest production. A better understanding of the defense industrial base would have resulted in better protection for the warfighter.

We live in an austere and unpredictable fiscal environment. However, even with the end of the war in Iraq and the drawdown of U.S. forces in Afghanistan, the United States needs to be prepared to confront future threats, including those that we cannot currently anticipate. Advanced fabrics play an important role in protecting U.S. military personnel. The ability of this sector of the defense industrial base to better serve the warfighter can be enhanced through closer coordination with policymakers and military planners, greater transparency and stability of DoD orders, and increased support to those domestic firms that sacrifice commercial business for the U.S warfighter's well-being.

ENDNOTES

1 DuPont, Kevlar® and Nomex® are registered trademarks of E.I. du Pont de Nemours and Company.

2 Paul Wakefield, "Polymer Advances in the Interwar Period: The Impact of Science on World War II." *Army Logistician* 39, no. 2 (2007). http://www.almc.army.mil/alog/issues/Mar-Apr07/polymer_advan.html.

3 Department of Commerce, Bureau of Industry and Security (BIS), "National Security Assessment of the U.S. Aerial Delivery Equipment Industry" (Washington, D.C.: BIS, May 26, 2004). https://www.bis.doc.gov/defenseindustrialbaseprograms/osies/defmarketresearchrpts/air_delivery_final_may26-04_mac_printfinal.pdf.

4 Naval Sea Systems Command, Direction of Commander, "Chapter 582: Mooring and Towing," in *Naval Ships' Technical Manual* (S9086-TW-STM-010/CH-582R2, December 1, 2001). http://towmasters.files.wordpress.com/2009/05/nstm-chapter-582-mooring-towing.pdf.

5 Anthony H. Cordesman, Charles Loi, and Vivek Kocharlakota, "IED Metrics for Iraq: June 2003 – September 2010" (Washington, D.C.: Center for Strategic and International Studies, November 11, 2010). http://csis.org/files/publication/101110_ied_metrics_iraq.pdf.

6 "Operation Enduring Freedom," iCasualties.org, accessed March 18, 2013. http://icasualties.org/oef/.

7 Thom Shanker, "Makeshift Bombs Spread Beyond Afghanistan, Iraq," *The New York Times*, October 28, 2009. www.nytimes.com/2009/10/29/world/29military.html?ref=improvisedexplosivedevices.

8 TenCate Defender™ M is a trademark of Koninklijke Ten Cate N.V.

9 C. J. Chivers, "In Wider War in Afghanistan, Survival Rate of Wounded Rises," *The New York Times*, January 7, 2011. http://www.nytimes.com/2011/01/08/world/asia/08wounded.html?pagewanted=all.

10 U.S. Marine Corps, "Interceptor System Saves Lives," *U.S. Federal News Service*, March 24, 2005.

11 Twaron® is a registered trademark of Teijin Aramid.

12 Spectra® is a registered trademark of Honeywell International Inc.

13 Dyneema® is a registered trademark of Koninklijke DSM N.V.

14 William C. Smith, *High Performance and High Temperature Resistance Fibers* (Greer, SC: Industrial Textiles Association, September 21, 1999). http://www.intexa.com/downloads/hightemp.pdf.

15 Government Accountability Office, *Military Uniforms: Issues Related to the Supply of Flame Resistant Fibers for the Production of Military Uniforms*, by William Solis (Washington, D.C.: GAO-11-682R, June 30, 2011), 13. http://www.gao.gov/assets/100/97609.pdf.

16 Abrams™ V is a trademark of Milliken & Company.

17 Ibid., 11.

18 Adrian Wilson, "TenCate's Defender M Fabric a Popular Military Choice," *WTIN Intelligence*, April 16, 2012. http://ei.wtin.com/article/fVE9tCq8YSk/2012/04/16/company_profile_tencates_defender_m_fabric_a_popular_militar/.

19 Federal Registry 64, no. 88 (May 7, 1999), 24523-24524. http://www.gpo.gov/fdsys/pkg/FR-1999-05-07/pdf/99-11550.pdf.

The DNAD was reaffirmed in a written determination dated August 15, 2008, issued by John. J Young, Under Secretary of Defense for Acquisition, Technology, and Logistics. "Determination Under Section 807 of the Strom Thurmond National Defense Authorization Act for Fiscal Year 1999" (August 15, 2008). http://www.acq.osd.mil/dpap/dars/pgi/si_docs/225.70 - No.3- Aug 2008.pdf.

20 "1924: Rayon," DuPont. http://www2.dupont.com/Phoenix_Heritage/en_US/1924_a_detail.html.

21 "Avtex Fibers," Environmental Protection Agency, Superfund Accomplishments. http://www.epa.gov/superfund/accomp/success/avtex.htm.

22 "North American Rayon Corporation and American Bemberg Corporation," The Tennessee Encyclopedia of History and Culture. http://tennesseeencyclopedia.net/entry.php?rec=1005.

23 United States International Trade Commission (USITC), "Viscose Rayon Staple Fiber: Probable Effect of Modification of U.S.-Australia FTA Rules of Origin," *USITC Publication 4041* (Washington, D.C.: USITC, October 2008). www.usitc.gov/publications/332/pub4041.pdf.

24 "Lenzing To Expand Lenzing FR Production To Supply Fiber For Military Uniforms," *Textile World*, February 19, 2008. http://www.textileworld.com/Articles/2008/February_2008/Textile_World_News/Lenzing_To_Expand_Lenzing_FR_Production_To_Supply_Fiber_For_Military_Uniforms.html.

25 "Rayon Fiber," Fibersource, http://fibersource.com/f-tutor/rayon.htm.

26 Textiles Intelligence, *Flame Resistant Fibres and Fabrics* (Alderley House, UK: Textiles Intelligence Limited, 2011). ftp://textilesintelligence.com/web_shared/tilupload/TISPAM_00036_1671.pdf.

27 Government Accountability Office, *Military Uniforms: Issues Related to the Supply of Flame Resistant Fibers for the Production of Military Uniforms*, by William Solis (Washington, D.C.: GAO-11-682R, June 30, 2011), 19. http://www.gao.gov/assets/100/97609.pdf.

28 Ibid., 31.

29 Ibid., 23.

30 See http://www.troopsdeservethebestprotection.com/ for copies of the letters sent to Sen. Carl Levin, Sen. John McCain, Sen. Lindsey Graham, Rep. Ike Skelton, Rep. Buck McKeon, and Rep. John Spratt on April 12, 2010.

31 Eric Beidel, "New Fabrics Promise Better Fire Protection for IED-Battered Troops," *National Defense Magazine*, October 2011.

http://www.nationaldefensemagazine.org/archive/2011/October/Pages/NewFabricsPromiseBetterFireProtectionForIED-BatteredTroops.aspx.

32 Government Accountability Office, *Military Uniforms: Issues Related to the Supply of Flame Resistant Fibers for the Production of Military Uniforms*, by William Solis (Washington, D.C.: GAO-11-682R, June 30, 2011), 25. http://www.gao.gov/assets/100/97609.pdf.

33 "High Performance and High Temperature Resistance Fibers," Industrial Textiles Association. http://www.intexa.com/downloads/hightemp.pdf.

34 Luis Villarreal, "DLA Industrial Capability Overview," (Washington, D.C.: USA Symposium Defense Industrial Base Seminar and Workshops, June 15-16, 2010), 23. http://www.usasymposium.com/ibconference/Conference%20PDF/DLA%2016Jun%20PM%20Breakout%20Session%20LH1119/Villarreal%2016Jun%20PM%20Presentation%201%20LH1119.pdf.

Donna Pointkouski, Defense Logistics Agency, "DLA Troop Support Clothing & Textiles Warstopper Program" (2011), 79. https://www.wewear.org/assets/1/7/2011GCCFallPointkouski.pdf.

35 Textiles Intelligence, "Performance Apparel Markets: Business and Market Analysis of Worldwide Trends in High Performance Active Wear and Corporate Apparel" (4th Quarter, 2006). ftp://textilesintelligence.com/web_shared/pdfstore/TISPAM_00019.pdf.

36 Government Accountability Office, *Military Uniforms: Issues Related to the Supply of Flame Resistant Fibers for the Production of Military Uniforms*, by William Solis (Washington, D.C.: GAO-11-682R, June 30, 2011), 13. http://www.gao.gov/assets/100/97609.pdf.

37 Bulwark Protective Apparel, "Industry Update: Secondary Flame-Resistant Apparel, Volume 13." http://www.vfimagewear.com/images/public/Bulwark_Industry_Update_13.pdf.

38 William C. Smith, *High Performance and High Temperature Resistance Fibers* (Greer, SC: Industrial Textiles Association, September 21, 1999). http://www.intexa.com/downloads/hightemp.pdf.

39 U.S. Marine Corps, "Interceptor System Saves Lives," *U.S. Federal News Service*, March 24, 2005.

40 Jonathan Turley, "U.S. Soldiers Lack Best Protective Gear," *USA Today*, December 18, 2003, A23.

41 Kevin Graman, "Senate Approves Reimbursement for Soldiers' Own Body Armor Purchases," *McClatchy-Tribune Business News*, June 16, 2004.

42 Congressional Budget Office, "Direct Spending Estimate for H.R. 4200, Ronald W. Reagan National Defense Authorization Act for Fiscal Year 2005" (October 21, 2004), 8. http://www.cbo.gov/sites/default/files/cbofiles/ftpdocs/59xx/doc5957/hr4200pgo.pdf.

43 Government Accountability Office (GAO), *Defense Logistics: Actions Needed to Improve the Availability of Critical Items during Current and Future Operations*, by Richard G. Payne and John W. Lee (Washington D.C.: GAO, 05-275, 2005), 76.

44 U.S. Marine Corps, "Interceptor System Saves Lives," *U.S. Federal News Service*, March 24, 2005.

45 Jonathan Turley, "U.S. Soldiers Lack Best Protective Gear," *USA Today*, December 18, 2003, pA23.

46 Government Accountability Office (GAO), *Defense Logistics: Actions Needed to Improve the Availability of Critical Items during Current and Future Operations*, by Richard G. Payne and John W. Lee (Washington D.C.: GAO, 05-275, 2005), 76-77.

47 Spectra Shield® is a registered trademark of Honeywell International Inc.

48 Ibid., 80-81.

49 Zylon® is a registered trademark of Toyobo Co., Ltd.

50 Innegra™ S is a trademark of Innegra Technologies, LLC.

51 Serge Rebouillat, "Aramids," in *High Performance Fibers*, ed. J.W.S. Hearle (Cambridge, UK: Woodhead Publishing: 2004), 48.

52 Brendan Kearney, "Kevlar Plant to Help Weave Strong Future," *McClatchy-Tribune Business News*, October 17, 2011.

53 John W. McCurry, "Terrorist Threat Continues to Drive U.S. Fibre Developments," *Technical Textiles International* 14, no. 2 (2005), 23-27.

54 Janet Bealer Rodie, "Life-Saving Fabrics," *Textile World* 159, no. 3 (2009), 22-25.

55 Henry J. Holcomb, "Kevlar Protecting Profit for DuPont: The Company will Spend $500 Million to Increase Production 25% by 2010," *Philadelphia Inquirer*, September 20, 2007, C1.

56 United States International Trade Commission, "Aramid Fiber Formed of Poly Para-Phenylene Terephthalamine from the Netherlands," *Publication No. 3394* (2001).

57 "DuPont Starts Up $500 Million Copper River Kevlar Site," *Defense Daily* 252, October 6, 2011.

58 Serge Rebouillat, "Aramids," in *High Performance Fibers*, ed. J.W.S. Hearle (Cambridge, UK: Woodhead Publishing, 2004), 29-35.

59 Jan L. J. van Dingenen, "Gel-Spun High-Performance Polyethylene Fibres," in *High Performance Fibers*, ed. J.W.S. Hearle (Cambridge, UK: Woodhead Publishing, 2004), 70-71, 76, 81.

60 National Defense University, Industrial College of the Armed Forces, *Industry Study: Strategic Materials* (2007). http://www.ndu.edu/icaf/programs/academic/industry/reports/2007/pdf/icaf-is-report-strategic-materials-2007.pdf.

61 Government Accountability Office (GAO), *Defense Logistics: Actions Needed to Improve the Availability of Critical Items during Current and Future Operations*, by Richard G. Payne and John W. Lee (Washington D.C.: GAO, 05-275, 2005), 80.

62 John W. McCurry, "Terrorist Threat Continues to Drive U.S. Fibre Developments," *Technical Textiles International*, 14, no. 2 (2005), 23-27.

63 Janet Bealer Rodie, "Life-Saving Fabrics," *Textile World* 159, no.3 (2009), 22-25.

64 "Innegra S Fiber Available January 2012 From Innegra Technologies LLC," Innegra Technologies press release, December 1, 2011. http://0353b82.netsolhost.com/ing/wp-content/uploads/2012/03/Press-Release-3.pdf.

65 Innegra Technologies, *Technical Reference Sheet for Composite Fibers*. http://0353b82.netsolhost.com/ing/wp-content/uploads/2012/03/Fiber-Reference-Sheet-Composite-Fibers.jpg.

Brian Morin, Innegra Technologies, LLC, Composite Materials Including Amorphous Thermoplastic Fibers, US Patent 8168292, filed June, 15, 2006, and issued May, 1, 2012. (2012). http://www.freepatentsonline.com/8168292.html.

66 Innegra Technologies, *Innegra S High Performance Fiber: Tough, Ultra-lightweight, Cost-effective Fiber* (2010), 5. http://www.innegrity.com/mydocuments/innegra_s_high_performance_fiber_presentation_2010.pdf.

67 Innegra Technologies, *Ballistic Response of Soft Panels Including Innegra S Fibers*. http://www.innegrity.com/mydocuments/ballisticinnegras.pdf.

68 Innegra Technologies, *Innegra S High Performance Fiber: Tough, Ultra-lightweight, Cost-effective Fiber* (2010), 35. http://www.innegrity.com/mydocuments/innegra_s_high_performance_fiber_presentation_2010.pdf.

69 Alan Brown, "New Options in Personal Ballistic Protection," *Composites World*, March 1, 2003. http://www.compositesworld.com/articles/new-options-in-personal-ballistic-protection.

70 Robert J. Young and C. L. So, "Other High Modulus-High Tenacity (HM-HT) Fibres from Linear Polymers," in *High Performance Fibers*, ed. J.W.S. Hearle (Cambridge, UK: Woodhead Publishing, 2004), 106.

71 Brown, "New Options in Personal Ballistic Protection," *Composites World*, March 1, 2003. http://www.compositesworld.com/articles/new-options-in-personal-ballistic-protection.

72 U.S. Department of Justice, National Institutes of Justice, *Third Status Report to the Attorney General on Body Armor Safety Initiative Testing and Activities*, (Washington, D.C.: National Institute of Justice, August 24, 2005), 5. http://www.ojp.usdoj.gov/bvpbasi/docs/SupplementII_08_12_05.pdf.

73 Alan Brown, "New Options in Personal Ballistic Protection," *Composites World*, March 1, 2003. http://www.compositesworld.com/articles/new-options-in-personal-ballistic-protection.

74 "National Settlement Concerning Body Armor Containing Zylon Made by Second Change Body Armor," Tyobo press release, July 12, 2005. http://www.toyobo-global.com/seihin/kc/pbo/pdf/Toyobo_press_release_071205.pdf.

75 United States International Trade Commission, "Aramid Fiber Formed of Poly Para-Phenylene Terephthalamine from the Netherlands," Publication No. 3394 (2001), 14.

76 John J. Young, Undersecretary of Defense for Acquisition, Technology, and Logistics, "Determination Under Section 807 of the Strom Thurmond National Defense Authorization Act for Fiscal Year 1999" (August 15, 2008). http://www.acq.osd.mil/dpap/dars/pgi/si_docs/225.70 - No.3- Aug 2008.pdf.

77 United States International Trade Commission, "Aramid Fiber Formed of Poly Para-Phenylene Terephthalamine from the Netherlands," *Publication No. 3394* (2001), 11-12.

78 Henry J. Holcomb, "Kevlar Protecting Profit for DuPont: The Company will Spend $500 Million to Increase Production 25% by 2010," *Philadelphia Inquirer*, September 20, 2007, C1.

79 "DuPont Starts Up $500 Million Cooper River Kevlar Site," *Defense Daily* 252, October 6, 2011.

80 Government Accountability Office (GAO), *Defense Logistics: Actions Needed to Improve the Availability of Critical Items during Current and Future Operations*, by Richard G. Payne and John W. Lee (Washington D.C.: GAO, 05-275, 2005), 80-81.

81 Government Accountability Office, *Military Uniforms: Issues Related to the Supply of Flame Resistant Fibers for the Production of Military Uniforms*, by William Solis (Washington, D.C.: GAO-11-682R, June 30, 2011), 11. http://www.gao.gov/assets/100/97609.pdf.

82 John J. Young, Undersecretary of Defense for Acquisition, Technology, and Logistics, "Determination Under Section 807 of the Strom Thurmond National Defense Authorization Act for Fiscal Year 1999" (Washington, D.C." DLA, August 15, 2008). http://www.acq.osd.mil/dpap/dars/pgi/si_docs/225.70 - No.3- Aug 2008.pdf.

83 "DuPont Starts Up $500 Million Cooper River Kevlar Site," *Defense Daily* 252, October 6, 2011.

CHAPTER 12
TELECOMMUNICATIONS

EXECUTIVE SUMMARY

Military effectiveness requires up-to-date communications systems. Communications must remain secure from eavesdropping and interception to protect missions, objectives, and the lives of U.S. soldiers. However, digital networks can be disrupted in many ways, and the increased military use of wireless communication only increases these risks. The United States historically has been the world leader in telecommunications but recently lost its edge in certain sectors to foreign companies, including state-supported Chinese firms.

The rapid growth of the Chinese telecommunications industry means that foreign-produced hardware is used throughout global telecommunications networks. The U.S. military maintains the world's most secure communications networks; however, not all sensitive communications can take place via secure lines, especially when U.S. military forces are deployed globally and communicating wirelessly. Notably, the communications of many logistics providers upon whom the U.S. military depends are not secure, and are open to many potential breaches. (For example, surveillance devices could be planted or built into communications equipment, including routers and switches, while fiber-optic and wireless communications might be intercepted or jammed.) As more and more foreign equipment is used in the United States' communications infrastructure, the risk of interruption and interception increases accordingly.

Chinese telecommunications firms (such as Huawei and ZTE) have been able to undercut U.S. and European firms by selling equipment at steep discounts in different parts of the world. This tactic reflects the Chinese government's emphasis on global communications networks as an instrument of Chinese national defense. Undergoing downsizing and mergers to survive, U.S. firms have struggled to remain competitive.

Global market trends could pose threats to the integrity of U.S. defense-related communications, including communications between logistics providers that forward-deployed forces depend on. Foreign telecommunications firms cannot be held accountable by U.S. officials. With the elevated attention the Chinese military is placing on telecommunications, it cannot be assumed that equipment manufactured in China will be free of surveillance devices. If U.S. manufacturers continue to lose global market share, in time the U.S. military might lose confidence in its ability to ensure the integrity of its defense-related communications.

TELECOMMUNICATIONS
SECURE WARFIGHTER COMMUNICATION

MANUFACTURING SECURITY

Telecommunications enable and enhance U.S. military strength

UNMANNED AERIAL DRONES **GLOBAL POSITIONING SYSTEM** **COMBAT COMMUNICATIONS**

DOMESTIC SECTOR 🏢 = 100

1,500 companies comprise the telecom sector

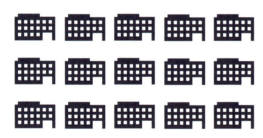

15
HUNDRED
COMPANIES

MARKET DOMINANCE

China dominates the manufacturing of telecom equipment

CHINA CONTROLS
71% OF THE
WORLD'S MOBILE
PHONE MARKET

DOD INNOVATIONS

DoD research and innovations have led to important advances in telecommunications and national security

GLOBAL POSITIONING **THE INTERNET** **CELLULAR TECHNOLOGY**

VULNERABILITIES

U.S. telecom systems face a number of vulnerabilities

COUNTERFEIT PARTS **UNSECURE COMMUNICATION** **NETWORK INTRUSION**

MITIGATING RISKS

Securing the U.S. telecommunications sector

SUPPORT NEW RESEARCH **IMPROVE CYBER SECURITY**

REVITALIZE CORE IT SECTOR **MONITOR SUPPLY CHAIN**

MILITARY EQUIPMENT CHART
SELECTED DEFENSE USES FOR TELECOMMUNICATIONS

DEPARTMENT	SYSTEMS
ARMY	■ AN/PRC-148 Multiband Inter/Intra Team radio (MBITR) ■ AN/VRC – 92A vehicular radio set ■ Ground-Based Midcourse Defense (GMD) ballistic missile defense system
MARINE CORPS	■ AN/VRC – 88C Single-Channel Ground and Airborne Radio System (SINCGARS) ■ AN/ARQ – 53 Shipboard Single-Channel Ground and Airborne Radio System (SINCGARS)
NAVY	■ Aegis air and missile defense system ■ AN/PRC-148 Multiband Inter/Intra Team Radio (MBITR) ■ Unmanned Underwater Vehicles (UUVs)
AIR FORCE	■ F-35 joint strike fighter communications systems ■ Unmanned Aerial Vehicles (UAVs) ■ AN/PSC – 2 radio

INTRODUCTION

"The PLA must develop the capability to fight and win local wars under informationized conditions."[1]

– Former President Jiang Zemin speech during the 16th Communist Party Congress, 2002.

No modern military can function without a sophisticated communication system. Communications—the gathering and dissemination of information and the coordination of actions or decisions—are at the core of modern warfare.

The U.S. telecommunications sector is massive and consists of hardware manufacturers, software designers, and service providers. The most dynamic subsector is mobile telecommunications, providing mobile telephone and data services. Telecommunications encompasses anything that involves exchanges of information over significant distances by electronic means.

A telecommunications network has multiple stations, each equipped with a transmitter and a receiver. The device is called a transceiver when both the transmitter and receiver are coupled into a single device. Transmissions may be wired (traveling over electrical or fiber-optic cables) or wireless (broadcast over the electromagnetic spectrum).[2]

Mobile telecommunications is the principal model of wireless transmission—a business sector that has undergone explosive changes. Today, mobile telecommunications is less about voice and increasingly about text and images. Smartphones (for example, iPhone[3] and cellphones using the Android[4] operating system) are, in fact, mini-computers used to access the Internet, make payments, listen to music, watch videos, and play games. The mobile telecommunications industry consists of several distinct segments. There are hardware manufacturing companies, software designers, sales and marketing outlets, and mobile service providers. This chapter focuses on two types of hardware manufacturers: those that produce mobile phone handsets and peripherals such as Bluetooth-based devices, and those that manufacture mobile network equipment.

In addition to manufacturing phones (frequently offshore in low-cost countries), mobile phone manufacturers control research and development (R&D), sales, and distribution of mobile phones. The global mobile phone market was worth $150 billion in 2011. Production of mobile handsets is concentrated in a handful of countries, including China and South Korea; the major players are Nokia, Samsung, LG, Motorola (owned by Google), and Sony Ericsson. Together these five companies account for 75 percent of world production of mobile phones. However, Motorola has suffered a steady decline, and its market share has shrunk, even in the United States. Motorola sold its mobile phone division to Google; whether Google will continue to operate Motorola's phone business remains to be seen.

The second type of hardware consists of network equipment. Network equipment vendors develop, manufacture, and install network equipment for mobile network operators. They produce routers, nodes, mobile switching centers, servers, storage, and radio access equipment (such as base stations, base transceivers, and base station controllers).[5] Because of the popularity of smartphones, tablets, and Internet Protocol (IP) interactions (including Voice over IP (VoIP) and video streaming services), the United States faces the twin issues of data overload and spectrum crunch. Federal and industry officials are

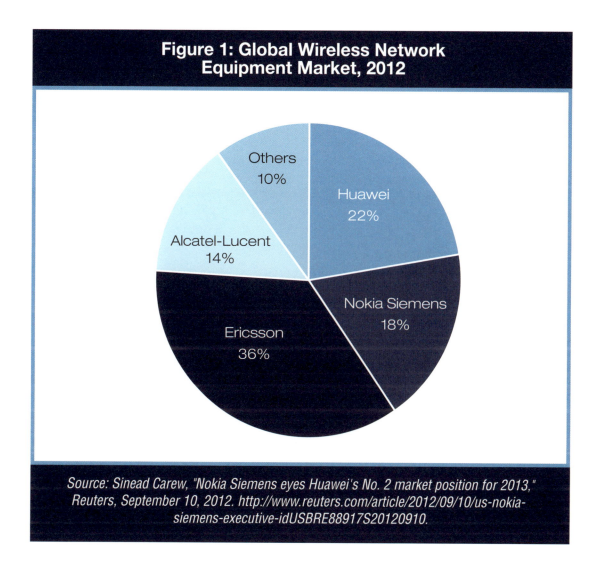

Figure 1: Global Wireless Network Equipment Market, 2012

Others 10%

Huawei 22%

Alcatel-Lucent 14%

Nokia Siemens 18%

Ericsson 36%

Source: Sinead Carew, "Nokia Siemens eyes Huawei's No. 2 market position for 2013," Reuters, September 10, 2012. http://www.reuters.com/article/2012/09/10/us-nokia-siemens-executive-idUSBRE88917S20120910.

increasingly concerned about how to manage the rapid growth of data traffic that is a result of the growing popularity of streaming video and other downloads, especially on mobile devices. Service providers and network equipment operators are exploring various strategies to increase the usability of the wireless spectrum quickly in order to expand network capacity. By current forecasts, wireless data usage will surpass capacity by 2014 if mobile network capacity is not expanded, forcing service providers to limit or reduce services.

Efforts to address this dilemma include expediting the implementation of more efficient fourth generation (4G) networks (for example, long-term evolution (LTE) and installing "small cells"—miniature cellular towers that can be installed almost anywhere—as well as encouraging the use of Wi-Fi-enabled devices to reduce data load. The most effective solution appears to be a combination of these improvements in conjunction with the auction of more bandwidth by the U.S. federal government for wireless data usage.

While the United States faces an imminent "bandwidth crunch" on the service side, the manufacture of network equipment and mobile handsets has moved offshore and is increasingly dominated by Chinese operators Huawei Technologies Co, Ltd.; ZTE Corporation; and—a distant third—China Mobile. Chinese companies have penetrated many markets (especially in developing countries) because they emphasize energy-saving, flexible, and configurable network equipment at low prices. Chinese operators have also secured orders from leading Western mobile network operators such as Vodafone, T-Mobile, and Verizon.

The success of Chinese network manufacturers is in large part due to the fact that they are "cost leaders" and deliver equipment (including modems, USB wireless devices, base stations, WiMAX, and third generation [3G] network equipment) at prices that Western vendors cannot match. As a result, the products made by Chinese manufacturers have acquired ever larger market shares.

How are Chinese manufacturers able to keep their prices so much lower while delivering quality that is more or less equal to Western network manufacturers? First, Chinese firms manufacture in a low-cost environment. Second the two main operators--Huawei and ZTE—receive sizable support from the Chinese state and military, which have prioritized upgrading and improving the Chinese telecommunication sector and invested billions of dollars in strengthening and facilitating Huawei and ZTE becoming key market players.[6] Since 2002, China's People's Liberation Army (PLA) has prioritized a new, historic mission, which involves supporting national economic development, expanding territorial interests, and providing military support to the Communist party. To that end it has committed to securing and protecting China's electromagnetic spectrum. Control over the electromagnetic spectrum strengthens China's national security, supports modernization, and adds another weapon to the PLA's toolkit.[7]

Key themes discussed in this chapter are:

■ The decline of the U.S. telecommunications sector as it has lost market shares in both the domestic and international markets.

■ The rise of Chinese telecommunications companies, which have close ties to the Chinese government and the military establishment.

■ The security and intelligence implications of a supply chain that is based in Asia and dominated by Chinese companies.

■ The loss of technological know-how and innovative capacity due to the decline of the U.S.-based telecommunications sector.

A NOTE ON CRITICALITY

The importance of telecommunications technologies to U.S. armed forces cannot be overstated. Deployed units must have access to a wide variety of operational, intelligence, logistical, and administrative communications that are resistant to interference or interception by enemy forces. Moreover, logistics in support of ongoing operations must be able to share and access massive amounts of data in real time, without risk of interference or interception. As the U.S. military fights on the network-centric battlefield, these reliability and security requirements are even more

DISRUPTED COMMUNICATION DURING A FUTURE CRISIS (a notional though realistic scenario)

In an attempt to stabilize tensions rising between China and Taiwan, the U.S. Navy deploys a carrier battle group to the waters off Taiwan. Shortly after the battle group arrives, Pacific Command's unclassified logisitics networks are disrupted. An inability to diagnose the cause of the disruption complicates efforts to restore full operations to the logistics network. Initially, it appears that the disruption is simply a system failure. Subsequent forensics determine that a computer network attack, focused on logistics contractor networks, caused the disruption. The attack penetrated telecommunications switches in the unclassified portions of the logistics network, leaving the battle group less ready to respond to a hostile outbreak.

essential. U.S. military mission success depends on access to the most advanced and secure telecommunications technology, manufactured in secure settings by trusted corporations. Vulnerability may be introduced by a single piece of equipment on a telecommunications network that malfunctions or has a backdoor built in to either its hardware or software. This type of weakness could jeopardize the integrity of all U.S. military assets and operations, and otherwise *incapacitate* the U.S. military.

The Chinese telecommunications industry has grown rapidly, with Chinese-manufactured telecommunications equipment spreading swiftly due to below-market prices supported by funding from the Chinese military. The widespread use of military-funded Chinese equipment in conjunction with the shrinking market share of trusted U.S. telecommunications firms increases the likelihood that kill switches or backdoors will be inserted into key communications infrastructure, jeopardizing the integrity of sensitive defense-related communications. Coincident with this threat, the U.S. domestic telecommunications industry

has shrunk considerably, with numerous mergers and takeovers, as domestic firms struggled to remain competitive with subsidized Chinese firms. This combination constitutes a *high* risk that reliable and secure telecommunications infrastructure and equipment will not be available if these trends remain unchecked.

BACKGROUND

The telecommunications industry has undergone convulsive changes driven by mass digitization. A growing number of households and businesses are becoming increasingly reliant on online services such as cloud storage and mobile payment systems. The widespread use of smartphones and tablets has given rise to an expanding market for mobile applications, as developers are given access to the devices underlying operating systems. Accordingly, the telecommunications sector is becoming more and more competitive, with a large number of new entrants seeking to obtain market share from more established firms.

The U.S. telecommunications equipment manufacturing industry includes about 1,500 companies with combined annual revenue of approximately $45 billion. Major telecommunications companies include Apple, Cisco, Motorola, and Qualcomm. The International Trade Centre values global trade in telecommunications equipment at about $280 billion, 80 percent of which is controlled by the 50 biggest firms. The leading equipment manufacturers are located in Asia, in countries including China, Hong Kong, and South Korea. Future growth will be in the emerging markets, especially in the growing economies of China, Mexico, and Brazil. Total telecommunications revenue is projected to rise from $2.1 trillion in 2012 to $2.7 trillion in 2017; however, most of the growth will be in Asia and Latin America rather than in the United States, and corresponds to an increasing demand for telecommunications (especially wireless infrastructure) in these regions resulting from the growing middle class.[8]

In the past, a single state company supplied phone services and many governments tightly regulated the telecommunications business to protect their public monopoly. In the 1990s, public monopolies fell out of favor, with a growing preference for private competition that touted the benefits of free markets. Subsequently, many European countries de-monopolized public utilities and telecommunications services. In the United States, deregulation came in several phases, anchored by the Telecommunications Act of 1996, which attempted to move all telecommunications markets toward competition. Prior to 1996, the telecommunications industry was subdivided into niche sectors, with service providers forced to focus on only one area. For example, cable television providers were unable to also provide telephone service, and local and long-distance telephone services were separated by regulations that barred competition between the two types of service providers. The 1996 Telecommunications Act removed these regulations, permitting companies to "bundle" television, data, and telephone services.

The Telecommunications Act was designed to increase competition throughout the industry. By removing the regulatory boundaries between service providers and hardware manufacturers, the legislation was expected to reduce consumer costs. By one estimate, approximately $550 billion would be saved between expected rate reductions for cable television, local telephone service, and (especially) long-distance telephone communications. However, these expectations were unfounded: both television and telephone rates increased. Despite higher rates for the most common services, the telecommunications industry stagnated, losing about half a million jobs and $2 trillion in value.[9]

The removal of regulatory barriers quickly prompted a series of mergers and acquisitions, resulting in the increasing centralization of the telecommunications industry into only a handful of major players. Smaller companies struggled to compete with immense conglomerates, and sought strategies of "leapfrogging" to new technologies to gain a competitive edge on the larger firms. Establishing telecommunications infrastructure requires significant capital investment, making it more realistic for smaller companies servicing small regions to install new technology or retrofit older technology, potentially giving them a technology advantage over larger companies. However, the costs of upgrading infrastructure proved burdensome for most small companies, who were often unable to recuperate their investment.

Once the 1996 Telecommunications Act was passed, increased competition over limited local markets and the rising demand for dial-up Internet access led multiple companies to invest in excess capacity in the same geographic area, in anticipation of continuous growth in demand for both service and telecommunications equipment (such as modems).[10] As consumer demand stabilized in the late 1990s, telecommunications entities faced falling profit margins due to surplus capacity that exceeded what was necessary to satisfy demand.

Fueled by irrational exuberance, the telecommunications bubble of the late 1990s eventually popped, resulting in the 2001–2003 telecommunications crash, which also happened to coincide with the dot-com crash. Funding for R&D declined. Historically, funding for telecommunications research has been supported by the National Science Foundation (NSF) and the Defense Advanced Research Projects Agency (DARPA); more recently, the NSF has focused its funding support more narrowly on networking, while DARPA has refocused its attention to more immediate military requirements, resulting in a decreased emphasis on telecommunications technology. During the 1980s, many DARPA-funded projects contributed to the development of the Internet, cellphones, and fiber-optics. These investments resulted in a rapidly changing telecommunications landscape. Likewise, major corporate R&D laboratories that supported applied research in telecommunications have also retrenched, contributing to the shrinking U.S. telecommunications research community. The National Research Council (NRC) called attention to the state of affairs in 2006. In *Renewing United States Telecommunications Research*, the NRC remarked, "the American position as a leader in telecommunications technology is at risk because of the recent decline in domestic support for long-term, fundamental telecommunications research."[11]

In recent years, the telecommunications industry has transformed from being primarily focused on wired telephone communications to being heavily reliant on fiber-optics and wireless communications. But that final phase—the wireless connection—is still evolving; in the sense the telecommunications sector is not yet fully mature, and is still characterized by major innovation and change. Of those U.S. companies that survived the earlier telecommunications bust, many have been content to meet short-term goals for investors rather than prioritizing the long-term future of their competitive strength. This outlook has resulted in the slowing down of investments and the delayed migration to 4G mobile broadband, and is partly why the United States will face a "bandwidth crunch." It could occur as early as 2014, especially when considering the rising use of smartphones, tablets, VoIP, and other IP-integrated activities over wireless networks.[12] Without additional resources for basic research, the United States' position of leadership in telecommunications is at risk. There are various solutions, many of which require corporate investment in R&D.[13]

TELECOMMUNICATIONS MANUFACTURING

Network equipment manufacturers must achieve economies of scale to support the massive fixed costs of R&D.[14] This achievement comes by disaggregating the manufacturing and assembly process and situating these different operations in regions with corresponding economic advantages. U.S. telecommunications companies have moved manufacturing,

Table 1: China's Production and Worldwide Share of Mobile Phones (2008-2010)

Production in 1000s				Worldwide Market Share %		
2008	2009	2010	(Year)	2008	2009	2010
559,640	619,520	993,000		44.7%	49.9%	71.3%

Adapted from Ed Pausa, "Continued Growth: China's Impact on the Semiconductor Industry, 2011 Update," Pricewaterhouse Coopers (November 2011). http://www.pwc.com/gx/en/technology/assets/china-semiconductor-report-2011.pdf.

packaging, testing, and distribution off-shore (see Figure 1). The savings obtained from moving offshore substitute for those that could potentially be achieved from technological improvements and innovative solutions; it is easier to cut costs by moving offshore than to gamble on improved technology or unproven new products.[15] Accordingly, U.S. telecommunications vendors are no longer the industry leaders. Rather, L.M. Ericsson (based in Sweden) has an R&D edge, while the Chinese firm Huawei (and to a lesser extent ZTE) has the most competitive pricing. By contrast, U.S. companies have been exiting the industry, unable to compete on either technology or cost.

Controlling some 27,000 patents and with a $35 billion revenue from telecommunications sales in 2011, Ericsson's revenue makes up roughly one-third of global revenue in the global mobile-infrastructure sector. Likewise, Huawei's market share has increased from 4.5 percent in 2006 to 15.6 percent in 2010, during a time when most other telecommunications firms were losing share, merging, being acquired, or otherwise going out of business. (Just behind Ericsson, Huawei saw $32 billion in revenue from telecommunications sales in

2011.)[16] For example, Alcatel and Lucent merged in 2006; Nokia and Siemens merged in 2009, and then acquired Motorola Networks in 2011 to gain control over its patents; and Nortel filed for bankruptcy in 2009.[17] Other firms, including Fujitsu, NEC, Hitachi, ADC Telecom, and ADTRAN have retrenched and now focus on local rather than global markets.[18] Despite increasing demand for mobile network infrastructure, the global telecommunications industry is now dominated by five international companies (Ericsson, Huawei, Alcatel-Lucent, NSN, and Cisco).[19]

Similar patterns apply to the mobile handset industry. Handset manufacturing is increasingly performed in China (see Table 1), although these figures cited include both Western-owned manufacturing plants situated in China to exploit lower labor costs as well as production contracted to Chinese factories, such as Foxconn's production of the iPhone for Apple. Chinese companies are increasingly producing handsets specifically intended for the growing Chinese market. This harsh business climate with increased competition from non-Western players persuaded Motorola to sell its handset division, Motorola Mobility, to Google for $12.5

billion, while Nokia, once the global leader in handsets, is rapidly losing market share.

THE DECLINE OF THE U.S. TELECOMMUNICATIONS INDUSTRY AND NATIONAL SECURITY IMPLICATIONS

The emergence of Chinese telecommunications companies has had a colossal impact on Western companies. Western companies—both niche and larger, vertically integrated—cannot compete against the prices and improved technological performance of Chinese companies. The rise of Chinese network equipment and cellphone handset manufacturers coincides with the decline of the U.S. sector, despite the increase in demand for upgraded and reconfigured mobile networks in support of the significant increase in data traffic. This situation has prompted considerable debate in U.S. government and industry circles, revolving around three general concerns.

First, telecommunications supply chains are lengthy, extremely diffuse, complex, and dispersed (see Figure 2), making it difficult to verify the authenticity of the purchased electronic equipment. The opacity of the supply chain creates a logistical vulnerability: counterfeit parts, unlicensed copies, and re-sold defective parts can corrupt the supply chain as companies attempt to squeeze out extra profits. (A similar issue bedevils semiconductors; see this report's chapter on semiconductors.) Nevertheless, if they are substandard and do not meet military specifications

(Mil-Spec), fake or faulty network routers can suddenly disable or debilitate communication systems at critical times.

Most often, U.S. telecommunications companies oversee the design and integration of various components, while production and assembly are outsourced to subcontractors, generic manufacturers, and separate packaging facilities. Because the Department of Defense (DoD) procures much of its hardware from commercial vendors (commercial-off-the-shelf [COTS]), it is exposed to any issues introduced throughout the geographically dispersed supply chain.[20]

Next, malicious hardware or software may be embedded in a product and used to intercept or interrupt the transmission of sensitive information. A supplier (or a party associated with that supplier) potentially could implant or build in special devices that would enable exfiltration of sensitive information. Foreign governments as well as non-state actors routinely attempt to intercept sensitive communications between U.S. governmental agencies and the military. Although the most sensitive U.S. government and military information is protected by tightly controlled, encrypted networks, a vast amount of unclassified yet sensitive information, including a substantial amount of logistics and administrative traffic, is transmitted on open unclassified telecommunications lines, making it vulnerable to information leakage or interception.[21] As more communications employ wireless networks, the threat of cyber attacks, intelligence-gathering, and hacking increases.

Finally, and perhaps most alarming, is that malicious activities could potentially disrupt or disable the entire Internet by manipulating routers and switches. The architecture of large networks is very

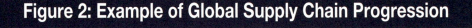

Figure 2: Example of Global Supply Chain Progression

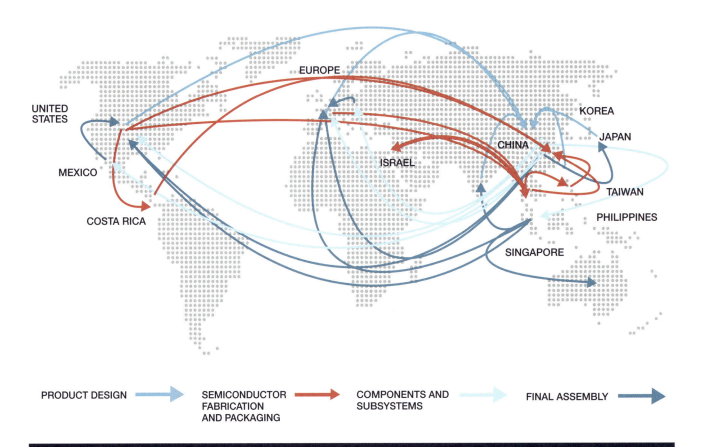

PRODUCT DESIGN ➡ SEMICONDUCTOR FABRICATION AND PACKAGING ➡ COMPONENTS AND SUBSYSTEMS ➡ FINAL ASSEMBLY ➡

Adapted from Mark T. Zetter, "Economic Drivers, Challenges Creating Regional Electronics Industry," IDC Manufacturing Insights, December 12, 2009. http://www.ventureoutsource.com/contract-manufacturing/ benchmarks-best-practices/executive-management/economic-drivers-challenges-creating-regional-electronics-

complex; data traffic could be disrupted if the integrated circuits that operate the routers and switches of the largest and most advanced networks were modified to permit that node to be destroyed or disabled remotely. These modifications could be relatively simple, such as undermining the connections that distribute signals and provide power to different integrated circuits. Yet such modifications could be designed to crash the system under specific conditions, such as when the load of traffic increases.[22]

Interception and sabotage become more likely as an increasing proportion of telecommunications equipment is manufactured and assembled in foreign countries by foreign companies. Because the supply chain is so complex and many tasks are distributed across many different facilities and regions, oversight of the assembly and production process is challenging. This dispersion creates numerous opportunities for tampering or interference, with minimal opportunity for detection.

Fiber-optic cables are also vulnerable to interception and interference by outsiders. Hacking into an optical network requires little more than bending a cable to allow a small amount of light to "leak" from the cable without actually breaking connections. An operator can then install couples (small photonic devices) at that bend to capture the light using an electro-optical converter that acts as an interface to a computer. In theory, although network engineers should be able to detect anomalies in the optical network, backdoors built directly into network hardware could aid eavesdroppers in circumventing detection, especially if network components or subcomponents are manufactured outside of the United States by actors with heightened incentives to "listen in." Components may contain malicious codes or malicious diagnostic tools exposing fiber-optic communications to third-party eavesdropping.

A dispersed and diffuse supply chain does not automatically become a target for hostile activities, tampering, and sabotage. What concerns defense analysts, policymakers, and industry experts is that much of the telecommunications equipment used by commercial and government entities alike is manufactured in China; that Chinese companies are gaining a global dominance; and that these global players in the telecommunications market are closely connected with the PLA and the Chinese government.

The possibility of a national telecommunications system that includes malicious components has prompted considerable political debate. Between February and June 2012, members of the House Select Committees on Intelligence met and corresponded with officials from Huawei and ZTE as part of an investigation into the threat Chinese telecommunication companies posed to U.S. national security. The Committees' concerns are twofold. First, they suspect that the two Chinese companies may be selling equipment to the United States that is designed to intercept information or "establish the ability to do cyber attacks," according to Representative Mike Rogers (R-MI), Chairman of the House Permanent Select Committee on Intelligence. This concern stems from, in part, significant subsidies from the Chinese government that makes this equipment less expensive and therefore more popular among users.[23]

Numerous Congressional letters to U.S. governmental agency heads have addressed concerns over the relationship between national security and Chinese-manufactured telecommunications. In 2011, then-Senator Jon Kyl (R-AZ) wrote to the Chairman of the Federal Communications Commission (FCC) inquiring as to what the FCC might do to mitigate this vulnerability.[24] Also in 2011, Senators Kyl, Tom Coburn (R-OK), James Inhofe (R-OK), and Jim DeMint (R-SC) as well as Representative Sue Myrick (R-NC) contacted both the Secretaries of Defense and Energy, concerned with the use of Huawei technology in a sensitive computer center.[25] The same group of legislators wrote to the Secretaries of Commerce and the Treasury with questions concerning the Committee on Foreign Investment in the United States' (CFIUS) review of Huawei's takeover of 3Leaf Systems.[26] Moreover, they contacted then-Secretary of State Hillary Clinton to ask about Huawei's relationship with Iran.[27] In light of the political interest in Capitol Hill, CFIUS has become involved, and has stressed repeatedly that Huawei, the larger of the two Chinese companies, has extremely close links to the PLA and to the Chinese state ministries, exposing the U.S. communication system to heightened security threats.[28] In an October 2010 letter to then- FCC Chairman Julius Genachowski, Senators Jon Kyl (R-AZ), Joseph Lieberman (I-CT), and Susan Collins (R-ME), and Representative Sue Myrick (R-NC) wrote:

> "We are very concerned that these companies are being financed by the Chinese government and potentially subject to significant influence by the Chinese military which may create an opportunity for manipulation of switches, routers, or software embedded in American telecommunications networks so that communications can be disrupted, intercepted, tampered with or purposely misrouted."[29]

The letter also expressed concern that equipment designed and manufactured in China "may be remotely accessed and programmed from that country," posing a national security threat to the United States.[30] Study after study has shown that many cyber attacks against Western companies and government agencies originate in China.[31] Yet the Chinese government has not committed itself to identifying and prosecuting the hacking operations coming from its country.

In October 2012, after a year of hearings, a bipartisan report drafted by the U.S. House of Representatives Permanent Select Committee on Intelligence concluded that both Huawei and ZTE should no longer be allowed to install phone and data networks in the United States because of national security risks. The committee made that decision after reviewing classified and unclassified information and after questioning officials from each company. Huawei and ZTE are both accused of attempting to extract

sensitive information from U.S. companies. Chairman of the House Permanent Select Committee Intelligence Representative Mike Rogers (R-MI) and Representative C. A. Ruppersberger (D-MD), the committee's ranking Democrat, said that the U.S. government should be barred from doing business with Huawei and ZTE, and that U.S. companies should avoid buying their equipment.[32] The list of alleged cyber-infractions committed by China is long, and includes allegations that Chinese agents have conducted "cyber-attacks" against the Australian Prime Minister and Members of Parliament, German Chancellor Angela Merkel, and senior French officials. British and Canadian corporations have been warned by their respective governments to be attentive when doing business with Chinese entities, while both South Korea and India attribute various intrusions to Chinese sources.[33]

In February 2013, a U.S. cyber-security company accused a PLA unit in Shanghai of engaging in cyber-warfare against U.S. corporations, organizations and government agencies, citing a growing body of digital forensic evidence that pointed to the involvement of a PLA unit in Shanghai, and alleging that U.S. intelligence officials were tracking the unit's activities.[34] According to various reports, a run-down neighborhood in Shanghai is the center of a growing corps of PLA-affiliated cyber-warriors.[35]

Aside from hacking, intelligence gathering, and espionage, the United States runs the risk of losing its capacity to keep up with revolutionary changes in information and communication technology (ICT). Many U.S. companies have shifted operations to China or other large market, low cost environments to control prices and retain market shares. As with semiconductors, the back-end assembly of telecommunications equipment is labor-intensive, and many U.S. operators have offshore packaging and testing, which often leads to the entire production process taking place outside of the United States. Investment in overseas manufacturing makes old technology more profitable, and reduces the incentive to switch to newer but more expensive technologies.

For example, the efficiency of most communications networks could be dramatically increased by removing the "switches" between the fiber-optic backbone of most large-scale networks and the electronic signals used to relay data between the fiber-optic hubs and the end-user. However, this change would require significant R&D, followed by investments in upgrading production facilities. Given that offshore manufacturing facilities increase the profitability of older equipment, switching to more efficient technologies would reduce profit margins (although this would likely aid in addressing the network capacity issues discussed earlier in this chapter). The U.S. telecommunications industry, after having invented the Internet and having been the first to commercialize cell-phones (Motorola), is now at risk of losing the race for leading the next generation of telecommunications innovation.

CHINA AND THE GLOBAL TELECOMMUNICATIONS INDUSTRY

In the 1980s and1990s, the PLA experienced difficulties communicating and coordinating between different military commands due to outdated and insufficient telecommunications infrastructure. To overcome this inadequacy, the PLA purchased technology abroad. Western companies were extremely accommodating and competed intensely for a share of this growing market, which is currently the largest in the world.

The Chinese military is a major player in China's telecommunications modernization, but its entry into the telecommunications business was due to a historic accident. The Communist state traditionally encouraged the creation of a separate infrastructure for the military in the interest of attaining military self-sufficiency: the PLA operates separate airports, seaports, and railways, and maintains an independent telephone network. In addition, citing national security, the PLA assumed control over significant portions of the electromagnetic broadcast spectrum.[36]

In the early 1980s, the PLA responded to deep cuts in its budget by commercializing its enterprises. The PLA diversified, entering into peripheral businesses such as hotels, transportation, and light industrial production. Commercializing the PLA's separate communications network provided an immediate source of income. For example, the PLA's phone system had excess capacity that could be leased to provincial authorities. Additionally, many of the bandwidths reserved for the military had been left unused. Rather than surrender control of these bandwidths to civilian authorities, the PLA made commercial use of them. Among the frequencies under PLA control was the 800-MHz band, which is well-suited for mobile cellular communications.

By the mid-1990s, Qualcomm, Motorola, Northern Telecom, Ericsson, and Lucent were competing to construct a nationwide cellular network in China after the Chinese government invited the companies to bid for contracts at the provincial and local levels. As foreign telecommunications companies worked with local authorities to construct this cellular network, technological know-how inadvertently was transferred to Chinese counterparts and agencies, including those close to the military.[37]

These economic interests resulted in corruption and stunted the PLA's transition into a modern, professional military. Many of these sectors were eventually divested in the early 2000s.

Nevertheless, the PLA still retained an indirect stake in China's telecommunications sector, through its connections with state-funded R&D institutes and their ties with the commercial telecommunications sector. The Chinese Ministry of Science and Technology promotes a selective list of "national champions," which include energy (power generation, oil, and coal industries), automobiles, information technology, telecommunications, construction, and ferrous and nonferrous metals. These national champions are selected based on their ability to compete internationally, and receive state support through land and energy subsidies, favorable tax policies, and below-market interest rate "loans" issued from state banks with reduced or no expectation of repayment. For example, Chinese banks have made massive loans to telecommunications companies' customers that have aided them in penetrating many emerging markets. In one such case, Chinese banks loaned $30 billion to a Brazilian company with an extended grace period and very low interest rates.[38] While these export-credit programs are common, the capital available to the customers of Chinese telecommunications companies is magnitudes of order greater than that

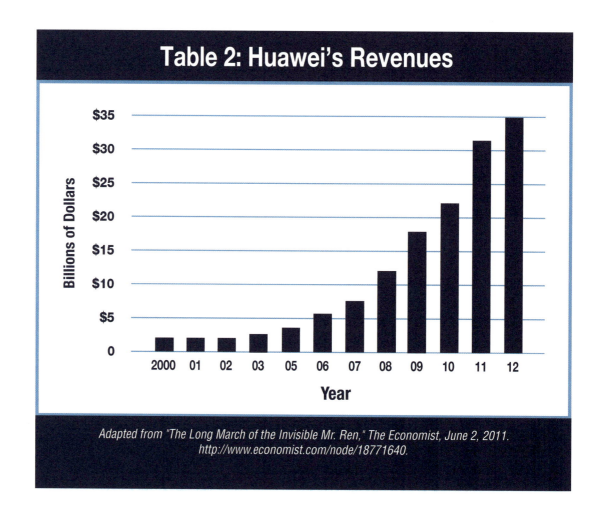

Table 2: Huawei's Revenues

Adapted from "The Long March of the Invisible Mr. Ren," The Economist, June 2, 2011. http://www.economist.com/node/18771640.

available elsewhere, and skews market competition.[39]

Much of the increased competitiveness of the Chinese telecommunications sector has resulted from funding, support, and direction from the Chinese government. The government has provided subsidies for infrastructure projects and licensing, and is insistent that the technology enabling China's wireless expansion be Chinese. For example, the construction of new roads has been transformed into an opportunity to install a high-volume fiber-optic network. Moreover, these cables are being installed regardless of population density, suggesting a forward-looking development strategy, where an area will already be wired prior to residents and businesses moving in.[40] Additionally, the Ministry of Science and Technology has recently awarded many Chinese telecommunications firms with "national laboratories," which are generally reserved for research universities and bring with them preferential access to funding.[41]

For all of these reasons, U.S., Australian, Indian, and European officials have expressed considerable alarm about the rise and success of Huawei, which reached sales of $28 billion in 2010 to take second position in the worldwide telecommunications market (after Ericsson at $30 billion). Importantly, Huawei controls the intellectual property rights to some 18,000 patents with 3,000 in overseas jurisdictions.[42]

Founded in 1988 by Ren Zhengfei, a former member of the PLA's Engineer Corps, Huawei ranks second among global telecommunications companies. Huawei's governance structure is opaque, and it has remained in private hands. Its corporate decisions are secretive, fueling suspicion that it is an appendage of the PLA or the Chinese state apparatus.[43] Repeatedly, Huawei has pledged to disclose more detailed financial information and information about its shareholders to dispel fears over suspected ties to the Chinese military, but officials in the United States and elsewhere and markets are still waiting to hear more about the company's unique ownership scheme.[44] Additionally, Huawei has grown rapidly as a result of offering network equipment at prices lower than established Western companies and selling its cheaper network equipment in the developing world.

ZTE, China's second largest telecommunications manufacturer, was founded in 1985. It has been less profitable than Huawei because it lacks Huawei's economies of scale. However, it has expressed ambitions to ramp up its production of telecommunications gear—a worrisome development for smaller vendors such as Nokia Siemens and Alcatel-Lucent. ZTE posted 2011 revenues of $13 billion, registering annual growth of 23 percent in 2010.

The entry of Chinese state-sponsored companies into markets historically friendly to the United States is particularly troubling. During the 1990s, Huawei expanded to Asian and African markets, where Western firms were uncompetitive due to high prices. Huawei also does considerable business in the Middle East, Southeast Asia, and Latin America. In the late 1990s, Huawei was accused of providing Iraq with improved fiber-optics for defense purposes.[45] In the early 2000s, China moved into the Argentine market after the United States withdrew. (It is important to note that Argentina controls a strategic longitudinal slot [81 degrees longitude] that is advantageous for satellite surveillance of all the Americas.) Huawei moved into the Argentine market after the Argentine economy crashed in 2001, and used that opportunity to sell equipment cheaply in order to capture market share. Chinese companies also took an active interest in developing Argentina's space program. Chinese companies sold equipment to the Argentine authorities at a steep discount and in return received a majority stake in the Argentine state satellite company ARSAT. Currently, China "owns" the Argentine space and telecommunications networks, a situation which itself is a concern for U.S. national security.[46]

Aside from its prices, Chinese telecommunications products also are attractive in developing markets because they have been leaders in introducing energy efficiency measures.[47] By using optical fiber cables, Huawei and ZTE were able to cut energy consumption by as much as a third.[48] Huawei currently operates in 140 countries and supplies components to nearly all of the world's large telecommunications companies. Huawei is a major player in Africa and the Middle East, having captured a large share of the telecommunications market in those regions.[49]

In sum, these two Chinese companies have made significant strides in dominating the global market and, especially in the case of Huawei, are thought to entertain close links with the military and the state. Observers point out that the telecommunications sector is a top priority for both the military and the Chinese government. Although Huawei's founder is a former military officer and the company is assumed to benefit from state funding, Huawei is listed as a privately-held company and does not divulge information about who sits on its board, how staff is promoted, or how it prices its products.

The lack of Western-style corporate governance norms coupled with the extremely low prices has generated an endless flow of rumors that both Huawei and ZTE receive substantial illegal aid from the Chinese government. Both companies deny that they engage in "dumping" (selling a product below its cost of production) and both claim that any subsidies they received were above board and legal. Years of investigation by the European Union (EU) has yielded evidence of unfair state subsidies and of Huawei's underpricing its telecommunications equipment, but the EU is hesitant to pursue a case with the World Trade Organization because of the possible repercussions for its larger trade relations with China.[50]

The monumental impact of Huawei's and ZTE's market behavior has been to drive down global telecommunications costs and bring about industry consolidation. Using aggressive price cuts (up to 50 percent) to acquire market share, Huawei and ZTE played a large role in forcing recent mergers and closings of other major telecommunications companies.[51]

RESPONSE BY THE U.S. GOVERNMENT

Telecommunications and information technologies, which undergird both commercial and military networks, occupy a no man's land between national security policies and national commercial, trade, and investment policies. The U.S. government response to the developments in the telecommunications market has been consistent, yet also *ad hoc*, without clear guidance about why the government has taken this position or how it seeks to address the falling competitiveness of U.S. telecommunications equipment and mobile set manufacturers. Over the last few years, lawmakers and regulatory agencies, as well as CFIUS, have taken steps to hinder or bar Huawei from entering the U.S. domestic market. CFIUS has reviewed many foreign investments in U.S. telecommunications businesses under its mandate to assess whether proposed foreign acquisitions of U.S. businesses threaten national security. CFIUS has imposed conditions on several acquisitions, while blocking others outright.[52]

In the last few years, Huawei has been in the news repeatedly as it tangled with authorities in its Western business markets. For example, Huawei neglected to report a relatively minor ($2 million) acquisition to CFIUS, creating concern over the Chinese company's opacity. Upon learning of this oversight, CFIUS advised Huawei to abandon several already completed acquisitions. In the face of American public outcry, Huawei has been prevented from acquiring other companies due to various national security concerns.

Huawei was also accused of corporate espionage in 2010, when Motorola claimed that the Chinese firm deliberately sought and acquired proprietary trade secrets. In 2002, several Motorola employees were discovered to be working for a Chinese firm that allegedly stole secrets, and later transferred them to Huawei.[53] Motorola's ensuing lawsuit against Huawei was later dropped in exchange for Huawei's dropping of an intellectual property case against Motorola. In 2004, Cisco sued Huawei for allegedly copying the company's router technology. This case was also settled out of court.[54]

As previously discussed, members of the U.S. House and Senate have expressed their concern to various U.S. government officials about Huawei's presence and business practices within the United States. Huawei and ZTE's aggressive pursuit of contracts to supply sensitive telecommunications infrastructure has raised red flags.

MITIGATING THE RISKS

Implement and enforce controls at each transaction point in the supply chain, but especially on delivery. New telecommunications products should be thoroughly tested prior to widespread adoption. Testing should focus on identifying small abnormalities in the circuitry of both end-systems and subcomponents. Additionally, DoD should request access to, and the right to inspect, any firmware or software needed for the hardware to function. Such a testing system should be proactive and build an active defense against tampering, sabotage, hacking, and infiltration. Since the supply chain is global, it makes sense for U.S. officials to cooperate with other nations to ward off cyber-attacks. Increased international cooperation to secure the integrity of the global IT system is a valuable long-term objective.

Refine CFIUS' mission and authority, to include formulating clear directions and guidelines as to when a technology investment or exchange constitutes a threat to national security. The manner in which CFIUS makes its determinations is unclear, and in some cases could jeopardize mergers, acquisitions, or joint ventures that are actually advantageous to the United States. Failure to clearly focus on national security implications could lead to allegations that the U.S. is politicizing trade policy, potentially reducing future foreign investment in the U.S. economy, or leading to trade disputes with foreign countries. Given the importance of China as a trade partner, the United States should make efforts to reassure China that these decisions truly are based in national security concerns rather than to gain an unfair economic edge.

Create incentives to ensure that a proportion of high-tech manufacturing of telecommunications equipment remains in the United Sates. The U.S. telecommunications industry has declined significantly in recent years. An appropriate package of tax and other incentives could reverse this trend. Telecommunications is a strategic sector, although lawmakers and DoD officials have neglected and ignored its decline in the United States, as well as its market challenges. The telecommunications sector is arguably as important as the aerospace sector, yet the latter is officially recognized as a key pillar of the defense industrial base. Aggressive action and targeted intervention are needed to revive the U.S. domestic telecommunications sector, including securing the more sensitive components of the U.S. IT infrastructure.

Rebuild the partnership between academia, the private sector, and defense to foster innovation and stimulate the ongoing commitment to design, manufacture, and assemble in the United States. For decades, U.S. industry thrived in an environment that encouraged basic and applied research and cemented close ties between academia, corporate labs, and DoD or DoE. An important step in the revival and survival of U.S. telecommunications sector would be to restore this triadic relationship between federal agencies, academia, and the private sector.

CONCLUSION

The telecommunications industry is a large and diverse sector that includes thousands of companies and hundreds of thousands of workers. However, and despite the fact that telecommunications is a strategic industry, the U.S. telecommunications sector is in decline. Its decline coincides with the rise of dynamic Chinese companies, allied with the Chinese military establishment and high-level Chinese research institutes, with a mandate to provide funds and resources to budding "national champion" firms in the telecommunications and information technology sectors. Thus, just as many U.S. network equipment manufacturers are throwing in the towel, Chinese companies are offering a cheap alternative. Aside from costing U.S. jobs and future opportunities to improve IT technologies, the ascendance of Chinese telecommunications companies also poses a threat to U.S. national security due to the growing risk of cyber-attacks directed at infiltration or disruption of the U.S. communications infrastructure. Such a scenario is all the more likely because even U.S. telecommunications companies, which still manufacture network equipment, rely extensively on global supply chains that include factories and distribution centers across the globe. We must forge a coordinated and comprehensive strategy to address the national security threat to this vital sector of our defense industrial base.

ENDNOTES

1   Daniel Hartnett, *Towards a Globally Focused Chinese Military: the Historic Missions of the Chinese Armed Forces* (Alexandria, VA: Center for Naval Analysis, June 2008). http://www.cna.org/sites/default/files/9.pdf.

2   A broadcast network (such as NBC or NPR), consisting of a single transmitting station and multiple receive-only stations, is considered a form of telecommunications. This chapter does not include this type of telecommunication.

3   iPhone is a registered trademark of Apple Inc.

4   Android is a trademark of Google Inc.

5   Routers are used to connect users between networks, while switches and hubs are used to connect users within a network. Many routers are now designed to perform the functions of switches and hubs as well as other security services, such as intrusion detection and prevention and antivirus scanning. Most networks are designed for redundancy and have multiple routers; in case of an outage the remaining routers are reconfigured. Routers, switches, and hubs are increasingly manufactured abroad.

6   One example came to light after the Belgian labor union and Option SA (a wireless wide-area network modem manufacturer) filed an official complaint with the European Union in 2010, alleging that both Huawei and ZTE received a combined $45 billion in credit from the China Development Bank on favorable terms, with a moratorium on interest payments. Matthew Dalton, "EU Finds China Gives Aid to Huawei, ZTE," *The Wall Street Journal*, February 3, 2011. http://online.wsj.com/article/SB10001424052748703960804576120012288591074.html.

7   Kevin McCauley, "The PLA's Three-Pronged Approach to Achieving Jointness in Command and Control,"

    China Brief Jamestown Foundation 12, no. 6 (March 15, 2012).

    Dean Cheng, "China's Military Role in Space," *Strategic Studies Quarterly* (Spring 2012), 55-77.

    Bryan Krekel, Patton Adams, and George Bakos (Northrop Grumman Corporation), *Occupying the Information High Ground: Chinese Capabilities for Computer Network Operations and Cyber Espionage* (U.S.-China Economic and Security Review Commission, March 7, 2012). http://www.uscc.gov/RFP/2012/USCCReport_Chinese_CapabilitiesforComputer_NetworkOperationsandCyberEspionage.pdf.

8   The Insight-Corporation, *Telecommunications Industry Review: An Anthology of Market Facts and Forecasts 2011-2016* (January 2012). www.insight-corp.com/reports/review12.asp.

9   Common Cause "The Fallout From the Telecommunications Act of 1996: Unintended Consequences and Lessons Learned," (Washington, D.C.: Common Cause Education Fund, May 9, 2005), 6. http://www.commoncause.org/atf/cf/%7BFB3C17E2-CDD1-4DF6-92BE-BD4429893665%7D/ FALLOUT_FROM_THE_TELECOMM_ACT_5-9-05.PDF.

10 Om Malik, *Broadbandits: Inside the $750b Telecom Heist* (Hoboken, NJ: John Wiley, 2003).

11 Robert W. Lucky and Jon Eisenberg, eds, *Renewing U.S. Telecommunications Research* (Washington, D.C.: The National Academies Press, 2006).

12 Theodore S. Rappaport, "Offshoring in the U.S. Telecommunications Industry," in *The Offshoring of Engineering: Facts, Unknowns, and Potential Implications* (Washington D.C.: The National Academies Press, 2008).

13 Many operators support six or more separate networks delivering different groups of services such as voice network, Internet/IP network, ATM, low-speed leased line, and high speed leased line. Combining all services into one operating system would be more effective.

14 Andrew Wheen, *Dot-Dash to Dot.Com: How Modern Telecommunications Evolved from the Telegraph to the Internet* (New York/Heidelberg: Springer Praxis Books, 2011).

15 Erica Fuchs and Randolph Kirchain, "Design for Location?: The Impact of Manufacturing Off-Shore on Technology Competitiveness in the Optoelectronics Industry," *Management Science* 56 (2010): 2323-2349.

16 "The Company That Spooked The World: The Success of China's Telecoms-Equipment Behemoth Makes Spies and Politicians Elsewhere Nervous," *The Economist*, August 4, 2012. http://www.economist.com/node/21559929.

17 "Two's Company: In An Industry with A Cost Leader And A Price Leader, Is There Room for Others?" *The Economist*, December 3, 2011.

18 Alan Weissberger, "Huawei, ZTE, and Ericcson to Dominate Telecom Infrastructure Equipment Market— Or Not?" *IEEE Communications Society* (February 14, 2011). http://community.comsoc.org/blogs/alanweissberger/huawei-zte-and-ericsson-dominate-telecom-infrastructure-equipment-market-or-no.

19 Nokia Siemens Networks and Cisco are more or less of equal size as measured by telecom equipment revenues. Ericsson and Huawei are considerably larger. "The Company That Spooked The World: The Success of China's Telecoms-Equipment Behemoth Makes Spies and Politicians Elsewhere Nervous," *The Economist*, August 4, 2012. http://www.economist.com/node/21559929.

20 Frank McFadden and Richard Arnold, "Supply Chain Risk Mitigation for IT Electronics," *IEEE (2010)*, 49-55. http://ieeexplore.ieee.org/stamp/stamp.jsp?arnumber=05655094.

21 John Grimes, "Secure Cellular Comms Could Continue to Be a Hero," *Defense Systems*, April 7, 2011. http://defensesystems.com/Articles/2011/03/29/Industry-Perspective-Secure-Wireless-Communications.aspx?Page=2&p=1.

22 Bryan Krekel, Patton Adams, and George Bakos (Northrop Grumman Corporation), *Occupying the Information High Ground: Chinese Capabilities for Computer Network Operations and Cyber Espionage* (U.S.-China Economic and Security Review Commission, March 7, 2012). http://www.uscc.gov/RFP/2012/USCCReport_Chinese_CapabilitiesforComputer_NetworkOperationsandCyberEspionage.pdf.

89.

23 The letters are found on the website of the U.S. House of Representatives, Permanent Select Committee on Intelligence. "Rogers and Ruppersberger Intensify Investigation of Huawei and ZTE," House of Representatives press release, June 13, 2012, http://intelligence.house.gov/press-release/rogers-and-ruppersberger-intensify-investigation-huawei-and-zte.

U.S. House of Representatives, Permanent Select Committee on Intelligence, letter to Charles Ding of Hauwei Technologies Co., LTD., June, 12, 2012. http://intelligence.house.gov/sites/intelligence.house.gov/files/documents/HuaweiCharlesDing12JUNE2012.pdf.

Jim Wolf, "China Telecom Firms May Be Subsidized: U.S. Lawmaker," Reuters, June 21, 2012. http://www.reuters.com/article/2012/06/22/us-china-usa-huawei-idUSBRE85L03G20120622.

24 Lewis Dowling, "US Blocks Huawei from Public Safety LTE Trial," Total Telecom, October 11, 2011. http://www.totaltele.com/view.aspx?ID=468333.

"Rogers and Ruppersberger Intensify Investigation of Huawei and ZTE," U.S. House of Representatives Permanent Select Committee on Intelligence press release, June 13, 2012. http://intelligence.house.gov/press-release/rogers-and-ruppersberger-intensify-investigation-huawei-and-zte.

25 Eli Lake, "Computer Lab's Chinese-Made Parts Raise Spy Concerns," The Washington Times, August 16, 2011. http://www.washingtontimes.com/news/2011/aug/16/computer-labs-parts-raise-spy-concerns/?page=all.

Todd Shields, "Locke Says Sprint's Chief Was Called about Huawei Bid Concerns," Businessweek, December 7, 2010. http://www.businessweek.com/news/2010-12-07/locke-says-sprint-s-chief-was-called-about-huawei-bid-concerns.html.

26 Senator Jon Kyl and Senator Jim Webb, letter to then-Secretary of the Treasury Timothy Geithner and U.S. Ambassador to China Gary Locke (February 10, 2011). http://myrick.house.gov/uploads/02102011%20letter%20to%20Geithner%20and%20Locke.pdf.

Claude Barfield, Telecoms and the Huawei Conundrum: Chinese Foreign Direct Investment in the United States. (Washington DC: AEI Economics Studies, November 2011), 3-4.

27 Jon Kyl, Sheldon Whitehouse, et al., Letter to then Secretary of State Hillary Clinton (December 22, 2011). http://myrick.house.gov/uploads/12222011_Huawei_Iran_Concerns1.pdf.

28 U.S.-China Economic and Security Review Commission (USCC), The National Security Implications of Investment and Products from the People's Republic of China in the Telecommunications Sector (Washington, D.C.: USCC, January 2011). http://www.uscc.gov/RFP/2011/FINALREPORT_TheNationalSecurityImplicationsofInvestmentsandProductsfromThePRCintheTelecommunicationsSector.pdf.

29 U.S. Senate Committee on Homeland Security and Governmental Affairs, "Congressional Leaders Cite Telecommunications Concerns with Firms That Have Ties with Chinese Government," October 19, 2010. http://www.hsgac.senate.gov/media/minority-media/congressional-leaders-cite-telecommunications-concerns-with-firms-that-have-ties-with-chinese-government.

30 Ibid.

31 See the U.S.-China Economic and Security Review Commission-sponsored reports Occupying the Information High Ground: Chinese Capabilities for Computer Network Operations and Cyber Espionage (2012) and The National Security Implications of Investment and Products from the People's Republic of China in the Telecommunications Sector (2011). Both reports rely on numerous outside sources, both in English and Chinese (Mandarin). They are available online at http://www.uscc.gov/.

32 Michael Schmidt, Keith Bradsher, and Christine Hauser, "U.S. Panel Cites Risks in Chinese Equipment, The New York Times, October 8, 2012. http://www.nytimes.com/2012/10/09/us/us-panel-calls-huawei-and-zte-national-security-threat.html?pagewanted=all&_r=0.

33 James Andrew Lewis, "Cyber Security and US-China Relations," China-US Focus, July 6, 2011. http://www.chinausfocus.com/peace-security/cyber-security-and-us-china-relations/.

34 David E. Sanger, David Barboza, and Nicole Perlroth, "Chinese Army Unit Is Seen as Tied to Hacking Against U.S.," The New York Times, February 18, 2013. http://www.nytimes.com/2013/02/19/technology/chinas-army-is-seen-as-tied-to-hacking-against-us.html?hp&_r=1&.

35 David Barboza, "China Says Army Is Not Behind Attacks in Report," The New York Times, February 20, 2013. http://www.nytimes.com/2013/02/21/business/global/china-says-army-not-behind-attacks-in-report.html.

David Sanger, David Barboza, and Nicole Perlroth, "Chinese Army Unit Is Seen as Tied to Hacking Against U.S.," The New York Times, February 18, 2013. http://www.nytimes.com/2013/02/19/technology/chinas-army-is-seen-as-tied-to-hacking-against-us.html?pagewanted=all.

36 James Mulvenon and Thomas J. Bickford, "The PLA and the Telecommunications Industry in China," in James C. Mulvenon and Richard H. Yang, The People's Liberation Army in the Information Age (Santa Monica, CA: RAND Corporation, 1999). http://www.rand.org/pubs/conf_proceedings/CF145/CF145.chap12.pdf.

37 James Mulvenon and Thomas J. Bickford, "The PLA and the Telecommunications Industry in Chin."

James Mulvenon, "The PLA in the New Economy: Plus Ça change, Plus C'est la Même Chose," in Civil-military relations in today's China: Swimming in a New Sea, eds. David M. Finkelstein and Kristen Gunness. (Armonk, NY: M.E. Sharpe, 2007).

38 London Interbank Offered Rate is the average interest rate leading London banks estimate that they would be charged if borrowing from other banks.

39 "Huawei's $30 Billion China Credit Opens Doors in Brazil, Mexico," Bloomberg News, April 24, 2011. http://www.bloomberg.com/news/2011-04-25/huawei-counts-on-30-billion-china-credit-to-open-doors-in-brazil-mexico.html.

40 Theodore S. Rappaport, "Offshoring in the U.S. Telecommunications Industry," The Offshoring of Engineering: Facts, Unknowns, and Potential Implications (Washington D.C.: The National Academies Press, 2008).

41 Claude Barfield, Telecoms and the Huawei Conundrum: Chinese Foreign Direct Investment in the United States. (Washington DC: AEI Economics Studies, November 2011), 6. http://www.aei.org/article/telecoms-and-the-huawei-conundrum/.

42 "The Long March of the Invisible Mr. Ren China's Technology Star Needs to Shine More Openly," *The Economist*, June 2, 2011. http://www.economist.com/node/18771640.

43 "The Long March of the Invisible Mr. Ren China's Technology Star Needs to Shine More Openly," *The Economist*, June 2, 2011. http://www.economist.com/node/18771640.

44 Kathrin Hille, "Huawei in Pledge to Disclose More Information," *The Financial Times*, January 21, 2013.

http://www.ft.com/intl/cms/s/0/4b0ab5ce-6398-11e2-84d8-00144feab49a.html.

45 William Jasper, "Chinagate All Over Again: The Disastrous Transfers of Military Technology to Communist China Continue Apace, Much as They Did under Clinton," *The New American*, March 10, 2003.

46 Janie Hulse, "China's Expansion into and U.S. Withdrawal from Argentina's Telecommunications and Space Industries and the Implications for U.S. National Security," *Strategic Studies Institute* (September 2007). http://www.strategicstudiesinstitute.army.mil/pubs/display.cfm?pubID=806.

47 "Up, Up and Huawei: China Has Made Huge Strides in Network Equipment," *The Economist*, September 24, 2009. http://www.economist.com/node/14483904.

48 Ibid.

49 With its competitive pricing and presence in Africa, Huawei consistently undercut Ericsson and Nokia by five to 15 percent. Knowledge@Wharton, *Huawei Technologies: A Chinese Trail Blazer in Africa* (Philadelphia, PA, Wharton Business School, April 20, 2009). http://knowledge.wharton.upenn.edu/article.cfm?articleid=2211.

50 Joshua Chaffin, "Beijing Faces Brussels Action on Telecoms Aid," *The Financial Times*, May 25, 2012. http://www.ft.com/intl/cms/s/0/876632ae-a689-11e1-aef2-00144feabdc0.html#axzz1z9YREQeY.

Joshua Chaffin and Gerrit Wiesmann, "EU Trade Officials Face China Dilemma," *The Financial Times*, September 2, 2012. http://www.ft.com/intl/cms/s/0/7873f2d2-f4e9-11e1-b120-00144feabdc0.html#axzz25tzShA22.

51 "Up, Up and Huawei: China Has Made Huge Strides in Network Equipment," *The Economist*, September 24, 2009. http://www.economist.com/node/14483904.

52 Warren G. Lavey, "Telecom Globalization and Deregulation Encounter U.S. National Security and Labor Concerns," *Journal on Telecommunications and High Technology Law* 6 (Fall 2007), 121-175. http://www.law.yale.edu/documents/pdf/cbl/Lavey_Telecom_Globalization.pdf.

53 Court documents indicate Motorola became aware of the alleged theft of trade secrets only after one of its employees was stopped during a routine U.S. customs check at Chicago's O'Hare airport on February 28, 2007. The employee carried a large amount of cash, and more importantly, she was in the possession of 1,300 electronic and paper documents belonging to Motorola, some of them marked confidential. Jamil Anderlini, "Motorola Claims Espionage in Huawei Lawsuit," *The Financial Times*, July 22, 2010. http://www.ft.com/cms/s/0/616d2b34-953d-11df-b2e1-00144feab49a.html#ixzz1wgVn2sCG.

54 Kathrin Hille and Paul Taylor, "Huawei Declares Truce with Motorola," *The Financial Times*, April 13, 2011. http://www.ft.com/cms/s/0/b6813068-65f6-11e0-9d40-00144feab49a.html#axzz1wkR1z7uD.

Stephanie Kirchgaessner, "Huawei U-turn on U.S. Deal Saves Blushes," *The Financial Times*, February 20, 2011. http://www.ft.com/intl/cms/s/2/28c1e442-3d20-11e0-bbff-00144feabdc0.html#axzz1wkR1z7uD.

CHAPTER 13
NIGHT VISION DEVICES

EXECUTIVE SUMMARY

Night vision devices (NVD) make use of image intensification (I2) technology to amplify light, allowing U.S. troops to operate in low-light environments. This technology is a significant advantage over opposing forces that lack similar capabilities, and often must curtail night operations for lack of visibility. State-of-the-art NVDs are defense-specific products, with only limited commercial demand; therefore NVD manufacturers are heavily reliant on military contracts. With increasingly constrained defense resources, and combat operations in Iraq and Afghanistan winding down, demand for NVDs is likely to decrease.

Maintaining a technological edge in night vision will require sustained investments in research and development (R&D). Military NVDs represent a niche market; military sales represent 70 percent of total sales for one major NVD firm. U.S. firms do invest in R&D, but a majority of funding comes from the Department of Defense (DoD).

NVD production requires access to the rare earth element (REE) lanthanum, which is predominantly supplied by Chinese firms. China provided 91 percent of lanthanum exports to the U.S. in 2010. China's near-monopoly of the REE market has prompted widespread allegations of price manipulation and artificial supply restrictions. In 2010, Japanese access to Chinese REEs was temporarily cut off following a diplomatic dispute. China (among other countries) has been actively pursuing NVD technology, and in 2007 was reported to have received access to U.S. night vision technology through illicit means. Dependence on China for inputs critical to NVDs creates a risk that China could withhold access to imports as a way to inhibit a U.S. technological advantage.

The United States must assure long-term R&D funding and demand for NVDs, and should identify either strategies to assure stable access to lanthanum or a practical alternative.

NIGHT VISION DEVICES
ENHANCING U.S. MILITARY EFFECTIVENESS

MANUFACTURING SECURITY

Night vision devices are an essential U.S. military capability

NIGHT-VISION DEVICES + **U.S. TROOPS** = **NIGHTTIME DOMINANCE**

SUPPLY VULNERABILITIES

China is the leading supplier of rare earth elements crucial to national security

CHINA IS THE SUPPLIER OF 97% **OF ALL REE**

The rare earth element lanthanum has special magnetic properties and is a critical input for advanced U.S. night vision devices

91% OF U.S. SUPPLY OF LANTHANUM COMES FROM **CHINESE COMPANIES**

FUNDING UNCERTANTIES

DoD funding for R&D is currently facing budget uncertainties

 + =

LOWER DOD INVESTMENTS + **BUDGET** UNCERTAINTY = **LOSS OF** CAPACITY

LEADING TECHNOLOGY

The U.S. is the world leader in night vision technology

U.S. LEADS THE WORLD IN **NIGHT VISION** TECHNOLOGY

SPREAD OF TECHNOLOGY

Export controls must prevent illicit technology transfer and enable cooperation with partners

PREVENT ILLICIT TECHNOLOGY **TRANSFERS**

ENABLE COOPERATION **WITH PARTNERS**

MITIGATING RISKS

Avoiding risks to U.S. night vision capabilities

STABLE DOD DEMAND   **FUNDING INNOVATION**   **SECURING SUPPLY CHAIN**

MILITARY EQUIPMENT CHART
SELECTED DEFENSE USES OF NIGHT VISION

DEPARTMENT	PLATFORMS AND WEAPON SYSTEMS	OTHER EQUIPMENT
ARMY	■ M24 Sniper rifle ■ AH-64 Apache helicopter ■ UH-60 Blackhawk helicopter	■ AN/AVS-6(V)3 aviator's night vision imaging system ■ AN/PVS-7 binocular night vision scope ■ AN/PVS-10 binocular night vision ■ AN/PVS-15 submersible night vision binoculars ■ AN/PVS-23 binocular night vision goggles ■ AN/AVS-7 Aviator's night vision imaging system ■ AN/GVS-5 laser rangefinder ■ AN/AAS-32 airborne laser tracker ■ AN/PVS-22 scope ■ Target Acquisition and Designation Sights (TADS)
NAVY	■ SH-60 Seahawk helicopter	■ AN/AVS-6(V)3 Aviator's night vision imaging system ■ AN/PVS-15 submersible night vision binoculars ■ AN/PVS-23 binocular night vision goggles ■ AN/AVS-7 aviator's night vision imaging system ■ AN/GVS-5 laser rangefinder ■ AN/AAS-32 airborne laser tracker ■ AN/PVS-22 scope
AIR FORCE	■ A-10 Thunderbolt II aircraft ■ MC-130 Combat Talon aircraft ■ AC-130U Spooky aircraft	■ Panoramic night vision goggles (PNVGs) ■ AN/PSQ-20 Enhanced night vision goggles (ENVG) ■ Fused multispectral weapon sight

INTRODUCTION

"Our night vision capability provided the single greatest mismatch in the [first Gulf] War."

–General Barry McCaffrey, Commanding General, 24th Infantry Division (Mechanized) during Operation DESERT STORM

U.S. technological superiority across a range of sophisticated capabilities is necessary for U.S. military dominance. As then Defense Secretary Leon Panetta said in a March 1, 2012, speech, "the force of the future" must be maintained with a "decisive, technological edge" in order to "defend this country and our global interests in the 21st century."[1] Among the capabilities that provide that edge are NVDs, a crucial tool that enhances U.S. military effectiveness in carrying out important operations.

NVDs allow U.S. troops to "own the night" by using image intensification (I2) technology to amplify available light from the moon, stars, or distant cities up to thousands of times.[2] The light is collected, amplified, and then projected onto a phosphor screen. This process creates an intensified and readily observable, green-tinted image visible to the naked eye.[3]

The operation that killed former al-Qaeda leader Osama bin Laden in Abbottabad, Pakistan, took place at night, and the elite U.S. Navy SEAL team that carried out the operation wore advanced night vision goggles.

The widespread adaptation of NVDs has made the United States more capable against its enemies, including foes such as the Taliban in Afghanistan. Afghan insurgents have attempted to gain their own night vision capabilities[4] but, as *Wired* magazine reported in 2011, "insurgents prefer daylight, mostly leaving the night to tech-savvy U.S. forces."[5]

Night vision capabilities are particularly crucial for U.S. special operations. The operation that killed former al Qaeda leader Osama bin Laden in Abbottabad, Pakistan, took place at night, and the elite U.S. Navy SEAL team that carried out the operation wore advanced night vision goggles when carrying it out.[6] As the global struggle against terrorists and violent extremists evolves, the United States' capability to carry out these kinds of operations likely will only grow in importance. Brookings Institution Senior Fellow Michael O'Hanlon has called night vision capabilities "essential" and "one of the real trump cards we have in the battle with al-Qaida."[7]

The United States possesses a critical edge in night vision technology. Maintaining this advantage, and preventing it from spreading to current or potential rivals, strengthens U.S. national security.[8] However, China's night vision capabilities are advancing, and China has attempted to acquire U.S. night vision technology.[9] The spread of night vision technology could diminish an important U.S. tactical advantage.

To ensure that the United States preserves this critical advantage, it is essential to understand the vulnerabilities associated with U.S. ability to produce NVDs, and to continue to successfully develop this innovative technology.

A NECESSARY CAPABILITY THAT MUST BE PRESERVED (a notional though realistic scenario)

U.S. Special Forces deploy near Kandahar, Afghanistan, to capture a senior al-Qaeda member. Escape vehicles must be disabled; civilian casualties must be minimized. Snipers deploy to an overwatch position. The al-Qaeda meeting occurs in darkness, and the team is equipped with a AN/PVS-27 PINNACLE Magnum Universal Night Sight® (MUNS™), a high-resolution clip-on weapon sight that adds night vision capabilities to an existing daytime scope.

An operative arrives at the appointed meeting time in a Toyota Land Cruiser. U.S. and Afghan troops emerge from their hide sites and approach the vehicle. The operative quickly reenters his vehicle and attempts to flee. The sniper takes his shot. Aiming at the engine compartment, he fires and misses. The operative escapes and speeds away, eluding capture. Even though the operation is unsuccessful, it would have been unthinkable without night vision capabilities.

Key themes discussed in this chapter are:

- Night vision is an important U.S. capability for a variety of military operations against enemies.

- The United States should take steps to retain its technological advantages over competitors in developing and producing night vision technology.

- The REE lanthanum, an important input for NVD production, is currently produced almost entirely in China.

A NOTE ON CRITICALITY

NVDs are important force multipliers that allow U.S. soldiers to operate in low-light environments. Night vision devices increase the effectiveness and safety of the U.S. warfighter. If NVDs were unavailable, this loss would represent an *isolated* impact on U.S. defensive capabilities and eliminate a significant tactical advantage.

While the United States remains the leader in night vision technology, it currently depends on China for the REE lanthanum, a critical input needed to manufacture NVDs (see Figure 1). Chinese domination of the production of this critical input potentially limits production capability, and constitutes an *extreme* vulnerability. Despite efforts to redevelop domestic REEs, U.S. lanthanum production is not projected to adequately meet domestic consumption needs in the near future.

BACKGROUND

Portable NVDs are made by the U.S. imaging and sensors industry, which produces devices for commercial, defense, and recreational applications.[10] U.S. firms lead the defense portion of the market. In the commercial portion, however, foreign firms are increasingly servicing global demand.[11] In recent years, NVD manufacturers such as ITT Exelis and L-3 Insight Technology saw increased sales due to demand associated with the wars in Afghanistan and

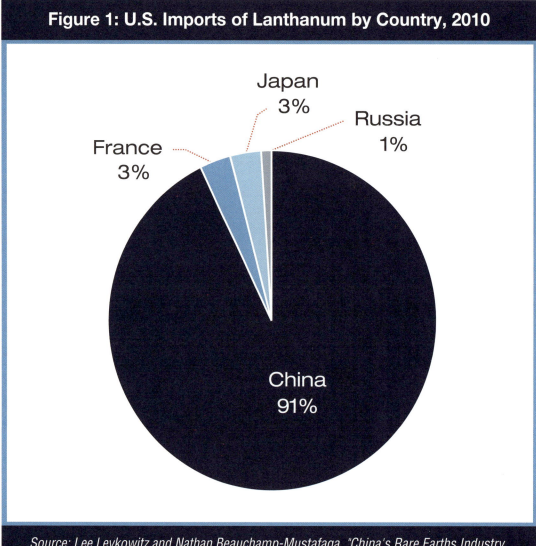

Figure 1: U.S. Imports of Lanthanum by Country, 2010

Japan
3%

Russia
1%

France
3%

China
91%

Source: Lee Levkowitz and Nathan Beauchamp-Mustafaga, "China's Rare Earths Industry and its Role in the International Market," U.S.-China Economic and Security Review Commission Staff Backgrounder, november 3, 2010. http://origin.www.uscc.gov/sites/default/files/Research/RareEarthsBackgrounderFINAL.pdf.

Iraq, although the U.S. government has NVD needs other than war-fighting.[12]

An in-depth 2005 U.S. Department of Commerce (DoC) survey of the sensors and imaging industry revealed that defense sales made up approximately two-thirds of all sales; the overall health of the industry depends heavily on DoD demand.[13]

The imaging and sensors industry has consolidated in the last decade, with larger firms buying up smaller firms. This mirrors larger trends in the defense industrial base. There are both positive and negative effects associated with consolidation in the U.S. defense industrial base. Consolidation can mean cost savings for the U.S. government, but it can also result in reduced competition, and may leave only a small number of firms able to supply a particular capability.[14]

U.S. NVD manufacturers were not significantly affected by the recent economic downturn, given the prominence of DoD demand relative to total demand. Revenue will likely shrink in the coming years, however, as defense spending declines and becomes more uncertain.

NIGHT VISION DEVICES AND U.S. DEFENSE CAPABILITIES

NVDs provide a powerful advantage over enemies with limited or nonexistent night vision capabilities. The U.S. military has pursued night vision capabilities since World War II. The first NVDs to use I2 technology were deployed during the mid-1960s. The technology has greatly advanced since then, achieving greater capacity in amplification, resolution, portability, and ease of use. U.S. forces have made extensive use of night vision over the last two decades. This was true, for example, of Operation Desert Storm in 1990 and 1991. General Barry McCaffrey said of Operation Desert Storm that U.S. "night vision capability provided the single greatest mismatch of the war."[15] Today, it is a critical capability for U.S. forces operating in hostile environments. NVDs enable success in a variety of missions, including:

- Targeting and tracking enemy forces during nighttime operations

- Protecting U.S. bases and facilities in hostile territory from nighttime attack

- Enhancing security of domestic facilities, including U.S. ports

The U.S. government currently recognizes three generations of NVDs (see Figure 2). Each provides progressively better capabilities. Certain NVDs are available commercially and are used recreationally or commercially. Advanced NVDs are used exclusively by the military; they are subject to export controls to prevent the technology from spreading to unauthorized countries.

ALTERNATIVES TO NIGHT VISION DEVICES

The main alternative to I2 technology is thermal imaging. While I2 devices amplify available light to attain greater visibility, thermal devices detect heat signatures[16] by detecting the infrared portion of the light spectrum, which objects emit as heat. Warmer objects emit more light than colder objects. For example, with the use of thermal imaging, a person or a moving vehicle will be more visible than a road or building.[17]

Thermal imaging devices and I2 devices have different advantages and disadvantages. For instance, a camouflaged enemy soldier would be more visible to a thermal imaging device than to an I2 device, because visual camouflage cannot disguise the heat signature of a person or a moving object. Thermal imaging also can see through fog and other environmental factors that decrease visibility. I2 devices, on the other hand, allow for better recognition of fine details, such as the contours of a human face. I2 devices are also cheaper and more compact.[18]

The helmet-mounted monocular Enhanced Night Vision Goggle (ENVG) integrates I2 and thermal sensors into a single device. These two capabilities are digitally blended

Figure 2: Generations of Night Vision Devices Using Image Intensification Technology

Generation-1	- Require approximate equivalent of one-half the light of a full moon to operate. - NVDs used for civilian applications are most often Generation-1. - Most do not require an export license. - Amplification: 1,000x
Generation-2	- Use more advanced electronics to provide greater clarity. - Utilize a micro-channel plate to add greater clarity. - Longer battery life than Generation-1. - Most Generation-2 NVDs require an export license. - Amplification: 20,000x
Generation-3	- Typically use a Gallium Arsenide photocathode. - More efficient and longer lasting. - Generation-3 NVDs require an export license. - Amplification: 30,000-50,000x

Source: U.S. Department of Commerce, Bureau of Industry and Security, Office of Strategic Industries and Economic Security, Defense Industrial Base Assessment: U.S. Imaging and Sensors Industry (Washington, D.C.: U.S. Department of Commerce, BIS/SIES, October 2006), p. ii-4. http://www.bis.doc.gov/news/2006/wholereportwithappendices10_12_06.pdf

to provide U.S. forces with enhanced detection and situational awareness. These goggles have been described as the "next logical iteration" of current U.S. Army I2 capabilities.[19]

Given the different advantages that these two capabilities provide in combat, thermal imaging cannot fully replace I2 for use on the battlefield. Thermal imaging and I2 capabilities are more complementary than interchangeable.[20]

RECENT DEVELOPMENTS

The U.S. Army procures NVDs through a series of multiyear Omnibus contracts.[21] The Army awarded the first, Omnibus I, in 1985. Omnibus VII was awarded in 2005, with an estimated potential value of $3.2 billion for up to 360,000 NVDs[22] in 2005.[23] Omnibus VIII was the most recent contract.

The unauthorized spread of night vision technology has long been a concern for the United States. The Department of Justice (DOJ) fined ITT Corporation[24] $100 million for sending sensitive technical data related to night vision to China, Singapore, and Britain in 2007.[25] In addition to the penalty, officers of the ITT Corporation pled guilty to two felony charges, and undertook steps to prevent future illegal transfers. According to U.S. Attorney for the Western District of Virginia John Brownlee, transfers of this kind threaten "to turn on the lights on the modern battlefield for our enemies and expose American soldiers to great harm."[26] The penalties required ITT to invest $50 million in the "acceleration, development and fielding of the most advanced" NVDs.[27]

Future cuts to Department of Defense procurement and research and development spending may affect the willingness and ability of firms to continue current levels of capital investment in NVD capabilities.

The U.S. imaging and sensors industry has gone through various high-level mergers and acquisitions in recent years, mostly with large manufacturers' purchasing of smaller, more specialized firms.[28] In 2002, for example, acquisitions worth almost $8 billion took place. In 2010, L-3 Communications, a major manufacturer, acquired NVD and electro-optical equipment maker Insight Technologies, a firm with approximately $290 million in sales that year.[29]

In 2010 the U.S. Army Research Development and Engineering Command ordered ENVGs from ITT Exelis, L-3 Insight Technology, and DRS Systems for testing in anticipation of wider fielding of the system. ITT Exelis had previously produced the system under a sole-source contract.[30]

The U.S. Army ordered 3,800 of the more advanced Spiral Enhanced Night Vision Goggles (SENVG) in May 2012. The device can incorporate video and data from outside sources, including potentially from unmanned aerial vehicles, according to the Vice President of Programs for ITT Exelis.[31]

U.S. companies have collaborated with foreign firms to supply NVD capabilities to partner nations. ITT Exelis formed a partnership with India-based Tata Advanced Systems Limited (TASL) in March 2012. ITT Exelis will supply TASL "with the latest

Gen 3 night vision image intensifier tubes, kits and other materials" so that TASL will be able to manufacture "high precision components and sub-assemblies of the devices" in India.[32] This partnership will not include the transfer of sensitive I2 tube technology.[33] ITT Exelis also recently signed a contract with the United Kingdom Ministry of Defense to supply Generation 3 NVDs.[34] Such partnerships are to be expected in an era of increasing economic and allied military interdependence. They also further globalize defense supply chains. The risks and benefits of such arrangements must be carefully evaluated to preserve the viability of this key sector of the U.S. defense industrial base and to prevent the transfer of defense technology to countries with interests that may conflict with those of the United States.

ISSUES AFFECTING NIGHT VISION DEVICE AVAILABILITY

DoD procurement and research and development (R&D) decisions. U.S. imaging and sensor firms that manufacture NVDs are highly dependent on the Pentagon for R&D and for sustained production orders. From 2003 to 2005, for example, DoD spent $350 million on manufacturer R&D, compared to $300 million of internal R&D by private firms in the United States.[35]

Future cuts to DoD procurement and R&D spending may affect the willingness and ability of firms to continue current levels of capital investment in NVD capabilities. ITT Exelis' Chief Strategy and Development Officer told *Aviation Week & Space Technology* that it was difficult for his company, a significant NVD manufacturer,

"to find areas where you can see growth in almost anything" due to tight budgets. Defense sales make up almost 70 percent of sales for the company, which projected declining revenues in 2012.[36] The *Washington Post* reported in April 2012 that ITT Exelis is increasingly moving into services while de-emphasizing manufactured products such as NVDs.[37] In February 2013, the company announced that it would be closing a factory in West Springfield, Massachusetts, that manufactures NVD power sources, laying off most of the factory's 235 workers. An ITT Exelis spokesman attributed the move to decreasing DoD demand for military NVDs.[38]

The future of U.S. combat operations. U.S. service members, including aviators and ground troops, will require night vision capabilities for combat missions. In addition to combat missions, key support missions such as search and rescue depend on night vision capabilities. Future demand for sophisticated NVDs will largely be determined by the Pentagon's assessment of their utility in a rapidly evolving global security environment.

REE trade disputes with China. The REE lanthanum is a crucial input for manufacturing NVDs. In 2012, the United States, together with Japan and the European Union, filed a complaint with the World Trade Organization (WTO) over Chinese trade policies, including export quotas that artificially limit the global supply of REEs.[39] Chinese trade policies, as well as lax environmental standards and other measures, have helped expand China's share of the REE market to approximately 80 percent (see this report's chapter on high-tech magnets).[40]

These trade disputes are part of a long-standing and broader challenge for the United States: contending with the questionable trade practices that have allowed for a Chinese near-monopoly on REE production. This issue has received high-level Congressional attention.[41]

The ongoing disputes in the WTO could take years to resolve, and the outcome is far from certain.[42] In the meantime, the ongoing discussion about REEs in political and national security circles will continue to affect how the United States insulates itself from supply disruptions and attendant cost increases.

VULNERABILITIES IN THE NIGHT VISION DEVICE SUPPLY CHAINS

Dependence on DoD demand in a time of constrained budgets. According to the Department of Commerce (DoC), the financial performance of firms that produce NVDs will depend largely on U.S. government demand.[43] Constrained and uncertain defense spending, the end of combat operations in Iraq, and the ongoing troop drawdown in Afghanistan all create uncertainty for military NVD producers about future demand. NVD maker L-3 cited decreasing U.S. government demand as a reason for declining sales in the fourth quarter of 2011—a trend that is likely to continue.[44]

Availability of the REE lanthanum.
Lanthanum, an REE used in the production of sophisticated lenses, is essential for U.S. military night vision systems. REEs are a group of seventeen elements with special magnetic and other physical properties that make them well suited for a variety of advanced manufacturing and electronics applications.[45] (For further

discussion, see this report's chapter on specialty metals.)

The United States previously performed all stages of rare earth extraction and processing, but those processes are now performed mostly in China.[46] China currently produces approximately 90 percent of REEs, even though the country only holds approximately 30 percent of global deposits.[47] In 2010, approximately 91 percent of U.S. lanthanum imports came from China.

Dependence on foreign suppliers.
Approximately two-thirds of firms surveyed by DoC procure inputs from foreign firms. Around seven percent of content used in night vision system devices and components between 2001 and 2005 was of foreign-origin.[48] The U.S. government currently lacks mechanisms for actively and consistently monitoring these supply chains, as well as assessing the extent to which they make U.S. firms reliant on sole foreign sources for crucial inputs.

According to the DoC study, the top three reasons for foreign sourcing are: [49]

- Cost compared to domestic alternatives

- The unavailability of certain items and services from domestic suppliers

- Higher quality provided by foreign sources

In an increasingly globalized world, as companies look for the best value and global suppliers aggressively compete, it is no surprise that more foreign content will be used. When it comes to producing defense goods, however, globalization of supply chains and offshoring of certain capabilities creates risks, especially during a crisis.

Besides the risk of supply cutoffs due to political conflict or purposeful manipulation (a low risk for most items, especially during peacetime), there is also a risk that unexpected market shifts (globally or within a particular country) or changes in national regulations could unexpectedly reduce the availability of a particular item. Furthermore, natural disasters or other major unexpected events can disrupt supply chains.

In most cases the market will work to correct disruptions in supply. However, there is a risk that lower-tier NVD suppliers will not be able to acquire a necessary material, service, or component (especially during political or economic crises), which could negatively affect U.S. production capability.

MITIGATING THE RISKS

The United States and DoD should take steps to preserve the U.S. advantage in night vision capabilities and its capacity to innovate in the future. Our key recommendations are:

DoD should take steps to strengthen awareness of NVD supply chains. DoD should strengthen and expand efforts to gain greater insight into the supply chains that support U.S. NVD capabilities. DoD's Sector-by-Sector, Tier-by-Tier (S2T2) effort will "gather industrial base data, map supplier relationships, and evaluate industrial capabilities."[50] According to Deputy Assistant Secretary of Defense for Manufacturing and Industrial Base Policy Brett Lambert, collecting this data will assist DoD in "getting out of the role of firefighter, waiting for a building to be on fire before we responded."[51] S2T2 will help identify instances of over-reliance on foreign suppliers, areas of limited competition, and potential "single points of failure."[52] S2T2 focuses mostly on large

combat platforms and weapons systems, and does not currently cover NVDs, but it may provide a model for future efforts to gain greater clarity about NVD supply chains.

Congress is taking an interest as well. In Section 854 of the FY2012 NDAA,[53] Congress instructed DoD to come up with a full assessment of the supply chains for NVD components to discover how dependent the United States is on foreign suppliers. The FY2012 NDAA also instructed DoD to:

"Identify and assess current strategies to leverage innovative night vision image intensification technologies being pursued in both DoD laboratories and the private sector for the next generation of night-vision capabilities."[54]

Greater awareness of defense industry supply chains and their relationships to other defense and civilian supply chains will improve U.S. defense industrial base policy, including decision-making to preserve key capabilities and strengthen competition.

Fund and encourage future innovation. To ensure that the U.S. military remains strong and flexible, DoD should provide incentives for NVD suppliers to innovate further. This innovation may mean the development of more sophisticated versions of current devices, such as I2 devices with better resolution that are better integrated into U.S. platforms and warfighter gear. It may also mean integrating I2 capabilities with other technologies, as is being done with the ENVG and SENVG.

Innovation is important for at least three major reasons:

- **Staying ahead of competitors.** Even as the United States innovates, other countries are acquiring their own capabilities. To ensure victory in a potential conflict, U.S. troops need to be better equipped than any potential rival. Innovation is important to address functional gaps in the technology. For example, while U.S. military NVDs are highly sophisticated and effective against modern foes, I2 devices still do not "see" as well as the human eye, and have a narrower field of vision. Different lighting conditions can make it appear as if objects are closer or further away than they actually are.[55]

- **Availability of inputs.** Current U.S. production of NVDs is dependent on Chinese supplies of the REE lanthanum. The development of NVD technologies that do not require foreign inputs, or that require fewer or different foreign inputs, would enhance U.S. self-sufficiency for this important capability.

- **Cost.** As defense funding decreases, delivering value for taxpayer dollars becomes a higher priority. Identifying cheaper alternatives that deliver the same level of performance and battlefield capability will free up resources for other critical U.S. defense needs.

Examine and rationalize trade rules.
The United States government and DoD must take the globalization of the defense market into account. In the words of former Under Secretary of Defense Jacques Gansler, the United States has to "understand and realize the benefits of globalization while of course mitigating its risks."[56] DoD must ensure that program managers, as well as prime and lower-tier contractors, are leveraging the global market in ways that do not create vulnerabilities for U.S. night vision capabilities.

I2 tubes are restricted for export under the International Traffic in Arms Regulations (ITAR).[57] However, these regulations do not always work perfectly. In addition to the instances of violations mentioned above, in 2009 the U.S. Government Accountability Office set up a fake company, certified that company as a distributor of night vision goggles, and were able to purchase an unrestricted quantity of export-controlled night vision goggles.[58]

The U.S. export control system should allow sales to, and collaboration with, U.S. allies and trusted partners in order to strengthen the capabilities of friendly countries and to provide commercial opportunities to U.S. companies. However, the system should also mitigate the risk of U.S. competitors or potential adversaries gaining access to unauthorized technology.

Communicate stable and predictable demand to industry. Given the importance of DoD demand for the companies that produce military NVDs, DoD should, to the extent possible, communicate its projected requirements for advanced night vision capabilities. At the beginning of 2012, ITT Exelis cut 75 salaried positions at its facilities in Roanoke, Virginia, after having already cut 300 hourly positions. ITT Exelis blamed the cuts on lower demand for night vision equipment.[59] (More recent cuts have been mentioned above). Reductions are to be expected in a more constrained fiscal environment for defense, but this adjustment should be balanced against the need to retain important technical skills in the defense industrial base. Stable and predictable demand will make it easier for the defense industrial base to retain its most skilled and important workers.

CONCLUSION

NVDs have proven their worth time and again in Iraq, Afghanistan, and many other combat scenarios, and they will continue to provide a key advantage against future foes. The United States must preserve its edge in advanced military technologies to hedge against an uncertain and evolving strategic environment. Maintaining the U.S. defense industrial base's ability to provide this capability will enhance U.S. national security and ensure that the United States' warfighters continue to win on the battlefield.

ENDNOTES

1 Leon Panetta, "Fighting for the American Dream," (Speech in Louisville, Kentucky, March 1, 2012. http://www.defense.gov/speeches/speech.aspx?speechid=1658.

2 "Night Vision Goggles NVG," GlobalSecurity.org (July 7, 2011). http://www.globalsecurity.org/military/systems/ground/nvg.htm.

3 DoC, Bureau of Industry and Security (BIS), Office of Strategic Industries and Economic Security (SIES), *Defense Industrial Base Assessment: U.S. Imaging and Sensors Industry* (Washington, DC: DoC, BIS/SIES, October 2006), II-3. http://www.bis.doc.gov/news/2006/wholereportwithappendices10_12_06.pdf.

4 Peter Eisler, "U.S. Foes Seek Edge in the Dark," *USA Today*, April 15, 2008. http://www.usatoday.com/news/military/2008-05-14-nightinside_N.htm.

5 David Axe, "Night Vision Tech Tangles Troops in Afghanistan," *Wired*, April 1, 2011. http://www.wired.com/dangerroom/2011/04/night-vision-tech-tangles-troops-in-afghanistan/.

6 Lynn Sweet, "Osama bin Laden Raid Took Months to Plan," *Chicago Sun-Times*, May 2, 2012. http://www.suntimes.com/news/nation/5147948-418/how-the-u.s.-found-killed-osama-bin-laden.

7 Amanda Onion, "Night Vision: How It Works, How It Helps," *ABC News*, April 7, 2012. http://abcnews.go.com/Technology/story?id=98233&page=1#.T4CJhY7WXz0.

8 U.S. Department of Justice, "Statement of U.S. Attorney John Brownlee on the Guilty Plea of ITT Corporation for Illegally Transferring Classified and Export-controlled Night Vision Technology to Foreign Countries" (March 27, 2007). www.justice.gov/nsd/pdf/itt_statement_by_usattorney.pdf.

9 Simon Cooper, "How China Steals U.S. Military Secrets," *Popular Mechanics*, January 10, 2009. http://www.popularmechanics.com/technology/military/news/3319656.

10 DoC, Bureau of Industry and Security (BIS), Office of Strategic Industries and Economic Security (SIES), Defense Industrial Base Assessment: U.S. Imaging and Sensors Industry (Washington, DC: DoC, BIS/SIES, October 2006), I-1. http://www.bis.doc.gov/news/2006/wholereportwithappendices10_12_06.pdf.

11 DoC, Bureau of Industry and Security (BIS), Office of Strategic Industries and Economic Security (SIES), Defense Industrial Base Assessment: U.S. Imaging and Sensors Industry (Washington, DC: DoC, BIS/SIES, October 2006), I-1. http://www.bis.doc.gov/news/2006/wholereportwithappendices10_12_06.pdf.

12 Ibid., III-3. http://www.bis.doc.gov/news/2006/wholereportwithappendices10_12_06.pdf.

In 2008 the increased NVD demand in connection with the Iraq War caused a shortage of NVDs for civilian medical helicopter pilots. Ryan Foley, "Iraq War Creates Shortage of Night Vision Gear in U.S.," *USA Today*, June 4, 2008. http://www.usatoday.com/news/nation/2008-06-04-4109775549_x.htm.

13 Institute for Defense Analyses (IDA), "Export Controls and the U.S. Defense Industrial Base," eds. Richard Van Atta et al. (Alexandria, VA: IDA, January 2007), 29. http://www.acq.osd.mil/mibp/docs/ida_study-export_controls_%20us_def_ib.pdf.

14 For a discussion on defense industry consolidation, see Barry D. Watts, *The U.S. Defense Industrial Base: Past, Present, and Future*, (Washington, D.C.: Center for Strategic and Budgetary Priorities, October 15, 2008), 2. http://www.csbaonline.org/wp-content/uploads/2011/02/2008.10.15-Defense-Industrial-Base.pdf.

15 U.S. Army Research, Development, and Engineering Command (RDECOM), Communications-Electronics Research, Development, and Engineering Center (CERDEC), "History of Night Vision." http://www.nvl.army.mil/history.html.

16 "Thermal Imaging," U.S. Army Research, Development, and Engineering Command, Communications-Electronics Research, Development, and Engineering Center, http://www.nvl.army.mil/thermal.html.

17 DoC, Bureau of Industry and Security (BIS), Office of Strategic Industries and Economic Security (SIES), Defense Industrial Base Assessment: U.S. Imaging and Sensors Industry (Washington, DC: DoC, BIS/SIES, October 2006), II-6. http://www.bis.doc.gov/news/2006/wholereportwithappendices10_12_06.pdf.

18 DoC, Bureau of Industry and Security (BIS), Office of Strategic Industries and Economic Security (SIES), Defense Industrial Base Assessment: U.S. Imaging and Sensors Industry (Washington, DC: DoC, BIS/SIES, October 2006), II-8. http://www.bis.doc.gov/news/2006/wholereportwithappendices10_12_06.pdf.

19 Barry Rosenberg, "PEO Soldier takes the lead on soldier situational awareness," *Defense Systems*, February 21, 2012. http://defensesystems.com/articles/2012/02/08/interview-bg-camille-nichols.aspx.

20 "Night Vision Gives U.S. Troops Edge, Through a Glass, Darkly," *Defense Industry Daily*, October 4, 2011. http://www.defenseindustrydaily.com/through-a-glass-darkly-night-vision-gives-us-troops-edge-06047/.

21 Ibid.

22 Jefferson Morris, "Army To Buy Up To 360,000 Night Vision Goggles," *Aviation Week*, September 19, 2005. http://www.aviationweek.com/aw/generic/story_generic.jsp?channel=aerospacedaily&id=news/NVIT09195.xml&headline=Army%20To%20Buy%20Up%20To%20360,000%20Night-Vision%20Goggles.

23 "Night Vision Gives U.S. Troops Edge, Through a Glass, Darkly," *Defense Industry Daily*, October 4, 2011. http://www.defenseindustrydaily.com/through-a-glass-darkly-night-vision-gives-us-troops-edge-06047/.

24 ITT Corporation spun off its defense arm, now known as ITT Exelis, in 2011. "ITT Exelis Begins Operations as a Publicly Traded Company," *ITT Exelis* press release, October 31, 2011. http://www.exelisinc.com/news/pressreleases/Pages/ITT-Exelis-Begins-Operations-as-a-Publicly-Traded-Company.aspx.

25 Mike M. Ahlers, "ITT Fined up to $100M for Illegal Night Vision Exports," *CNN Money*, March 27, 2007. http://money.cnn.com/2007/03/27/news/international/itt_export/index.htm.

26 "Statement of U.S. Attorney for the Western District of Virginia John Brownlee on the guilty plea of ITT Corporation for illegally transferring classified and export controlled night vision technology to foreign countries."

27 U.S. Department of Justice, "ITT Corporation to Pay $100 Million Penalty and Plead Guilty to Illegally Exporting Secret Military Data Overseas" (March 27, 2007). http://www.justice.gov/opa/pr/2007/March/07_nsd_192.html.

28 U.S. Department of Commerce, Bureau of Industry and Security (BIS), Office of Strategic Industries and Economic Security (SIES), *Defense Industrial Base Assessment: U.S. Imaging and Sensors Industry* (Washington, DC: U.S. Department of Commerce, BIS/SIES, October 2006), IV-6. http://www.bis.doc.gov/news/2006/wholereportwithappendices10_12_06.pdf.

29 "L-3 Agrees to Acquire Insight Technology, Expands Electro-Optical and Night Vision Business," L-3 Communications press release, February 19, 2010. http://www.l-3com.com/press-releases/l-3-agrees-to-acquire-insight-technology-expands-electro-optical-and-night-vision-business.html.

30 Tamir Eshel, "Update: U.S. Army Awards Enhances NVG Contracts to ITT, DRS and L-3," *Defense Update*, August 19, 2010. http://defense-update.com/20100819_envg.html.

31 Paul McLeary, "Clearer, and in Color: ITT Exelis' Latest Night-Vision Goggles Provide Better Images," *Defense News* (May 7, 2012). http://www.defensenews.com/apps/pbcs.dll/article?AID=2012305070005.

32 "ITT Exelis and Tata Advanced Systems Limited partner to manufacture Generation 3 night vision devices in India," ITT Exelis press release, March 29, 2012. http://www.exelisinc.com/News/PressReleases/Pages/ITT-Exelis-and-Tata-Advanced-Systems-Limited-partner-to-manufacture-Generation-3-night-vision-devices-in-India.aspx.

33 Sandra Erwin, "Night Vision Technology Arms Race Accelerates," *National Defense*, April 9, 2012. http://www.nationaldefensemagazine.org/blog/Lists/Posts/Post.aspx?ID=740.

34 "ITT Exelis awarded $33 million to provide night-vision devices to U.K. Ministry of Defence," ITT Exelis Press Release, March 30, 2012. http://www.exelisinc.com/News/PressReleases/Pages/ITT-Exelis-awarded-$33-million-to-provide-night-vision-devices-to-U.K.-Ministry-of-Defence.aspx.

35 U.S. Department of Commerce, Bureau of Industry and Security (BIS), Office of Strategic Industries and Economic Security (SIES), *Defense Industrial Base Assessment: U.S. Imaging and Sensors Industry* (Washington, DC: U.S. Department of Commerce, BIS/SIES, October 2006), IX-5. http://www.bis.doc.gov/news/2006/wholereportwithappendices10_12_06.pdf.

36 Joseph C. Anselmo, "Cut Loose, Exelis Faces Tough Times On Its Own," *Aviation Week & Space Technology*, April 16, 2012.

37 Marjorie Censer, "Contractors Try to Find Right Services, Products Balance," *Washington Post*, April 9, 2012. http://www.washingtonpost.com/business/capitalbusiness/exelis-boosts-services-business-lowering-margins-but-maintaining-sales/2012/04/06/gIQARxud4S_story.html.

38 Patrick Johnson, "ITT Exelis to Close West Springfield Manufacturing Plant, Lay Off 200; Cites Expected Cuts in U.S. Defense Spending," *The Republican* February 13, 2013. http://www.masslive.com/business-news/index.ssf/2013/02/itt_exelis_to_close_west_springfield_man.html.

39 Eric Martin and Sonja Elmquist, "U.S. to File WTO Complaint Over China Rare-Earth Export Caps," *Bloomberg*, March 13, 2012. http://www.bloomberg.com/news/2012-03-13/u-s-will-ask-for-wto-s-help-to-fight-chinese-curbs-on-rare-earth-exports.html.

40 Emily Coppel, "Rare Earth Metals and U.S. National Security," *American Security Project* (February 1, 2011), 2. http://americansecurityproject.org/wp-content/uploads/2011/02/Rare-Earth-Metals-and-US-Security-FINAL.pdf.

41 U.S. House of Representatives, Committee on Foreign Affairs, Subcommittee on Asia and the Pacific, "China's Monopoly on Rare Earths: Implications for U.S. Foreign and Security Policy" (September 21, 2011). http://foreignaffairs.house.gov/hearing_notice.asp?id=1352.

42 Keith Bradsher, "Specialists in Rare Earths Say a Trade Case Against China May Be Too Late," *The New York Times*, March 13, 2012. http://www.nytimes.com/2012/03/14/business/global/rare-earth-trade-case-against-china-may-be-too-late.html?pagewanted=all.

43 DoC, Bureau of Industry and Security (BIS), Office of Strategic Industries and Economic Security (SIES), *Defense Industrial Base Assessment: U.S. Imaging and Sensors Industry* (Washington, DC: DoC, BIS/SIES, October 2006), I-7. http://www.bis.doc.gov/news/2006/wholereportwithappendices10_12_06.pdf.

44 "L-3 Announces Fourth Quarter 2011 Results," L-3 Communications press release, January 31, 2012. http://www.l-3com.com/media-center/press-releases.html?pr_id=1654500.

45 Lee Levkowitz and Nathan Beauchamp-Mustafaga, "China's Rare Earths Industry and its Role in the International Market," *U.S.-China Economic and Security Review Commission* (November 3, 2010), 1. http://www.uscc.gov/researchpapers/2011/RareEarthsBackgrounderFINAL.pdf.

46 Government Accountability Office (GAO), Rare Earth Metals in the Defense Supply Chain, by Belva Martin (Washington, DC: GAO, 10-617-R, April 2010), 13. http://www.gao.gov/new.items/d10617r.pdf.

47 "China Rare Earths to Last 15-20 Years, May Import," *Bloomberg.com,* October 16, 2010. http://www.bloomberg.com/news/2010-10-16/china-says-its-medium-heavy-rare-earth-reserves-may-last-only-15-20-years.html.

48 DoC, Bureau of Industry and Security (BIS), Office of Strategic Industries and Economic Security (SIES), *Defense Industrial Base Assessment: U.S. Imaging and Sensors Industry* (Washington, DC: DoC, BIS/SIES, October 2006), IV-9. http://www.bis.doc.gov/news/2006/wholereportwithappendices10_12_06.pdf.

49 Ibid., I-4.

50 Department of Defense, Office of Manufacturing and Industrial Base Policy, "Sector-by-Sector, Tier-by-Tier (S2T2) Industrial Base Review," (January 1, 2012). http://www.acq.osd.mil/mibp/s2t2.shtml.

51 U.S. House of Representatives, Armed Services Committee, "The Defense Industrial Base: The Role of the Department of Defense" (November 1, 2011). http://armedservices.house.gov/index.cfm/2011/11/the-defense-industrial-base-the-role-of-the-department-of-defense.

52 Brett Lambert, "Presentation to National Defense Industrial Association" (August 2011), p. 4. http://www.ndia.org/Advocacy/Resources/Documents/NDIA_S2T2_Briefing_AUG11.pdf

53 The NDAA is the legislation that sets defense spending priorities for the coming year as well as certain policies governing issues such as acquisitions see Nathan Hodge, "Congress (Finally) Passes Defense-Policy Bill," *The Wall Street Journal*, December 22, 2010. http://blogs.wsj.com/washwire/2010/12/22/congress-finally-passes-defense-policy-bill/.

54 "Congress Orders Pentagon To Assess Health Of Defense Industrial Base, Consider A Rare-Earths Stockpile, and Find Out Why U.S. Is Dependent on Foreigners for Night Vision Systems," *Manufacturing and Technology News*, December 30, 2011.

55 "Night vision Goggles (NVG)," GlobalSecurity.org. http://www.globalsecurity.org/military/systems/ground/nvg.htm.

56 Jacques Gansler, "Testimony to the Senate Armed Services Committee Subcommittee on Emerging Threats and Capabilities" (May 3, 2011). http://armed-services.senate.gov/e_witnesslist.cfm?id=5163.

57 Code of Federal Regulations Title 22 Part 121.1 Category XII. http://pmddtc.state.gov/regulations_laws/documents/official_itar/ITAR_Part_121.pdf.

58 Government Accountability Office (GAO), *Military and Dual-Use Technology: Covert Testing Shows Continuing Vulnerabilities of Domestic Sales for Illegal Export*, by Gregory D. Kutz (Washington, DC:, 09-725R, June 4, 2009). http://gao.gov/new.items/d09725t.pdf.

59 "ITT Exelis Trims 75 Salaried Workers in Latest Set of Layoffs," *The Roanoke Times*, January 27, 2012. http://www.roanoke.com/business/wb/304037.

CHAPTER 14 • MACHINE TOOLS

EXECUTIVE SUMMARY

Machine tools cut, mill, grind, and drill metals and other materials. The mother of all industrial manufacturing machines, they enable the fabrication of custom components for commercial and defense products. Almost all manufactured goods contain at least one component fabricated by a machine tool, and the tools also produce precision components for the most sophisticated U.S. defense goods. The five-axis simultaneous control machine tool is the most advanced system; its three linear axes and two rotary axes accurately shape complex parts and components. Because of their complexity, machine tools are significant investments. Off-the-shelf units easily cost half a million dollars, while a custom machine tool, as is often required for advanced weapons systems, can cost several million dollars.

The U.S. machine tool sector represents a small portion of the global market, with just five percent in 2012. Roughly 70 percent of machine tools used in U.S. manufacturing are imported, mostly from Germany and Japan. The machine tool sector is highly susceptible to economic cycles. Downturns cause companies to cut back spending and repeated business downturns and global recessions since the 1980s have led to a decline in the U.S. machine tool industry. Although sales picked up again from 2011 to 2012 by a modest seven percent, U.S. machine tool manufacturers witnessed a 60 percent decline in sales in 2009. The U.S. machine tool sector ranks seventh after China, Japan, Germany, Korea, Italy, and Taiwan, respectively, and is hampered by U.S. export control legislation.

The decline of the U.S. sector does not directly constitute a risk to the defense industrial base, because machine tools manufactured in foreign countries are readily available to U.S. firms wishing to purchase them. However, the sector's decline is indicative of a broader trend in U.S. manufacturing. It also points to a gap in capabilities between U.S. manufacturers and foreign firms that could pose longer-term risks to the defense industrial base. With a much larger share of the market, foreign firms will drive innovation and are likely to prioritize their domestic markets over U.S. demand. New technologies may be made available to U.S. firms only after being offered to foreign competitors. Defense innovations driven by manufacturing innovations abroad (and perhaps with a significant time lag) could threaten U.S. technological advantages, corresponding with a broader loss of innovation, research and development (R&D), and capacity across the entire defense industrial base.

MACHINE TOOLS
PRODUCING U.S. DEFENSE INPUTS

MANUFACTURING SECURITY

Machine tools are necessary for the production and maintenance of U.S. military platforms and other systems

U.S. NAVAL VESSELS   **U.S. MILITARY AIRCRAFT**   **ADVANCED WEAPONS**

JOBS SUPPORTED

 = 4,000

The U.S. machine tool sector alone supports approximately

40
THOUSAND U.S. JOBS

ENABLING AN ADVANCED ECONOMY

Machine tools produce inputs used in advanced military and civilian systems

MACHINE TOOLS   **MILITARY SYSTEMS**   **CIVILIAN TECHNOLOGY**

DOMESTIC TRENDS

DOMESTIC CAPACITY

DOMESTIC DEMAND

VULNERABILITY

Most machine tools in the United States are imported

 70% FROM FOREIGN SOURCES

MITIGATING RISKS

Avoiding uncertainties in the U.S. machine tool supply

RE-EXAMINE EXPORT CONTROLS   **EDUCATION AND VOCATIONAL TRAINING**   **SUPPORT FOR SMALL/MEDIUM SIZED BUSINESSES**

MILITARY EQUIPMENT CHART
SELECTED DEFENSE USES FOR MACHINE TOOLS

DEPARTMENT	WEAPON SYSTEMS	PLATFORMS
ARMY	■ M777 155mm Howitzer ■ AGM-114 HELLFIRE air-to-surface missile ■ M1014 combat shotgun ■ M252 81mm mortar ■ High Mobility Artillery Rocket System (HIMARS)	■ M1 Abrams main battle tank ■ Humvee ■ UH-60 Blackhawk helicopter
MARINE CORPS	■ M777 155mm Howitzer ■ M16 Assault rifle ■ M110 Sniper rifle ■ M1014 Combat shotgun ■ M252 81 mm mortar ■ High Mobility Artillery Rocket System (HIMARS)	■ V-22 Osprey aircraft ■ AAV-7A1 Assault Amphibious Vehicle ■ M1 Abrams main battle tank ■ F-35B Joint Strike fighter
NAVY	■ BGM-Tomahawk ■ 57 Mk110 naval gun system	■ F-35C Joint Strike fighter ■ Nimitz-class nuclear-powered aircraft carrier ■ Littoral Combat Ship (LCS) ■ SSN-774 Virginia-class nuclear-powered attack submarine
AIR FORCE	■ AIM-9 Sidewinder air-to-air missile ■ BGM-Tomahawk	■ HH-60 Pave Hawk helicopter ■ F-15 Eagle fighter ■ F-22 Raptor fighter ■ F-35A Joint Strike fighter ■ C-130 military transport aircraft

Mitigating these risks requires strengthening the U.S. machine tool industry against economic downturns by regaining significant shares of both domestic and foreign machine tool markets. Many machine tool producers are highly specialized and relatively small. They need more support in coping with market failures, in gaining access to long-term capital, and in navigating export licenses. In addition, the machine tool sector relies on skilled labor and craftsmen. The United States has failed to invest sufficiently in the formation of a skilled manufacturing labor force that will increase the productivity and competitiveness of the U.S. sector.

INTRODUCTION

Advanced U.S. military platforms, weapons systems, and devices require thousands of precisely manufactured components to function properly. Sophisticated machine tools, used to cut and sculpt metal, are essential to manufacturing many of these components. A company's capability to produce reliable systems for the U.S. military therefore relies partially on access to high-performance machine tools. Five-axis simultaneous control machine tools have the ability to manufacture products or parts (most often metallic). They are used for abrading, cutting, drilling, forming, grinding, nibbling, or shaping of a piece of metal or other material, such as a plastic or ceramic.

In June 2011, the President's Council of Advisors on Science and Technology reported that while other nations are investing heavily in manufacturing, U.S. investments have fallen behind. The council concluded that the erosion of domestic manufacturing capabilities had national security implications. The report urged the U.S. government to undertake an Advanced Manufacturing Initiative (AMI), bringing together elements of the Departments of Commerce, Defense (DoD), and Energy (DoE) under the coordination of the White House.[1] The overall objective is to foster innovation in manufacturing technology by reforming tax and business policy to provide more consistent support for basic research, and for training and education of a highly skilled workforce. The 2010 Quadrennial Defense Review issued a similar warning: "In the mid to long term, it is imperative that we have a robust industrial base with sufficient manufacturing capability and capacity to preserve our technological edge and provide for the reset and recapitalization of our force."[2]

These official recommendations notwithstanding, U.S. manufacturing declined from 27 percent in 1957 as a fraction of U.S. GDP to approximately 12 percent by 2012. Manufacturing employment declined from 17.6 million jobs in 1998 to around 12 million jobs at the end of 2012.[3]

The machine tool industry plays a key role in preserving the international competitiveness of the U.S. industrial base.

The 2010 Quadrennial Defense Review issued a similar warning: "In the mid to long term, it is imperative that we have a robust industrial base with sufficient manufacturing capability and capacity to preserve our technological edge and provide for the reset and recapitalization of our force."[a]

– Government Accountability Office (GAO)

Table 1: Leading Global Exporters of Machine Tools ($ Millions)

Country	2002	2003	2004	2005	2006	2007	2012
Germany	$2,144	$2,481	$2,979	$3,295	$3,795	$4,382	$10,410
Japan	$1,590	$2,182	$2,599	$3,168	$3,476	$3,846	$11,565
Taiwan	$549	$654	$904	$1,046	$1,231	$1,545	$4,236
Italy	$829	$922	$1,122	$1,081	$1,247	$1,516	$4,434
Switzerland	$547	$533	$621	$727	$827	$877	$2,773
United States	$594	$522	$570	$599	$760	$740	$2,088
South Korea	$133	$215	$285	$284	$364	$478	$2,551
Czech Republic	$142	$150	$231	$267	$292	$438	$823
Spain	$239	$266	$251	$279	$321	$400	$983
China	$104	$129	$149	$185	$258	$400	$2,750
Total	$7,971	$9,407	$11,206	$12,641	$14,368	$17,170	$42,613

Adapted from U.S. Department of Commerce Bureau of Industry and Security, "Critical Technology Assessment: Five Axis Simultaneous Control Machine Tools," July 2009, 15. http://www.bis.doc.gov/defenseindustrialbaseprograms/osies/defmarketresearchrpts/final_machine_tool_report. pdf. 2012 data from Gardner Research, The World Machine Tool Output & Consumption Survey (February 2013). http://www.gardnerweb.com/cdn/cms/uploadedFiles/2013wmtocs_SURVEY.pdf.

Furthermore, machine tools are essential for reproducing the technological innovations required to sustain a competitive industrial economy. Currently, approximately 80 percent of U.S. machine tools are imported from Japan, Germany, Taiwan, Italy, and Switzerland (see Figure 1). The U.S. machine tool sector represents only five percent of global production in 2012, down from 7.5 percent in 2002 (see Table 1).[4] Following the 2007 economic slowdown, U.S. domestic consumption of five-axis simultaneous control machine tools fell by over 60 percent.[5] The current state of the U.S. domestic machine tool industry does not itself constitute a vulnerability to American national security—provided that U.S. firms can buy machines from other countries without hindrance. However, there are three reasons to be concerned about the erosion of the domestic machine tool sector. First, U.S. companies may find that they do not have prompt access to the newest machines because foreign manufacturers give preference to their domestic customers or have tailored the performance of the machine to the needs of a compatriot customer. U.S. customers also could experience delays with special orders or gaining access to the newest models.

Second, studies have analyzed the relative cost to the industrial knowledge base that can arise when an advanced technological sector relies predominantly on imports for inputs and innovation. These studies have found that proximity to other sources of innovation is important for sustaining expertise and ensuring innovative progress. Even with advances in communications, the exchange of data, ideas, and findings works better when machine tool designers, for example, enjoy routine personal interaction with the downstream of the supply chain—offering end-users access to better performance.[6]

Much of the U.S. advantage in military capabilities comes from its superior

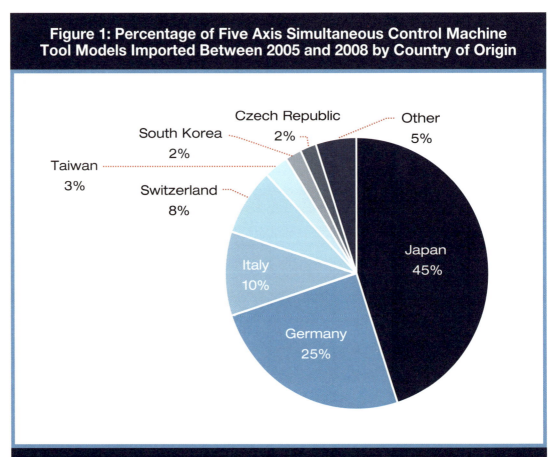

Figure 1: Percentage of Five Axis Simultaneous Control Machine Tool Models Imported Between 2005 and 2008 by Country of Origin

Other 5%

Czech Republic 2%

South Korea 2%

Taiwan 3%

Switzerland 8%

Italy 10%

Japan 45%

Germany 25%

Source: Jennifer Watts, Jason Bolton, and Ashley Miller, "Critical Technology Assessment: Five Axis Simultaneous Control Machine Tools," U.S. Department of Commerce Bureau of Industry and Security (July 2009). http://www.bis.doc.gov/defenseindustrialbaseprograms/osies/defmarketresearchrpts/final_machine_tool_report.pdf.

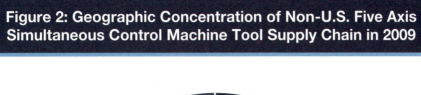

Figure 2: Geographic Concentration of Non-U.S. Five Axis Simultaneous Control Machine Tool Supply Chain in 2009

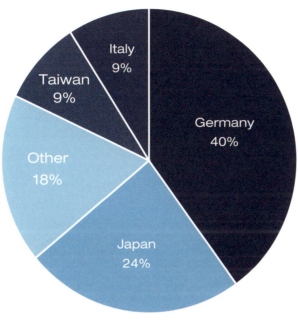

Italy
9%

Taiwan
9%

Germany
40%

Other
18%

Japan
24%

Jennifer Watts, Jason Bolton, and Ashley Miller, "Critical Technology Assessment: Five Axis Simultaneous Control Machine Tools," U.S. Department of Commerce Bureau of Industry and Security (July 2009). http://www.bis.doc.gov/defenseindustrialbaseprograms/ osies/defmarketresearchrpts/final_machine_tool_report.pdf.

industrial base and knowledge-based economy. The decline in production of the most sophisticated machine tools is indicative of broader trends that may eventually compromise U.S. military superiority. Over time, U.S. economic and political leaders have ignored the symbiotic relationship between basic research and manufacturing technology. In advanced industrialized economies, R&D and manufacturing processing are optimally intertwined. "Design for manufacturing" is important for the ability to compete internationally.[7]

The gradual erosion of the manufacturing of the most advanced precision machine tools in the United States has been accompanied by a loss of engineering, computer, and manufacturing talent and manufacturing experience and skills. The United States has lost skilled workers, machinists, engineers, and production workers who represent a reservoir of talent needed to generate innovations and technological advances. This skilled labor force is particularly important for the U.S. industrial base because every manufacturing sector (including pharmaceuticals, medicine, nuclear science, automobiles, information technology, green energy, oil and gas, mining, and aviation) depends on precision machine tools.

Third, observers talk of a Third Industrial Revolution, which follows the success of assembly-line mass production.[8] This revolution will be digital and based on 3-D printing.[9] Whereas current machine tools gradually strip away subsequent layers of a material to create an end-product, 3-D printing gradually builds up layers of material (often plastic, ceramic, or metal) to create an end-product. A mock-up of the desired object is built using computer-aided design (CAD) software, which can then be "printed" in slices.[10] This printing technology allows for the efficient fabrication of many items with minimal waste. The technique largely depends on innovations in CAD software. Such technology enables the rapid prototyping of new products, increases the level of customization available to manufacturers, and permits more innovative designs.[11, 12] Estimates claim that this market will grow rapidly and be worth about $6 billion by 2019.

Recognizing the possibilities, the Obama Administration has identified 3-D printing as one means of bringing manufacturing back to the United States. A new program allocates $100 million to research 3-D printing and direct digital manufacturing.[13] As the United States competes to lead the next wave of manufacturing innovation, to be successful the private sector will need a supportive business environment that includes tax incentives, special rules, and access to a skilled labor force.

DoD has expressed great enthusiasm about the emergence of 3-D printing, which is very well suited to the small-batch, custom manufacturing needed for many defense products, especially in expeditionary operations. The printing technique may provide DoD's suppliers with an inexpensive means of producing customized components of new, more technologically advanced weapons systems. Moreover, miniature 3-D printers could become portable repair units, allowing deployed units to construct replacement units for damaged equipment quickly, even in battlefield environments.[14]

Due to the gradual decline of the conventional five-axis machine tool sector in the United States, U.S. scientists, corporations, and laboratories may be unable to translate cutting-edge innovations such as 3-D printing into commercially viable products that can compete with those produced in Japan, Germany, or other countries. Although the United States is almost always at the forefront of new waves of technological innovation, it often falls behind when bringing such innovations to the global marketplace.

The U.S. machine tool industry is relatively small compared to other domestic industries, such as automotive, aircraft, construction, and energy. In 2008, the machine tool industry encompassed approximately 180 companies, with an estimated 35,000 to 40,000 employees, and with total revenues from the sale of metal-cutting and metal-forming machines of $4.2 billion.[15] However, given that so many other industries depend upon machine tools to manufacture their own products, the size of this sector fails to adequately indicate its importance.

The most advanced system is the five-axis simultaneous control machine tool. The multiple axes can be operated simultaneously with the help of a computer numerical control (CNC) machine that directs and synchronizes the movements of the axes, and is programmed automatically. These machine tools are used in a variety of manufacturing processes to cut, grind, and shape materials and can produce parts with complex shapes and angles with great accuracy. Some machines produce micro parts used in small engines, while others have large-sized milling machines and can shape bigger aircraft parts. There are hundreds of different models of three-, four-, and five-axis machine tools, as is needed to produce thousands of different parts. Defense contractors also special order custom five-axis machine tools to produce special components. The average price of an off-the-shelf machine is $300,000; a custom machine can cost well over $1,000,000.

Machine tools are used for many commercial applications, but the most advanced precision machine tools are required for manufacturing the components of modern weapons systems. Every kind of advanced weapon system relies on precision parts such as propellers, turbines, other blades, gyroscopes, and engine parts. Machine tools are essential in producing components for stealth technology, nuclear weapons technology, night vision devices, and radar and sonar domes.[16]

The decline of the U.S. machine tools sector was initially a result of competitive pressure from Asia, beginning with Japan and then with added pressure from newly industrialized countries such as Korea and Taiwan. The first CNC machine was designed at the Massachusetts Institute of Technology (MIT) Servomechanisms Lab in 1959; by the early 1980s, Japan had assumed the leading position in commercializing new machine tool technology.[17] Competition has now become so globalized that the United States now imports 70 percent of all machine tools currently in use and runs a $3.7 billion deficit in the machine tools balance sheet.[18] While the economy has recovered from a deep

recession, the increased demand for machine tools is mostly met by an increase in imports of close to 30 percent in 2012.[19] Under certain scenarios, U.S. national security eventually could be compromised if U.S. domestic defense manufacturers continue to increase their reliance on advanced manufacturing technology designed and fabricated abroad.

Key themes discussed in this chapter are:

- The ability to develop and produce state-of-the-art machine tools and compete against emerging and advanced industrialized countries

- The decline of the U.S. advanced machine tool sector in contrast to the sustained viability of this sector in Japan and Germany, two other advanced industrialized countries with high standards of living and thus high labor costs

- The eventual emergence of new competitors such as China

- The less than optimal environment for fostering a highly innovative small-to-medium-sized manufacturing sector, which may impede the commercialization of the next wave of advanced manufacturing driven by 3-D printing

A NOTE ON CRITICALITY

The ability to develop and produce state-of-the-art machine tools within the United States is at risk. Since the 1980s, the U.S. machine tool industry has declined steadily, losing ground to European and Asian competitors who increasingly have entered the American market. Because it enables nearly all other manufacturing industries, the machine tool industry is highly sensitive to market forces. It essentially can come to a halt in difficult economic times, because other manufacturers, struggling to survive, are unlikely to upgrade outdated equipment. The current state of prolonged economic hardship further diminishes the United States' competitive advantage relative to foreign manufacturers. Losing the ability to design and construct machine tools domestically would be detrimental to the health of the U.S. industrial base. However, it is important to note that foreign production is predominantly located in Germany and Japan, both of which have longstanding positive relations with the United States. For these reasons, the risk that machine tools will become unavailable to the U.S. defense industrial base is *moderate*.

The importance of machine tools to the health and integrity of the defense industrial base cannot be understated. Machine tools are the basic enabler of virtually all other manufacturing, and are essential in the fabrication of custom components necessary for the fabrication of advanced weapons systems and military vehicles. Losing domestic design and production capability for machine tools creates the potential that technicians using machine tools to design and fabricate other defense products will not have access to the most advanced programs and machinery. An inability to acquire machine tools in general would cripple the defense industrial base, eliminating the ability to produce uncountable essential inputs to a broad spectrum of defense systems. This loss would constitute an *incapacitating* impact to U.S. military capabilities.

BACKGROUND

The U.S. machine tool sector traditionally has endured extreme cycles of booms and busts. During successive periods of global economic recession (1970, 1975, 1982, and 2003), the decline in demand and attendant drop in revenue has driven out firms that were not cost-competitive. The cyclical nature of the industry has made many manufacturers so cautious that they have failed to adequately prepare for future upturns after an extended slowdown, leaving them off-guard and unready to capitalize when global demand rises again.

President Ronald Reagan declared machine tools to be a core industry required to sustain a superpower-level military and economy, and implemented measures to bolster domestic machine tool production by limiting imports from foreign manufacturers.[20] Agreements were reached with Japan and Taiwan, two leading exporters of machine tools, to freeze their share of the U.S. market at 1986 levels for five years.[21] Additionally, the Reagan administration developed a domestic action plan to supplement the industry's own modernization efforts.

In 1986, a team of scientists and economists working for the MIT Commission on Industrial Productivity launched a study called *Made in America: Regaining the Productive Edge*.[22] Published in 1989, the report examined the U.S. machine tool industry and highlighted several reasons for its decline. Among these reasons were costly delays in introducing new product lines and new production methods to keep pace with foreign producers. As a result, Japanese manufacturers had been able to take over the low end of the global market, not least because of targeted measures and policies (including standardization) pursued by the Japanese government to enhance export competitiveness. Thanks to better engineering, application, and advanced machine technology, German manufacturers squeezed U.S. builders from the high end of the market.

The relative inertia of the U.S. machine tool industry also was influenced by an attitude of complacence towards innovation among U.S. producers; many end-users had become accustomed to being dominant. A shakeout of the industry in the 1980s encouraged the growth of large conglomerates. Instead of providing wider incentives for modernization and innovation in the industry as a whole through access to capital and advances in R&D, the conglomerates sought to maximize their own growth and profits. When the industry faced another recession in the early 1990s, the largest companies failed to offer capital investments to smaller firms, and the less competitive companies were unable to survive.

In the mid-1990s, the machine tool sector experienced further decline. The White House Office of Science and Technology Policy commissioned a report on the status of the machine tool industry in 1994.[23] The report repeated the main findings of the earlier Reagan-era study, reiterating that the U.S. machine tool industry was overly dependent on a declining U.S. market and that machine tool companies were responding inefficiently to recessions and hoarding orders during boom periods to deliver finished products during recessions. These artificial delays provided space for foreign exporters to compete by delivering machines in weeks instead of months. The report also pointed to reasons for the rise in Japan's share of the market, including combining new product technology (standardized CNC tools) with major process innovations (modular productions

to lower the costs of high-volume production) (see Figure 2).

While the report listed other incidental factors such as unfavorable exchange rates, the main finding was that the U.S. machine tool industry failed to recognize the arrival of a global economy and the rise of new global competitors. Ultimately, there were too many small firms struggling to access scarce financing for new investments, and they were unable to find common ground for standard computer-driven control systems that operated five-axis machines. Finally, the report noted that U.S. manufacturers faced an inadequate export infrastructure.

In the late 1990s, the Senate Committee on Commerce, Science, and Transportation's Subcommittee on Manufacturing and Competitiveness held a hearing to review the challenges confronting the U.S. machine tool industry.[24] The hearing covered the decline of the industry in the U.S. Midwest and the rise of Chinese tool producers, and emphasized this small sector's important contribution to maintaining a strong defense industrial base.

MACHINE TOOLS AND U.S. DEFENSE CAPABILITIES

There is general consensus that machine tools are critical for defense capabilities. In 2004, some members of Congress expressed great concern about the decline of U.S.-made precision machines. Subsequently, the Pentagon issued new regulations to encourage the purchase of domestic machine tools, reflecting a program in the FY2004 National Defense Authorization Act (NDAA) that provided preferential treatment to defense contractors that purchased U.S. manufactured machine tools.[25] The corresponding rule in the Defense Federal Acquisition Regulation Supplement (DFARS) stipulated that defense contractors should "purchase and use capital assets (including machine tools) manufactured in the United States."[26] Language in section 822 of the NDAA further supported this rule.

Nevertheless, the U.S. machine tool sector continued to decline. For example, in 2005, the Department of Commerce's Bureau of Industry and Security (BIS) reported that imported machine tools contained only a small fraction (an average of three percent) of U.S. produced components. In contrast, custom-built U.S. machine tools contained an average of 84 percent domestic components.[27] The survey additionally indicated that the strength of the U.S. machine tools sector is in the area of service/support (see Table 2). However, U.S. users of five-axis machine tools listed superb engineering, accuracy, precision, and speedy delivery as the most important considerations in acquisitions, criteria that tend to favor foreign-manufactured machine tools over domestic models. Therefore, a majority of U.S. buyers do not own any U.S. manufactured machine tools. A majority (64 percent, according to BIS) of commercial five-axis machine tools used to fulfill government contracts are imported.

EXPORT CONTROL

The largest demand for machine tools comes from China and India. However, the Department of Commerce has introduced an export licensing process for machine tools, listing them as "dual-use" (military and commercial). BIS reports that many American machine tool firms lose customers due to long delivery delays or the possibility of license denial. Often,

Table 2: Categories of U.S. Competitive Advantage and Disadvantage

U.S. Competitive Advantage	U.S. Competitive Disadvantage
Machine rigidity and/or durability Lifespan of machine Thermal stability and control Service and support Precision/repeatability	CNC IS/IT network/interface Precision/repeatability CNC rotary tables* Materials of construction Spindle speed/durability

The table shows the responses listed in descending order of frequency as mentioned by surveyed manufacturers of machine tools.
* CNC Rotary Tables can be used to make (form) arcs and circles in manufacturing processes. Such components are widely used in industrial robots, fiberoptics and photonics, machine tools, semiconductor equipment, and medical equipment. Jennifer Watts, Jason Bolton, and Ashley Miller, "Critical Technology Assessment: Five Axis Simultaneous Control Machine Tools," U.S. Department of Commerce Bureau of Industry and Security (July 2009). http://www.bis.doc.gov/defenseindustrialbaseprograms/osies/defmarketresearchrpts/final_machine_tool_report.pdf.

European or Asian manufacturers are able to deliver on orders as much as twice as fast as U.S. manufacturers, largely due to less restrictive export controls.[28] Moreover, most of the companies in the countries facing export licensing controls have developed their own domestic industries independent of the U.S. industry. BIS identified 45 companies that manufacture five-axis machine tools in Brazil, China, India, Russia, and Taiwan. None of them use U.S. technology, parts, components, or materials, yet advanced machine tools are readily available on the international market. When facing U.S. export controls, countries that would otherwise purchase a U.S. machine tool simply turn to another supplier. Although export controls do exist for Japanese and many European manufacturers, they are not nearly as cumbersome as those in the United States. Between 2004 and 2007, the sales of 148 U.S. machines were authorized; however, only 34 were actually delivered. In most cases, during the several months of delay in obtaining an export license, foreign consumers purchased and received a foreign model, resulting in cancellation of their original order from the United States. In one case, a Chinese firm cancelled its order after waiting seven months for an export license without response.

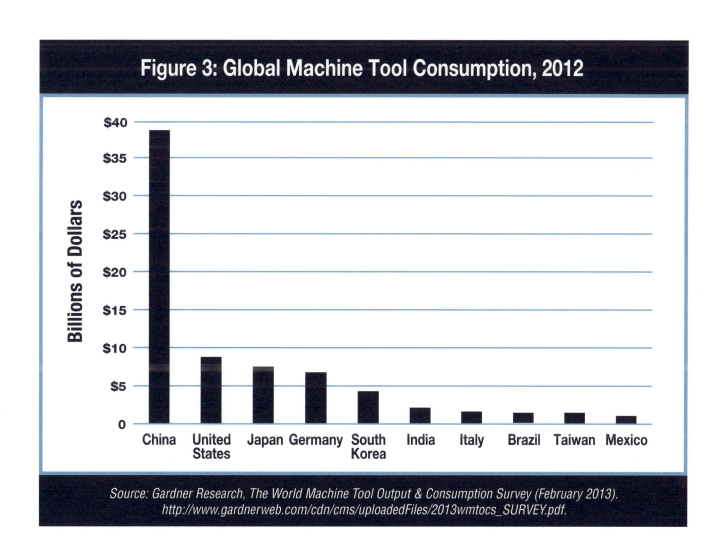

Figure 3: Global Machine Tool Consumption, 2012

Source: Gardner Research, The World Machine Tool Output & Consumption Survey (February 2013).
http://www.gardnerweb.com/cdn/cms/uploadedFiles/2013wmtocs_SURVEY.pdf.

In the global market, the U.S. machine tool industry is simply not competitive. The more advanced European and Asian markets are more than capable of supplying their own machine tools, while export controls render U.S. manufacturers uncompetitive in less developed markets.[29]

RECENT DEVELOPMENTS

The Association of Manufacturing Technology (AMT) sponsored a special report on the rise of the Chinese machine tools market in 2006. The report's conclusion says that "Chinese buyers are eager to buy U.S.-built machines, but competition is stiff, especially with Japanese and German models."[30] However, U.S. expansion into this market remains unlikely due to the difficulty in granting export licenses and business visas. However, AMT concludes that creating the regulatory conditions that allow for U.S. manufacturers to successfully expand into the Chinese market is vital to the success and growth of the U.S. machine tool industry.

While sales of U.S. -made high-precision five-axis machine tools are declining sharply, countries such as China and Taiwan are gaining market share, often due to favorable exchange rates and governmental policies. The sale of U.S.-made five-axis machine tools declined by 19 percent between 2005 and 2008, and collapsed by 60 percent in 2009.[31] It recovered in 2012, but U.S. users/buyers ended up importing a large volume of machine tools.

Moreover, China has at least 20 domestic five-axis machine tool manufacturers and Taiwan has 22, compared to only six manufacturers in the United States. Chinese production has expanded significantly in recent years, and was reported to have doubled between 2005 and 2007. China now imports only about 10 percent of the machine tools used in the country. The Chinese military is now fully supplied by domestically manufactured machine tools.[32] AMT has told U.S. authorities for years that Chinese factories have access to "whatever five-axis machine tools they want to acquire for its high-technology industry." The U.S. export control system appears to have no impact on China's ability to acquire machine tool technology, largely because U.S. manufacturers only account for six percent of world machine tool production.[33]

HUMAN CAPITAL

Keeping up with the newest trends in manufacturing requires skilled workers. The U.S. educational system and vocational schooling have not been able to ensure the continuous upgrading of labor skills.[34] The skills that workers need are often learned on the job and in the workplace. Yet U.S. employers have expressed a repeated reluctance to invest time and resources in training workers to improve relevant skills. The dilemma is a classic collective action problem: employers refuse to invest in training without a guarantee that the worker will stay on and continue to work for the firm. As all employers express the same reservations, nobody gets any training. Usually, an impartial third party, namely a government agency or a local authority, solves a collective action problem. Nevertheless, spending on education has been continually reduced, just when more spending is needed to close the skills gap.

According to a 2011 survey by the Manufacturing Institute (conducted by Deloitte), many firms struggle to hire machinists, operators, technicians, and craft workers—precisely the workers needed for advanced manufacturing sectors.[35] The Manufacturing Institute reported that hired workers often lacked the necessary skills to operate complex machinery. At the same time, workers reported that their employers were often unwilling to invest in expanding training opportunities or implement an apprenticeship system to obtain the requisite talent.[36, 37]

MITIGATING THE RISKS

The following recommendations are designed to mitigate risks to the defense industrial base by encouraging a more robust domestic machine tools industry:

Simplify and streamline the export licensing controls. U.S. manufacturers have been losing international business due to the lengthy and cumbersome process of obtaining export licenses for many high-volume consumers of machine tools. As a result, the United States has virtually no presence in foreign machine tools markets; instead, international customers purchase from countries not burdened by export restrictions. Although many export restrictions can be necessary in ensuring U.S. security, the licensing process itself should be streamlined. For example, foreign consumers cannot obtain an export license prior to submitting an order, translating to a several-month delay in delivery when ordering from U.S. suppliers. Due to that risk of time lost, many foreign consumers do not even consider U.S. suppliers. Additionally, the domestic security benefits some export restrictions provide are unclear.

Short of an outright suspension of export licensing for machine tools, allowing foreign customers to obtain pre-approved export licenses would help ensure that U.S. machine tools were at least considered by foreign consumers. At the very least,

this change would allow U.S. machine tool manufacturers the opportunity to be globally competitive. Although some officials worry about the proliferation of nuclear technology, in reality, since the 1990s, emerging markets such as China, Brazil, and India have been able to purchase the most advanced manufacturing technology on international markets. The export controls for machine tools are largely irrelevant, in part because the United States is a minor player in the global advanced manufacturing technology sector.

Simplify and streamline the process for issuing business visas to potential clients and individuals attending trade conferences. Since the U.S. is coping with a shortage of engineers, computer scientists, and other scientists, it should consider recruiting engineers from other parts of the world as a temporary stopgap.

Invest in cross-curriculum programs that provide high school students with opportunities to enter vocational training programs tailored to the needs of the advanced manufacturing sector. U.S. firms largely have abandoned training and apprenticeship programs. Many firms are unwilling to invest time and resources to train workers because those workers may use those new skills to find new jobs. One solution is to build partnerships between local community colleges and manufacturing firms to develop training programs. If firms donate old manufacturing equipment to the colleges, students can receive firsthand training on manufacturing equipment, increasing the pool of skilled machinists available to companies. Such a partnership between local authorities and the private sector will benefit all firms, improving the industrial commons as well as the defense industrial base.

Shorten the capital-investment depreciation schedule to more accurately reflect the degree to which the machinery is up-to-date. The standard depreciation schedule for capital-intensive investments, such as machinery, permits write-offs for portions of a capital-intense purchase over an extended time period, most often five years. This period is much longer than the "life" of many machines, and discourages upgrading to the most current and efficient machinery.

Temporary measures following the 2007-2008 economic slowdown changed capital investment depreciation schedules, allowing new capital-intense investments to be either 50 percent or 100 percent tax deductible in the purchase year, rather than depreciating deductions over an extended period. Under the temporary bonus depreciation policy, companies were encouraged to update machinery with more current and efficient models because they were able to fully expense the cost of new machinery in the purchase year.

Support U.S. machine tools capacity. DoD has been a principal driver in major breakthroughs in basic research. The U.S. federal government has supported basic research by making a long-term commitment to innovation, which ultimately has supported private research and projects. Other governments are intent on gaining advantages in the growth of new technology that underpins manufacturing. DoD should sustain machine tool capacity in areas such as the aerospace industry and other high-tech sectors.

CONCLUSION

Five-axis simultaneous control machine tools are the most advanced machinery that produces other machines. While many of the original innovations took place in the U.S. in the 1950s and 1960s, Japanese and German firms have been able to sustain their global competitiveness in spite of high labor costs because they have been singularly focused on preserving their industrial base. Ultimately, the decline of the most advanced machinery illustrates the decline of U.S. manufacturing prowess. Can this situation be reversed? We believe that the U.S. government and federal agencies should remain committed to the creation of fertile grounds for innovation, which will enable the United States to provide the overall best environment for business. We believe this growth can be accomplished through tax and business policy, robust support for basic research, and training and education of a highly skilled workforce. U.S. companies tend to falter when they need to translate new technologies and design methodologies into commercial competitive ventures; however, here, too, we can ensure that we possess the proper infrastructure to promote innovation.

ENDNOTES

a. Government Accountability Office (GAO), *Defense Logistics: Actions Needed to Improve the Availability of Critical Items during Current and Future Operations GAO-05-275*, by Richard G. Payne and John W. Lee (Washington, D.C.: GAO, April 2005). http://www.gao.gov/assets/250/245974.pdf.

1 John P. Holdren and Eric Lander, *Report to the President on Ensuring American Leadership in Advanced Manufacturing* (Washington, DC: June 2011). http://www.whitehouse.gov/sites/default/files/microsites/ostp/pcast-advanced-manufacturing-june2011.pdf.

2 Department of Defense, *Quadrennial Defense Review Report*, (Washington, DC: Department of Defense, February 2010). p. 103. http://www.defense.gov/qdr/images/QDR_as_of_12Feb10_1000.pdf.

3 U.S. Department of Labor, Bureau of Labor Statistics, *Employment, Hours, and Earnings from the Current Employment Statistics survey (National)*. http://data.bls.gov/pdq/SurveyOutputServlet.

4 Bureau of Industry and Security, U.S. Department of Commerce, *Critical Technology Assessment: Five Axis Simultaneous Control Machine Tools* (Washington, DC: BIS, 2009). http://www.bis.doc.gov/defenseindustrialbaseprograms/osies/defmarketresearchrpt/final_machine_tool_report.pdf. For recent figures, see Gardner Research, *The World Machine Tool Output & Consumption Survey* (February 2013). http://www.gardnerweb.com/cdn/cms/uploadedFiles/2013wmtocs_SURVEY.pdf.

5 Richard A. McCormack, "U.S. Precision Machine Tool Industry is No Longer a Global Competitive Force," *Manufacturing and Technology News*, March 5, 2010. http://www.manufacturingnews.com/news/10/0305/fiveaxis.html.

6 Gary Pisano and Willy Smith, "Restoring American Competitiveness," *Harvard Business Review*, July 1, 2009.

7 Gregory Tassey, "Rationales and Mechanisms for Revitalizing U.S. Manufacturing R&D Strategies," *The Journal of Technology Transfer* 35, n. 3 (2010): 283-333.

8 The first revolution began in Britain and centered on textiles and steam power. The second industrial revolution was ushered in by Ford, mass production, and the assembly line. The third revolution is digital.

9 Peter Marsh, "Manufacturing and Innovation: A Third Industrial Revolution," *The Economist*, April 21,2012. http://www.economist.com/blogs/schumpeter/2012/04/special-report-manufacturing-and-innovation.

10 "Layer by Layer: How 3D printers work," *The Economist*, April 21, 2012. http://www.economist.com/node/21552903.

11 Peter Marsh, "Programmed for Personalization," *The Financial Times*, June 14, 2012. http://www.ft.com/intl/cms/s/0/433a1172-b584-11e1-ab92-00144feabdc0.html.

12 Peter Marsh, "Democracy Made with Personalised Products," The Financial Times, June, 14, 2012. www.ft.com/intl/cms/s/0/aacde32a-b56d-11e1-ab92-00144feabdc0.html#axzz266ilWT8g.

13 The White House, Office of the Press Secretary, "We Can't Wait: Obama Administration Announces New Public-Private Partnership to Support," *The White House* (August 16, 2012). http://www.whitehouse.gov/the-press-office/2012/08/16/we-can-t-wait-obama-administration-announces-new-public-private-partners.

14 "U.S. Ready to Bet $60 Million on 3D Printing," *Innovation News*, May 9, 2012. http://www.innovationnewsdaily.com/1129-bet-3d-printing.html.

15 Albert Albrecht, *The American Machine Tool Industry: Its History, Growth & Decline - A Personal Perspective*. (Book Factory: 2009), 2.

16 Bureau of Industry and Security, U.S. Department of Commerce, *Critical Technology Assessment: Five Axis Simultaneous Control Machine Tools* (Washington, DC: BIS, 2009). http://www.bis.doc.gov/defenseindustrialbaseprograms/osies/defmarketresearchrpt/final_machine_tool_report.pdf.

17 David Noble, *Forces of Production: A Social History of Industrial Automation* (New York: Oxford University Press, 1986). See also MIT's Servomechanisms Laboratory vision of "three-axis continuous path control." http://museum.mit.edu/150/86.

18 Gardner Research, *The World Machine Tool Output & Consumption Survey* (February 2013). http://www.gardnerweb.com/cdn/cms/uploadedFiles/2013wmtocs_SURVEY.pdf.

19 Ibid.

20 Ronald Reagan, "Statement on the Revitalization of the Machine Tool Industry" Online by Gerhard Peters and John T. Woolley, *The American Presidency Project*, December 16, 1986. http://www.presidency.ucsb.edu/ws/?pid=36815.

21 Paul Freedenberg, "Are Machine Tools Important to our National Security?" *American Machinist*, August 5, 2005. http://americanmachinist.com/government-matters/are-machine-tools-important-our-national-security.

22 Michael L. Dertouzos, Richard K. Lester, Robert M. Solow and The MIT Commission, *Made in America: Regaining the Productive Edge* (Cambridge, MA: MIT Press, 1989).

23 David Feingold, et al., *The Decline of the U.S. Machine-tool Industry and Prospect for its Sustainable Recovery* (Santa Monica, CA: RAND: Critical Technologies Institute [MR-479/1-OSTP], 1994). http://www.rand.org/content/dam/rand/pubs/monograph_reports/2006/MR479.1.pdf.

24 *Challenges Confronting the Machine Tool Industry: Hearing before the Subcommittee on Manufacturing and Competitiveness of the Committee on Commerce, Science, and Transportation*, U.S. Senate, 106th Cong., 10 (1999) (S. Hrg. 106-1064). http://www.gpo.gov/fdsys/pkg/CHRG-106shrg74874/html/CHRG-106shrg74874.htm.

25 Paul Freedenberg, "Are Machine Tools Important to Our National Security?" *American Machinist*, August 5, 2005. http://americanmachinist.com/government-matters/are-machine-tools-important-our-national-security.

26 See DFARS 216.470 and *Federal Registry* 71(54), 14108-14110 (March 21, 2006).

27 U.S. Department of Commerce, Bureau of Industry and Security, *Critical Technology Assessment: Five Axis Simultaneous Control Machine Tools* (July 2009). http://www.bis.doc.gov/defenseindustrialbaseprograms /osies/defmarketresearchrpts/final_machine_tool_report.pdf.

28 U.S. Department of Commerce, Bureau of Industry and Security, *Critical Technology Assessment: Five Axis Simultaneous Control Machine Tools* (July 2009), 10. http://www.bis.doc.gov/defenseindustrialbaseprograms/osies/defmarketresearchrpts/final_machine_tool_report.pdf.

29 U.S. Department of Commerce, The International Trade Administration, "Industry Assessment: Machine Tools and Metalworking Equipment,"by Dawn Kawasaki , (February 2009). http://trade.gov/static/doc_Assess_MachineTools_Metalwork.asp.

30 Sergei Anisimov et al., *The China Machine Tools Market Analysis* (September 22, 2006), 19. http://www.maherlink.com/wp-content/uploads/2011/03/China-Machine-Tool-Market-Analysis-AMT-2006.pdf.

31 Richard A. McCormack, "U.S. Precision Machine Tool Industry is no Longer a Global Competitive Force," *Manufacturing and Technology News*, March 5, 2010. http://www.manufacturingnews.com/news/10/0305/fiveaxis.html.

32 U.S. Department of Commerce, Bureau of Industry and Security, *Critical Technology Assessment: Five Axis Simultaneous Control Machine Tools* (July 2009), 10. http://www.bis.doc.gov/defenseindustrialbaseprograms/osies/defmarketresearchrpts/final_machine_tool_report.pdf.

33 Paul Freedenberg, "We told you so," *American Machinist* (September 14, 2009). http://www.americanmachinist.com/304/Issue/Article/False/84785/Issue

34 Michael Hirsh and Fawn Johnson, "Desperately Seeking Skills: How an Unqualified Workforce is Prolonging America's Economic Misery," *National Journal*, August 3, 2011.

35 Tom Morrison *et al*, "Boiling Point? The Skills Gap in U.S. Manufacturing," (Washington, D.C.: Deloitte and the Manufacturing Institute, October, 2011). http://www.themanufacturinginstitute.org/~/media/A07730B2A798437D98501E798C2E13AA.ashx.

36 Anthony P. Carnevale and Stephen J. Rose, "The Undereducated American," (Washington, D.C.: Georgetown Center on Education and the Workforce, 2011).

37 Alan Berube, *Growth through Innovation: Identifying and Responding to the U.S. Metropolitan Employment Crisis* (Washington, D.C.: Brookings Institution, 2011).

CHAPTER 15
BIOLOGICAL WEAPONS DEFENSE

EXECUTIVE SUMMARY

The 2001 anthrax terror attacks in the United States underscored the importance of planning responses to and defense against biological attacks. This incident also highlighted critical shortages of available medical countermeasures (MCMs), including shortages of stockpiled vaccines that inoculate Americans against a range of biological threats. The U.S. government spent more than $22 billion on biodefense programs between 2001 and 2006; currently, however, no major pharmaceutical firms are developing MCMs or new broad-spectrum antibiotics.

Given the sheer number of possible biological threats for which vaccines may be needed and the fact that no new MCMs or broad-spectrum antibiotics are under development, the United States is underprepared for the possibility of a large-scale biological attack.

The Biomedical Advanced Research and Development Authority (BARDA) reported in 2011 that the United States has no new vaccines in development and does not plan to add new broad-spectrum antibiotics or antivirals to the stockpile (which currently is only prepared for anthrax and smallpox-based biological agents). The U.S. current smallpox vaccine stockpile would be inadequate in the event of a coordinated nationwide attack. Additionally, the current anthrax vaccine is over 50 years old, has documented and serious side effects, and has not been approved for use in children. All MCMs have limited shelf lives and must be periodically replaced as they reach expiration.

Preparedness for a biological attack requires high initial costs for research and development (R&D) and funding for stockpile maintenance. Most U.S. firms are hesitant to invest in developing vaccines and other MCMs, which are expensive products with unpredictable demand that offer no guarantee of revenue or profit. Instead, firms prefer to focus on daily-use medications that generate funding for future research. Public funding likely will be necessary in order to stimulate development of any new MCMs.

BIODEFENSE: MEDICAL COUNTER MEASURES
PROTECTING AGAINST BIOLOGICAL ATTACK

MANUFACTURING SECURITY

Investments to counter biological threats protect American lives

PRIVATE RESEARCH + **GOVERNMENT SUPPORT** = **SAFER AMERICANS**

COSTS OF PREPARATION

The average cost of a new vaccine is very high

$1 BILLION PER NEW VACCINE

 **THE U.S. HHS** MAINTAINED 10,000,000 **ANTHRAX VACCINES** BETWEEN 2006 AND 2011

U.S. INVESTMENTS

 = $1,000,000,000

Since 2001, the U.S. has invested roughly $22 billion in programs and products to defend against the threat of bioweapons

$22 BILLION

VULNERABILITY

 ANTHRAX ATTACKS IN 2001 COST OVER $100 MILLION

PERISHABLE PROTECTION

Vaccinations maintained by the U.S. are perishable and must be continually replaced and stockpiled

VACCINATIONS MUST BE CONTINUALLY REPLACED

The U.S. may face continuous bio-attacks from a capable opponent wielding bioweapons

MITIGATING RISKS

Robust U.S. biodefenses require a variety of steps to confront a possible bio-attack

DEVELOP NEW VACCINES   **INVEST IN RESEARCH**   **EDUCATE POLICYMAKERS**

INTRODUCTION

"A study of Disease—of Pestilences methodically prepared and deliberately launched upon man and beast—is certainly being pursued in the laboratories of more than one great country. Blight to destroy crops, Anthrax to slay horses and cattle, Plague to poison not armies but whole districts—such are the lines along which military science is remorselessly advancing."[1]

– Winston Churchill, September 1924

Biological weapons have been a part of warfare and terror campaigns since well before the field of microbiology formally was established. Since 1874, international treaties have prohibited the use of biological weapons. However, these arrangements lack real enforcement mechanisms and have not prevented the development and use of these weapons by a number of countries and non-state actors. Even prior to the U.S. Civil War, there have been a number of instances when biological weapons were used in warfare (see Table 1).

Biological weapons and the threat of their use by terrorists were not a particularly high priority for the U.S. government or public until 2001, when envelopes containing anthrax spores were mailed to several offices in the U.S. Congress. The anthrax

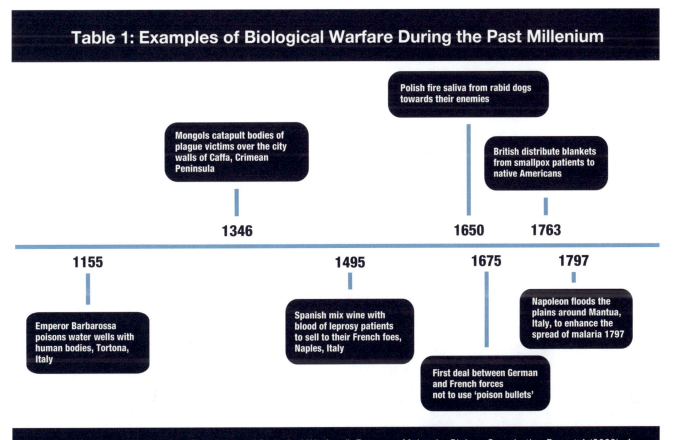

Table 1: Examples of Biological Warfare During the Past Millenium

Polish fire saliva from rabid dogs towards their enemies

Mongols catapult bodies of plague victims over the city walls of Caffa, Crimean Peninsula

British distribute blankets from smallpox patients to native Americans

1346 — 1650 — 1763

1155 — 1495 — 1675 — 1797

Emperor Barbarossa poisons water wells with human bodies, Tortona, Italy

Spanish mix wine with blood of leprosy patients to sell to their French foes, Naples, Italy

Napoleon floods the plains around Mantua, Italy, to enhance the spread of malaria 1797

First deal between German and French forces not to use 'poison bullets'

Source: Friedrich Frischknecht, "The History of Biological Warfare," *European Molecular Biology Organization Report 4 (2003), 4.*

scare revived governmental and public awareness of how dangerous and relatively easy to use biological weapons could be, including for non-state actors. The U.S. government began to fund research for new, more effective vaccines, and began stockpiling existing vaccines to prepare for a potential biological attack. However, with no signs of an imminent threat, the priority and funding for these programs declined over the course of the past decade.

A National Biodefense Science Board (NBSB) was created under the authority of the Pandemic and All-Hazards Preparedness Act, signed into law on December 19, 2006. According to its charter,

> "The NBSB was established to provide expert advice and guidance to the Secretary of the U.S. Department of Health and Human Services (HHS) on scientific, technical, and other matters of special interest to HHS regarding activities to prevent, prepare for, and respond to adverse health effects of public health emergencies resulting from chemical, biological, nuclear, and radiological events, whether naturally occurring, accidental, or deliberate."[2]

In March 2010, the NBSB published a report that urged the U.S. government to develop MCMs to protect against biological threats.[3] With the exception of first-generation anthrax and smallpox vaccines, the U.S. lacks vaccine stockpiles for any other agents the Department of Homeland Security (DHS) considers to be potential threats (as identified on the so-called material threat list). It is also problematic that U.S. R&D funding has primarily been awarded to biotech firms outside of the United Sates. While more recent funding went to some small U.S.

firms, no large U.S. pharmaceutical manufacturer has shown interest in developing MCMs. The March 2010 report summarized the challenge thusly: "If achieving national MCM goals is likened to climbing a mountain, then most of the mountain remains to be climbed."[4]

This chapter examines the degree to which the U.S. government has responded to the recommendations of the March 2010 report by increasing funding for R&D by U.S. firms. It also assesses the status of development of vaccines against the most prominent threats and examines whether efforts to stockpile vaccines have improved. This chapter also provides recommendations for creating and enhancing MCM capabilities commensurate with rising threats despite constraints on resources. It concludes with a discussion of why solutions to U.S. shortages of MCMs will depend on closer collaboration between government and industry.

Key themes discussed in this chapter are:

- The development and stockpiling of MCMs, as a public good and security imperative, cannot be left to market forces and requires special U.S. government investment.

- The current standing of the U.S. biodefense industrial base must be improved.

- The security of the U.S. military and public is at risk due to the lack of sufficient MCMs to protect against biothreats.

- U.S. policymakers must address the need for accelerated research, investment, and stockpiling of MCMs, notwithstanding fiscal constraints.

ANTICIPATING A THREAT ON THE MODERN BATTLEFIELD (a notional though realistic scenario)

A U.S. Army platoon is dug in at a forward position. An enemy convoy, located miles away, slows its movement and releases a spray. The enemy had previously determined that the prevailing wind would carry the disease-laden aerosol mist in the direction of the platoon. Unlike a chemical attack, which would have caused an almost immediate reaction, the biological agent utilized in this case requires an incubation period that can range from 24 to 48 hours after exposure before symptoms appear. This delay can also increase the likelihood of additional infections, as the exposed members of the platoon come into contact with others.

The platoon, even after realizing what has happened, lacks sufficient stores of medical countermeasures to treat every soldier. The platoon returns to base for treatment, but the biological agent has already caused significant negative health effects for most of the soldiers, placing lives at risk and reducing U.S. presence in the area.

A NOTE ON CRITICALITY

The United States is largely unprepared for a biological weapons attack. While an attack on U.S. troops or civilians by a bio-weapon is a "low probability event," currently the existing supply chains would be unable to quickly produce enough vaccine to effectively respond to such an attack. Moreover, despite the dire consequences of a successful biological attack, the development of new broad-spectrum anti-biotics and antiviral mediations has been slow, due to the fact that production and testing of new drugs would require large capital investments in R&D, and because demand for these drugs would be limited and inconsistent. Because of these and other considerations, the risk of a shortage of critical medical countermeasures in the event of a biological attack is *high*.

Given the likely shortage of adequate MCMs, the impact of a well-executed biological attack against a strategic military target would be *significant*. Although it is unlikely that a large-scale biological attack could be successfully executed across the entire military structure, strategic military assets (including in-theatre command headquarters, aircraft carriers, and nuclear submarines in port) could be targeted and rendered inoperable. Additionally, it could take several days to transport MCMs to forward operating assets, during which time in-theater operations would be severely compromised. For example, if a biological attack rendered an aircraft carrier battle group ineffective for an extended period, U.S. readiness would be significantly diminished.

BACKGROUND

The most significant examples of the use of biological weapons in modern warfare occurred during the World Wars I and II. Biological weapons were used by Germany during World War I and by Japan during World War II. The German attacks were relatively small scale and considered unsuccessful, relying on covert operations using both anthrax and glanders (an infectious disease carried by horses, donkeys, and mules) to infect animals or to contaminate animal feed in several countries.[5] During World War II, the Japanese army poisoned more than 1,000 water wells in Chinese villages in an attempt to spark cholera and typhus outbreaks. Japanese aircraft also dropped plague-infested fleas over Chinese cities and distributed bioagents by means of saboteurs in rice fields and along roads. Some of the epidemics caused by these Japanese attacks persisted for years. By 1947, long after the

Table 2: Crucial Biological Agents
(Centers for Disease Control, Atlanta, Georgia, USA)

Disease	Pathogen	Abused
Category A (major public health hazards)		
Anthrax	Bacillus antracis	First World War
		Second World War
		Soviet Union, 1979
		Japan, 1995
		USA, 2001
Botulism	Clostridium botulinum	-
Hemorrhagic fever	Marburg virus	Soviet bioweapons program
	Ebola virus	-
	Arenaviruses	-
Plague	Yersinia pestis	Fourteenth-century Europe
		Second World War
Smallpox	Variola major	Eighteenth-century N.America
Tularemia	Francisella tularensis	Second World War
Category B (public health hazards)		
Brucellosis	Brucella	–
Cholera	Vibrio cholerae	Second World War
Encephalitis	Alphaviruses	Second World War
Food poisoning	Salmonella, Shigella	Second World War
		USA, 1990s
Glanders	Burkholderia mallei	First World War
		Second World War
Psittacosis	Chlamydia psittaci	–
Q fever	Coxiella burnetti	–
Typhus	Rickettsia prowazekii	Second World War
Various toxic syndromes	Various bacteria	Second World War

Source: Friedrich Frischknecht, "The history of biological warfare," European Molecular Biology Organization Report 4 (2003).

Japanese surrender, the death toll in China was more than 30,000 people.[6]

The study of biological warfare and biodefense accelerated during the Cold War in the United States and Soviet Union. The United States had a small-scale defense and testing program, originally created to counter Japanese weapons. President Nixon cancelled this program after controversies over testing methods and the use of biological agents during the Vietnam War. In 1972, Nixon signed the Biological and Toxin Weapons Convention to ban the development, production and stockpiling of biological and toxin weapons. The agreement improved upon the 1925 Geneva Protocol (which only banned the use of these weapons) and also called for the destruction of stockpiles of such weapons. The agreement currently has 165 signatories. Although efforts have been made to create ways to verify countries' compliance, there is no way to enforce against those who are determined to violate the agreement. Some skeptics allege that shortly after signing the treaty, the Soviet Union had over 50,000 people employed at various biological weapons research facilities, producing and stockpiling tons of smallpox and anthrax, while also working to develop drug-resistant plague bacteria and various hemorrhagic fevers.[7]

Not all bioterror attacks are perpetrated by sovereign nation-states. An example of an attack committed by non-state actors took place in a small Oregon community in 1984.[8, 9, 10] "A religious sect tried to poison a whole community by spreading *Salmonella* in salad bars to interfere with a local election. The sect, which ran a hospital on its grounds, obtained the bacterial strain from a commercial supplier. Similarly, a laboratory technician tried to get hold of the plague bacterium from the American Tissue Culture Collection, and was only discovered after he complained that the procedure took too long."[11]

Organized, determined groups and individuals (including American citizens) have demonstrated they can gain access to biological materials to pursue violent objectives. The 2001 anthrax attacks in the United States killed five people and cost an estimated $100 million due to the ensuing panic and attempts to ensure the safety of the affected facilities. The attacks also lead to the overuse of antibiotics, which is thought to have contributed to higher levels of drug resistance among those who overmedicated.[12]

While the United States has not ignored the problems outlined in this report, efforts to invest in countermeasures or sustain a domestic production capacity for biological weapons defense have not kept pace with the threats.

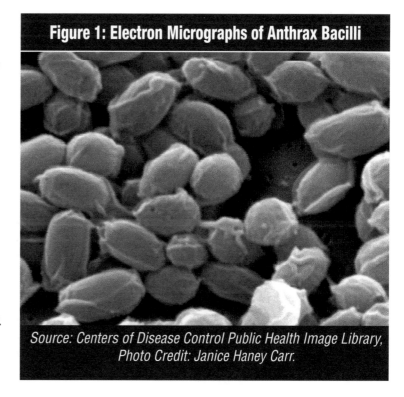

Figure 1: Electron Micrographs of Anthrax Bacilli

Source: Centers of Disease Control Public Health Image Library, Photo Credit: Janice Haney Carr.

STANDARDS AND BEST PRACTICES

Unlike other defense requirements, such as shipbuilding or aircraft manufacture, there are no commonly accepted standards for gauging the amount of vaccines required for stockpiling, the length of time in which particular vaccines are expected to expire, or how effective different kinds of vaccines or antibiotics might be to protect against a biological attack. In 2004, Congress enacted the Project Bioshield Act outlining some of these guidelines.[13] In addition to funding the purchase of additional vaccines, the Act:

- Relaxes procedures for some chemical, biological, radiological, and nuclear (CBRN) terrorism-related spending, including hiring and awarding research grants

- Guarantees a federal government market for new CBRN medical countermeasures

- Permits emergency use of unapproved countermeasures

When Soviet "virologist Nikolai Ustinov died after injecting himself with the deadly Marburg virus, his colleagues, with the mad logic and enthusiasm of bioweapon developers, re-isolated the virus from his body and found that it had mutated into a more virulent form than the one that Ustinov had used."[a]

—Friedrich Frischknecht

The Act also includes guidance for the Department of Health and Human Services (HHS) about the kinds and quantities of MCMs that should be stored in the Strategic National Stockpile (SNS).

"All medicines, including those added to the Strategic National Stockpile through Project Bioshield, must have explicit expiration dates and are not approved for use after this expiration date. As a consequence, HHS is required to procure a number of doses greater than that which is stored in the SNS at any given time. For example, HHS had to buy 29 million doses of anthrax vaccine to maintain a stockpile of at least 10 million doses from 2006 to 2011."[14]

The Act stipulates that HHS must keep the stockpile in a state of "consistent readiness," although it did not provide a definition or clear guidelines of what this entails.

The Food and Drug Administration (FDA) also is involved in approving and testing vaccines. However, the 2010 report of the NBSB called for the FDA to streamline its processes, stating that, "At present, MCM developers believe that the standards adopted by the FDA for regulation and review of CBRN MCMs are too often unclear, confusing, and inconsistently applied."[15]

VULNERABILITIES IN THE BIODEFENSE DEFENSE INDUSTRIAL BASE

Recent U.S. government and media studies have concluded that the development of MCMs designed to counter biological weapons in the United States is inadequate, even as other reports indicate that hundreds of millions of dollars have been spent on vaccine R&D. According to a 2005 report in the *National Journal*:

"U.S. government civil-biodefense spending has totaled about $22 billion since fiscal 2001, climbing from $414 million that year to a requested $7.6 billion in 2005, about an 18-fold increase, according to an analysis by Center for Biosecurity research analyst Ari Schuler. Most of the money has gone for research and development, and for stockpiling of vaccines, to counter six classic bio-warfare diseases (anthrax, botulism, plague, smallpox, tularemia, and viral hemorrhagic fevers like Ebola); for detectors to identify an attack using such agents; and for construction or upgrading of dozens of biodefense research-and-development laboratories across the country."[16]

A report released in 2004 by the University of Pittsburgh Medical Center and the Sarnoff Corporation based on interviews of 30 leading experts in the field of biomedicine concluded that there were serious shortfalls in the U.S. readiness level to protect against a biological attack.[17]

Unfortunately, evidence indicates that U.S. policymakers tasked with enhancing U.S.

''Between 50 and 60 pounds of freeze-dried tularemia produced in our production facility would eliminate about 60 percent of the population of London, England.''[b]

– Wil S. Hylton, The New York Times Magazine

biodefense remain unaware of the true extent of the threat.

Congress initially proposed allocating significant funding in support of Project Bioshield, but a 2009 Congressional Research Service report indicated that a large percentage of that funding was allocated elsewhere, which could result in a substantial reduction in the amount of Project Bioshield money available for CBRN countermeasures.[18] Additionally, because there is no consistent demand for these types of vaccines, and because the initial cost of developing them is so high (with an average of $1 billion to develop a single vaccine), the money allocated to Project Bioshield ($5.6 billion) was not enough to attract large pharmaceutical manufacturers.[19]

A December 2011 study of Project Bioshield concluded:

"[The United States has] enough smallpox vaccine in the stockpile to inoculate every United States citizen; enough anthrax vaccine to respond to a ‹three-city attack›; and a variety of therapeutic drugs to treat the infected…many other goals are incomplete. After spending hundreds of millions of dollars, for example, to develop a new vaccine for anthrax that would replace the controversial formula developed fifty years ago by the Army—which is known to have serious side effects and has never been approved for children—there is still no new vaccine. There also are no new broad-spectrum antibacterial drugs in the stockpile and no new antivirals."[20]

Perhaps the most serious biodefense risk arises from disagreements among policymakers tasked with creating an adequate, self-sustaining biodefense infrastructure. The director of BARDA admits that, due to inadequate government funding, there are no new vaccines in the pipeline. Instead of prevention, BARDA is focusing on therapeutic drugs designed to treat the infected.[21] While this focus is worrying, it is equally troubling that BARDA apparently believes there are only two biothreats—anthrax and smallpox—for which vaccines are needed. On the other hand, the lead of the Science and Technology Directorate at DHS views vaccines as essential, arguing that if there is a bioterror attack, citizens will want their children vaccinated against a "reload" or follow-up attack.[22]

"With biological weapons, we're talking about acquiring the ability to produce weapons. So if you acquire the ability to produce 100 grams of anthrax, you can keep doing that. You really have to think about biology as potentially the subject of a campaign, where somebody keeps attacking, rather than a one-shot incident."[c]

– Former Secretary of the Navy Richard Danzig

A "reload" arguably makes bioterrorism a greater threat than use of nuclear weapons by non-state actors. A nuclear attack would be disastrous, but would represent a single event, while a bioattack would be sustained.

Brett Giroir, a former official at the Defense Advanced Research Projects Agency who is now Vice Chancellor for Strategic Initiatives at Texas A&M University, reiterated the importance of developing vaccines: "Vaccines are critical components of a biodefense posture, and anybody who thinks they're not isn't thinking seriously about how we approach this."[23]

MITIGATING THE RISKS

The MCM program fundamentally addresses both public health and national security.[24] While the approximately $12 billion needed to develop new vaccines is a staggering amount when thought of in terms of a public health project, it represents less than a quarter of what is spent on some major defense projects.[25] Once an enemy state or non-state actor develops the ability to manufacture biological weapons, their ability to "reload" becomes one of the greatest threats to U.S. security, its citizens, and its warfighters.

Given market realities that create a disincentive for large companies to develop lower-demand, lower-profit vaccines, the Department of Homeland Security should incentivize the defense industrial biodefense sector through continued investment in smaller start-up businesses.

The NBSB, while laying out damning evidence of current U.S. readiness in MCMs, also has made extensive recommendations for rectifying these shortcomings.

Arguably, the most important is allowing the Secretary of the Department of Health and Human Services (HHS) to grant an operational MCM leader the authority to synchronize the efforts of HHS agencies, with end-to-end oversight.[26] Ideally, this appointment would provide a solution to the ongoing disconnect between policymakers about whether therapeutic measures are enough or whether a focus on vaccines is necessary. Given the nature of the biological weapons threat, the latter course of action appears wisest.

The following are recommendations to address the vulnerabilities presented by the lack of a U.S. defense industrial base capacity to address biological weapons attack:

Address the urgency of the threat with policymakers. Experts in and out of government largely agree that the United States is unprepared to address the threat of a biological weapons attack, and that the capacity to develop MCMs is inadequate. However, funding is lacking as well. Industry and government should focus on building a viable biodefense sector that can sustain research and development for advanced vaccines and other MCMs.

Recognize that the market will not, on its own, spur sufficient investments in vaccine research. Without government support, the barriers to creating a new vaccine are prohibitive to all but the largest drug-makers. Only a worst-case scenario of a bioterror attack will cause demand for a vaccine to be high enough for the market to drive innovation, by which point it will be too late. Without public support, MCM and vaccine innovation and production will lag, putting U.S. warfighters and citizens at grave risk.

CONCLUSION

The risks to U.S. national security posed by a biological weapons attack remain high due to a combination of misaligned incentives for industry, inadequate innovation, underfunding of U.S. MCMs, and a lack of policymaker understanding of the risks. Given these factors and the shortage of adequate MCMs (including vaccine stockpiles), the impact of a carefully executed biological attack against the U.S. would be significant.

Public funding is necessary if the United States is to prepare adequately for potential bioterror attacks. Public funding to spur research would not only benefit national security, but also be a boon to the economy because it could potentially lead to the creation and support of new high-tech research companies and would drive the United States to be a world leader in MCM development and production.

ENDNOTES

a   Friedrich Frischknecht, "The history of biological warfare," *European Molecular Biology Organization Report* 4 (2003).

b   Wil S. Hylton, "Warning: There's Not Nearly Enough of This Vaccine to Go Around," *The New York Times*, October 30, 2011, 34.

c   Ibid.

1   Winston Churchill, "Shall We All Commit Suicide?" *Pall Mall*, September 1924, reprinted in *Thoughts and Adventures* (1932), 250.

2   The Secretary of Health and Human Services, "Charter: National Biodefense Science Board," (Washington, D.C.: Health and Human Services, 2012). http://www.phe.gov/Preparedness/legal/boards/nbsb/Documents/2012-renewed-charter.pdf.

3   U.S. Department of Health and Human Services, National Biodefense Science Board, *Where Are the Countermeasures?* (Washington, D.C.: U.S. Department of Health and Human Services, 2010). http://www.phe.gov/Preparedness/legal/boards/nbsb/meetings/Documents/nbsb-mcmreport.pdf.

4   Ibid., 5.

5   Glanders is an infectious disease caused by a bacterium that directly affects animals and humans, but primarily affects horses. (See http://www.cdc.gov/glanders/ for more details.) Mark Wheelis, "Biological Sabotage in World War I," in *Biological and Toxin Weapons: Research, Development and Use from the Middle Ages to 1945, eds. Erhard Geissler and John Ellis van Courtland Moon,* (New York: Oxford University Press, 1999), 35-62.

6   Friedrich Frischknecht, "The History of Biological Warfare," *European Molecular Biology Organization Report* 4 (2003), p. S47; and S. Harris, "Japanese Biological Warfare Research on Humans: A Cause of Microbiology and Ethics," *Annals of the New York Academy of Sciences,* (1992), 21-52.

7   K. Allbek and S. Handelman, *Biohazard* (New York: Random House, 1999).

8   L.A. Cole, "The Specter of Biological Weapons," *Scientific American* 275 (1996): 30-35.

9   T.J. Torok, et al., "Large Community Outbreak of Salmonellosis Caused by Intentional Contamination of Restaurant Salad Bars," *JAMA* 278 (1997), 389-395.

10 M. Miller and W. Broad, *Germs, Biological Weapons and America's Secret War* (New York: Simon & Schuster, 2002).

11 Friedrich Frischknecht, "The History of Biological Warfare," *European Molecular Biology Organization Report* 4 (2003), 50.

12 M. Leitenberg, "Biological Weapons in the Twentieth Century: A Review and Analysis," *Critical Reviews in Microbiology* 27 (2001): 267-320.

13 Public Law 108-276 (July 21, 2004).

14 Frank Gotton, *Project Bioshield: Purposes and Authorities* (Washington, DC: Congressional Research Service, July 2006), 9.

15 Leigh Sawyer, "Where Are the Countermeasures? Protecting America's Health from CBRN Threats: A Report from the National Biodefense Science Board*,"* *Biosecurity and Bioterrorism: Biodefense Strategy, Practice, and Science* 8, no. 2 (2010): 203-7.

16 David Ruppe, "Threat-Mongering?" *The National Journal* 37 (2005): 17.

17 Staff, "U.S. Unprepared on Bio-terror, Report Says," *Homeland Security & Defense* 41, no. 3 (2004): 5.

18 Frank Gotton, *Project Bioshield: Purposes and Authorities* (Washington, DC: Congressional Research Service, July 2006), 2.

19 Ibid.

20 Wil S. Hylton, "Warning: There's Not Nearly Enough of This Vaccine to Go Around," *The New York Times*, October 30, 2011, 28.

21 Ibid., 33.

22 Wil S. Hylton, "Warning: There's Not Nearly Enough of This Vaccine to Go Around," *The New York Times*, October 30, 2011, 28.

23 Also see Hylton's discussion of the disturbing issues the Tularemia virus presents: as of now the United States has no vaccine against it, and the head of BARDA does not feel that the country needs one. Ibid., 34.

24 Wil S. Hylton, "Warning: There's Not Nearly Enough of This Vaccine to Go Around," *The New York Times*, October 30, 2011, 24.

25 Ibid., 37.

26 Leigh Sawyer, "Where Are the Countermeasures? Protecting America's Health from CBRN Threats: A Report from the National Biodefense Science Board," *Biosecurity and Bioterrorism: Biodefense Strategy, Practice, and Science* 8, no. 2 (2010): 83.

CHAPTER 16 • CONCLUSION

"… (T)he defense industry is second only to our people … our defense industry is what makes us a great military power."[1]

– Deputy Secretary of Defense Ashton Carter

The United States' national security rests solidly upon the shoulders of the men and women in uniform who defend this country. But our national security rests equally upon the shoulders of the millions of Americans in coveralls, lab coats, and business suits who man the stations of our defense industrial base. Our soldiers, sailors, airmen, and Marines share the privilege of defending the United States with their brothers and sisters throughout the U.S. defense industrial base who develop the technologies and build the weapons and systems necessary for victory on the battlefield.

Accordingly, this report examined a series of defense industrial base sectors that are vital to U.S. security. Some are especially important and require immediate attention to prevent critical loss of supply or production capacity, constituting an immediate threat to national security. Some sectors are vulnerable to immediate disruption arising from excessive or misaligned foreign dependency, while others face longer-term challenges. In general, the risks posed to many sectors of the defense industrial base may prove very difficult to fix, because they are a part of powerful, prevailing trends in the international technology market and the global economy.

> **Just as we demand strategic thinking about the problems confronting our armed forces on the battlefield, we demand strategic thinking about the problems confronting the defense industrial base.**

All of these challenges demand our best strategic thinking about how to prevent or contain significant and potentially dangerous risks to national security. This report investigates those risks and is a call to action to mitigate them.

In the late 1990s, Andrew Marshall, then-Director of the Pentagon's Office of Net Assessment, published a list of 20 critical defense technologies, the so-called "crown jewels" of U.S. defense. Now, in 2013, it is past time to update Marshall's list to address 21st century defense requirements. Nevertheless, no matter how comprehensive or up-to-date, no list will drive a real strategy unless the United States agrees upon and implements the highest priority measures for preserving an enduring and effective defense industrial base to guide the nation's security into the next century.

We need a defense industrial base strategy that serves our most important security requirements. We need to review that strategy continually to ensure that it keeps pace with rapidly shifting global trends and endures the test of time. If we are to preserve the United States' current advantages and ensure dominance on future battlefields, we need to identify the sectors that are and will be the most strategically important. Not only must we produce superior weaponry for today's warriors, we must preserve our technological edge to ensure that future generations can rely on U.S. ability to meet its defense commitments fully.

It is hard to overstate the difficulties inherent in making the right choices about competing priorities, just as one cannot over-emphasize the importance of being sure that we choose wisely. Deputy Secretary of Defense Ashton Carter discussed this dilemma recently, when he urged that we act decisively to protect well-performing weapons systems and eliminate those that are inadequate, ineffective, or simply obsolete.[2] Partnership between government and industry will be essential to ensure success in making these tough choices. In Carter's words, "We want to work together with the industry upon which we depend so much, so that they make the transition with us, and that they're here to make the greatest military in the world 10, 20 years from now."[3]

CRITICALITY – WHAT TO ADDRESS FIRST?

The issues the U.S. defense industrial base faces are complex and diverse. The challenges discussed in this report will at some point require the attention of policymakers, but they first must be prioritized according to relative risk, impact, and urgency. This report examines sectors that already are being exposed to risks that we judge to have the potential for disrupting timely production of essential combat systems. However, these risks are not evenly distributed across the defense industrial base; some are more imminent than others. The criticality matrix (Figure 1 below) enables a quick comparison of the severity and short-term likelihood of a supply disruption across the nodes discussed in this report. The top-right corner (extreme risk, incapacitating impact) represents the most urgent challenges. Supply chain disruptions to nodes in the bottom-left corner (moderate risk, isolated impact) are low probability in the short term and will not cripple key capabilities. Notably, we do not judge that any of the nodes we studied fell into the latter, relatively benign, category.

"Ours is a business of anticipation, not reaction. There is nothing magical about it. To meet tomorrow's crisis or conflict requires continuous investment today to ensure we can deliver capability critical to our nation and economic security."[a]

– Mike Petters, President and CEO of Huntington Ingalls Industries

FIGURE 1: THE U.S. DEFENSE INDUSTRIAL BASE CRITICALITY MATRIX FOR KEY NODES

DISRUPTION OF DOMESTIC SUPPLY CHAIN

RISK

IMPACT		Moderate	High	Extreme
Incapacitating: Widespread loss of capability		• Steel Armor Plate • Machine Tools	• Fasteners • Semi-conductors • Titanium • Telecommunications	• High-Tech Magnets
Significant: Specific weapons systems become unavailable			• Biodefenses • Cu-Ni Tubing	• HELLFIRE Missile Propellant • Specialty Metals • Platinum Group Metals
Isolated: Advantage reduced or lost			• Advanced Fabrics • Li-ion Batteries	• Night vision

The U.S. defense industrial base is a critical asset for U.S. national security. The supply chains that support the U.S. industrial base face an array of vulnerabilities, leaving U.S. military capabilities at risk. The Criticality Matrix evaluates the vulnerabilities to key defense industrial base nodes based on the likelihood of disruption, and a disruption's likely effect on U.S. national security.

More detailed and sophisticated strategies are required to guide defense priorities in a time of declining and uncertain budgets. Each and every defense industrial base node cannot be a priority, especially not an immediate priority. Civilian and military leaders currently lack the means for setting these priorities, though policymakers appear to recognize that more must be done in this regard. By making comparisons and setting priorities across defense industrial base nodes and sectors, the United States can preserve core defense capabilities while developing essential future capabilities.

MITIGATING THE RISKS

The U.S. government and industry already are undertaking important measures to mitigate risks. According to several senior Pentagon officials, including Under Secretary of Defense for Acquisition, Technology, and Logistics Frank Kendall, the defense industrial base should be considered part of the United States military force structure—and rightfully so. Managing the defense industrial base with as much care as we manage the branches of the armed services makes sense. This report makes 10 recommendations to reduce U.S. dependence on imported products vital to our national security. These recommendations are based on the premise that the U.S. defense industrial base is a vital national asset that is no less critical to our national security than our men and women in uniform. This report's recommendations call for the following actions:

1. **Increasing long-term federal investment in high-technology industries, particularly those involving advanced research and manufacturing capabilities.** Measures aimed at the health of the industrial base cannot be limited to the production of inputs or hardware. The distinguishing attribute of the U.S. defense industrial base is technological innovation. Collaboration among government, industry, and academia has had a leading role in developing the United States' 21st century economy and its world-class defense industrial base. As foreign nations manufacture an ever-larger share of U.S. defense supplies, the risk grows that the United States will have a diminished capacity to design and commercialize emerging defense technologies. To help ensure that our armed forces dominate the future battlefield, Congress should provide long-term funding for U.S. manufacturers to develop and implement advanced process technologies. On February 14, 2013, Brett Lambert, Deputy Assistant Secretary of Defense for Manufacturing and Industrial Base Policy, announced plans for the Department of Defense (DoD) to use public and private funding to start up two institutes comprising leading manufacturing companies and research universities. Lambert said, "[d]efense has to be a catalyst for market need. Once we are able to form these institutes, then DoD becomes the customer, not the provider of funds."[4] DoD's progress in this area is encouraging, but there is much more work to do. Our defense industrial base's capacity to adapt and develop new technologies must be nurtured.

2. **Properly applying and enforcing existing laws and regulations to support the U.S. defense industrial base.** Domestic source preferences already enacted into law, such as those that apply to steel and titanium under the Specialty Metals Clause, must be retained to ensure that important defense capabilities remain secure and available for the U.S. armed forces. Moreover, competition to reduce costs and achieve efficiency is always welcome, but only as long as the playing field is level. Here, the Executive Branch and Congress must aggressively enforce regulations aimed at ensuring fair competition.

3. **Developing domestic sources of key natural resources that our armed forces require.** Right now the United States relies far too heavily on foreign nations for certain key metals and other raw materials needed to

manufacture weapons systems and other military supplies. For example, most rare earth elements—essential components of many modern military technologies—currently must be purchased from China. As a solution to this dependency, the U.S. government and industry must stockpile these vital raw materials, strengthen efforts to resume domestic mining and processing of the materials, improve recycling to make more efficient use of current supplies, and identify alternate materials. The FY2013 National Defense Authorization Act (NDAA) gives Defense Logistics Agency Strategic Materials important new tools to diagnose and mitigate supply chain vulnerabilities for specialty metals and other critical raw materials. The NDAA must be effectively used to avoid supply disruptions such as those that have already caused procurement delays and caused the decline of key sectors of the defense industrial base. However, implementing these new tools will require greater coordination and collaboration across federal agencies to reach agreement on which materials are termed critical and strategic.

4. **Developing plans to strengthen our defense industrial base in the U.S. National Military Strategy, National Security Strategy, and the Quadrennial Defense Review process.** We must develop a strategy to identify and remedy the health of the most important and vulnerable sectors. Past iterations of these key strategic review documents have addressed the defense industrial base, but we must accord higher priority to these efforts, incorporate the defense industrial base into our national and defense strategy, and allocate resources accordingly.

5. **Building consensus among government, industry, the defense industrial base workforce, and the military on the best ways to strengthen the defense industrial base.** These sectors must work collaboratively to successfully address the concerns of all defense industrial base stakeholders. As important as it is to analyze and understand particular risks to the industrial base or the desirability of alternative mitigation strategies, creating consensus about the nature of the challenges and choice of remedies is even more important. No effective collaboration between industry and government is feasible without consensus. Solving supply chain problems will require the concerted efforts of prime contractors, original equipment manufacturers, and government. Defense-related firms depend upon increasingly global and complex supply chains. Government and industry managers need effective tools to detect supply chain risks, to determine the scope of those risks, and to address persistent problems (such as conformance and counterfeit issues) aggressively.

6. **Increasing cooperation between federal agencies and industry to build a healthier defense industrial base.** The health of the defense industrial base must not be solely the business of DoD. As part of U.S.

national strategy, assuring the health of the defense industrial base requires the coordinated efforts of a number of executive departments and agencies, including among others the Departments of Treasury, Energy, Commerce, Homeland Security, and State. If the United States is to succeed in sustaining its defense industrial base, we need to improve transparency and routine cooperation among government agencies and between government and industry.

7. **Strengthening collaboration between government, industry, and academic research institutions to educate, train, and retain people with specialized skills to work in key defense industrial base sectors.** The loss of U.S. manufacturing jobs has reduced the size of the workforce skilled in research and development, as well as advanced manufacturing processes. As then-Under Secretary of Defense Carter said in May 2012, the Pentagon's focus on selected skill sets is "an example of something we didn't do in [FY] '13 … [but] as we put together the [FY] '14 budget … [we] definitely want to look at those holes [and] make those kind of investments."[5] Our defense industrial base workforce is itself a national security asset, and must be nurtured accordingly.

8. **Crafting legislation to support a broadly representative defense industrial base strategy.** Congress and the Administration must collaborate on economic and fiscal policies that budget for enduring national security capabilities and sustain the industrial base necessary to support them.

9. **Secure defense supply chains should be modernized and secured through networked operations that provide regular communications among defense procurement agencies, prime contractors, and the supply chains upon which they depend.** By mapping the supply chain at levels below that of the prime defense contractors, DoD's Sector-by-Sector, Tier-by-Tier (S2T2) program offers great promise to DoD and industry managers alike. S2T2 allows DoD and industry managers to focus on and document the important role that lower tier defense industrial base firms play in sustaining U.S. defense capabilities. However, S2T2 cannot be expected to inform managers about how to prioritize efforts or address recurring problems such as non-conforming or counterfeit products. Ongoing, secure communications linking the different tiers of the defense supply chain, patterned on the networked operations of U.S. military forces around the world, would help managers identify and solve recurring problems involving military supplies.

10. **Identifying potential defense supply chain chokepoints and single points of failure, and planning to prevent disruptions.** This recommendation requires determining the scope of foreign control over critical military supply chains and finding ways of restoring U.S. control. In addition to the need to map the lower tiers of the supply chain, there is an urgent need to determine the scope of foreign control over critical supply chains, such as those for high-tech batteries or HELLFIRE missile propellant. Foreign control of defense supply chains makes U.S. defense capabilities vulnerable, especially in times of crisis. This dependence potentially enables foreign suppliers to leverage supply in return for concessions. Supply chain disruptions

are not solely a result of foreign exploitation. To plan for potential disruptions, the United States needs to understand the details and dynamics of the most critical defense supply chains. The United States should not wait for the next Fukushima-style disaster or a coup in a supplier nation that will necessitate urgent and perhaps very difficult countermeasures.

A CALL TO ACTION

Like our foreign competitors, the United States must take a long view in sustaining the health of its defense industrial base. We know that the future battlefield requires technologies that currently are in the laboratory or on the engineer's drawing board. We know that our defense industrial base is a pillar of U.S. prosperity and security. And we know that we need to address important vulnerabilities to today's defense supply chains, as well as to safeguard tomorrow's defense industrial capacity.

Now is the time to address the risks to our defense industrial base comprehensively. Concerted action on the part of government and industry is essential to success, particularly as defense spending declines. The effects of globalization and offshoring tell us that the invisible hand of the market is not sufficient to ensure that U.S. defense capabilities will be available in the future.

Without a healthy and technologically advanced defense industry, we cannot provide the weapons that our future generation of warriors must have to defend the United States now and into the future.

Nothing less is at stake in sustaining our defense industrial base than our nation's survival.

ENDNOTES

a   Mike Petters, "How Long Would It Take the Shipbuilding Industry to Grow Capacity and Throughput if the Nation Faced a Naval Crisis or Conflict?" *Information Dissemination*, June 5, 2012. http://www.informationdissemination.net/2012/06/how-long-would-it-take-shipbuilding.html.

1   Deputy Secretary of Defense Ashton Carter, quoted in Emilie Rutherford, "Carter: DoD to Protect Vital Industry Skillsets in Next Year's Budget," *Defense Daily*, May 31, 2012. http://www.defensedaily.com/free/17916.html.

2   Emilie Rutherford, "Carter: DoD to Protect Vital Industry Skillsets in Next Year's Budget," *Defense Daily*, May 31, 2012. http://www.defensedaily.com/free/17916.html.

3   Ibid.

4   Marcus Weisgerber, "DoD to support Advanced Manufacturing Projects," Defense News, February 18, 2013. http://www.defensenews.com/article/20130217/DEFREG02/302170009/DoD-Support-Advanced-Manufacturing-Projects.

5   Deputy Secretary of Defense Ashton Carter , quoted in Emilie Rutherford, "Carter: DoD to Protect Vital Industry Skillsets in Next Year's Budget," *Defense Daily*, May 31, 2012. http://www.defensedaily.com/free/17916.html.

EXPERTS CONSULTED

GERALD ABBOTT, PH.D.	Director Emeritus of Industry Studies, National Defense University
PETER AMBLER	Legislative Assistant, Office of Rep. Gabrielle Giffords (D-AZ)
JOHN ARNETT	Associate, Kelley Drye & Warren LLP
JOHN BARTO	President, Ansonia Copper & Brass
EILEEN P. BRADNER	Senior Director and Counsel, Federal Government Affairs, Nucor Corporation
CHRISTOPHER BRADISH	Consultant, The Grossman Group
TOM DEAN	Director, Armor Plate, ArcelorMittal USA
PETE GOYETTE	Vice President, BGM Fastener Co., Inc.
MICHAEL GREATHEAD	President, National Bronze & Metals, Inc.
JEFFREY A. GREEN	President, J.A. Green & Company
GARY GULINO	Project Administrator, Defense Logistics Agency
ROBERT J. HARRIS	Managing Director, Industrial Fasteners Institute
JAY HOROWITZ	President, Lewis Brass & Copper Company, Inc.
JIM HOUSTON	Retired Commander, U.S. Navy
ALAN KESTENBAUM	Executive Chairman, Global Specialty Metals, Inc.
ART KRACKE	Vice President, R&D and Business Development, ATI Allvac
LARRY LASOFF	Associate, Kelley Drye & Warren LLP
PATRICK MEADE	Manager, Aerospace Products Division, Industrial Fasteners Institute
TOM MILLER	General Manager, Nucor Fastener Division, Nucor Corporation
KELLY MISHKIN	Legislative Assistant, Office of Rep. Raul Grijalva (D-AZ)
CARL MOULTON	Senior Vice President, Allegheny Technologies Incorporated

PAUL O'DAY	President, American Fiber Manufacturers Association, Inc.
NEAL ORRINGER	Director of Manufacturing, Office of the Secretary of Defense
MARLIN PERKINS	Vice President of Sales, Globe Metallurgical Inc.
JENNIFER PHILLIPS	Contractor, Defense Logistics Agency
DAVID PINEAULT	Economist, Defense Logistics Agency
VICKY PLESKO	General Secretary, National Center for Defense Manufacturing and Machining
GRANT REEVES	President, PBI Performance Products, Inc.
JENNIFER REID	Vice President and Partner, The Laurin Baker Group, LLC
RALPH RESNICK	President and Executive Director, National Center for Defense Manufacturing and Machining
MARK RITACCO	Legislative Assistant, Office of Rep. Chris Murphy (D-CT)
ERIC SAYERS	Military Legislative Assistant, Office of Rep. J. Randy Forbes (R-VA)
MATTHEW SEAFORD, PH.D.	Deputy Director, Production Act Title III, Office of the Secretary of Defense
CHRIS SLEVIN	Economic Policy Director, Office of Sen. Sherrod Brown (D-OH)
PAULA STEAD	Deputy Administrator, Office of Strategic Materials, Defense Logistics Agency
DEBRA WAGONER	Director, Global Government Affairs, Corning Incorporated
CAPT. NEVILLE WELCH, USMC	Legislative Assistant, Office of Rep. Tim Ryan (D-OH)
ELAINE WILSON	Legislative Assistant, Office of Rep. Donald Manzullo (R-IL)
DOUGLAS WOODS	President, Association for Manufacturing Technology
JOEL S. YUDKEN, PH.D.	Principal, High Road Strategies, LLC

ACRONYMS AND ABBREVIATIONS

▶ **AGM** – Air-to-ground missile

▶ **AGMS** – Air-to-ground missile system

▶ **AIA** – Aerospace Industry Association

▶ **AMI** – Advanced manufacturing initiative

▶ **AMT** – Association of Manufacturing Technology

▶ **ANL** – Argonne National Laboratory

▶ **AMRDEC** – U.S. Army Aviation and Missile Research, Development & Engineering Center

▶ **AT&L** – Office of the Under Secretary of Defense for Acquisition, Technology, and Logistics

▶ **BAA** – Buy American Act

▶ **BIS** – Bureau of Industry and Security

▶ **BT** – Butanetriol

▶ **BTTN** – Butanetriol trinitrate

▶ **CFIUS** – Committee on Foreign Investment in the United States

▶ **CNC** – Computer numerical control

▶ **COTS** – Commercial-off-the-shelf

▶ **CRS** – Congressional Research Service

▶ **Cu-Ni** – Copper-nickel

▶ **DFARS** – Defense Federal Acquisition Regulations Supplement

▶ **DIB** – Defense industrial base

▶ **DLA** – Defense Logistics Agency

▶ **DoD** – Department of Defense

▶ **DoE** – Department of Energy

▶ **DPA** – Defense Production Act

▶ **DPAC** – Defense Production Act Committee

▶ **EoP** – Executive Office of the President

▶ **EV** – Electric vehicle

▶ **FCS** – Future Combat Systems

▶ **FLIR** – Forward looking infrared

▶ **FQA** – Fasteners Quality Act

▶ **FR-ACU** – Flame-Resistant Army Combat Uniform

▶ **FR Rayon** – Flame-resistant rayon

▶ **GAO** – Government Accountability Office

▶ **I2** – Image intensification

▶ **IED** – Improvised explosive device

▶ **IFI** – Industrial Fasteners Institute

▶ **INL** – Idaho National Laboratory

▶ **ISO** – International Organization for Standardization

▶ **JAGM** – Joint Air-Ground Missile

▶ **JDAM** – Joint Direct Attack Munition

▶ **JLTV** – Joint Light Tactical Vehicle

▶ **LCD** – Liquid crystal display

▶ **LCS** – Littoral combat ship

▶ **LED** – Light-emitting diode

▶ **Li-ion** – Lithium-ion

- **LNBL** – Lawrence Berkeley National Laboratory
- **LOI** – Limiting oxygen index
- **LPD** – Landing platform dock
- **ManTech** – Department of Defense Manufacturing Technology Program
- **Mil-Spec** – Military specifications
- **MRAP** – Mine-Resistant Ambush-Protected vehicle
- **NATO** – North Atlantic Treaty Organization
- **NAVSEA** – Naval Sea Systems Command
- **NC** – Nitrocellulose
- **NDAA** – National Defense Authorization Act
- **NdFeB** – Neodymium iron boron
- **NDIA** – National Defense Industrial Association
- **NDS** – National Defense Stockpile
- **NG** – Nitroglycerin
- **NREL** – National Renewable Energy Laboratory
- **NSWC** – Naval Surface Warfare Center
- **NVD** – Night vision device
- **OD** – Outside diameter
- **OECD** – Organization for Economic Cooperation and Development
- **OEM** – Original equipment manufacturer
- **ONR** – Office of Naval Research
- **ORNL** – Oakridge National Laboratory

- **PASGT** – Personal Armor System for Ground Troops
- **PBI** – Polybenzimidazole
- **PGM** – Platinum group metals
- **PLA** – People's Liberation Army
- **R&D** – Research and development
- **RDT&E** – Research, development, testing, and evaluation
- **RE** – Rare earth
- **REE** – Rare earth element
- **REM** – Rare earth magnet
- **SAPI** – Small arms protective insert
- **SBIR** – Small business innovation research
- **S2T2** – Sector-by-Sector, Tier-by-Tier Industrial Base Review
- **SMC** – Specialty Metals Clause
- **SmCo** – Samarium-Cobalt
- **SMPB** – Strategic Materials Protection Board
- **SMSP** – Strategic Material Security Program
- **SNL** – Sandia National Laboratory
- **SRM** – Solid rocket motor
- **SWIR** – Short-wave infrared
- **TARDEC** – U.S. Army Tank Automotive Research, Development, and Engineering Center
- **UAV** – Unmanned aerial vehicle
- **USGS** – U.S. Geological Survey
- **UUV** – Unmanned underwater vehicle
- **WTO** – World Trade Organization

GLOSSARY

Office of the Under Secretary of Defense for Acquisition, Technology, and Logistics (AT&L) – The branch of the Department of Defense that deals most directly with defense industrial base issues. It contains the Office of Manufacturing and Industrial Base Policy.

Advanced Manufacturing Partnership (AMP) – A U.S. government initiative to bring together various stakeholders in the manufacturing sector to support emerging technologies. They aim to create high-quality domestic jobs and enhance U.S. economic competitiveness.

Berry Amendment – A section of the U.S. code (10 U.S.C. 2533a) that requires the Department of Defense to procure certain items, including food and textiles, from domestic sources. The Berry Amendment previously governed specialty metals acquisitions as well, but that portion was later removed. Domestic sourcing restrictions for specialty metals are currently found in the Specialty Metals Clause (10 U.S.C. 2553b).

Biological weapon – A weapon that uses a living organism or a virus to harm or kill a target. Most countries are party to the Biological Weapons Convention, which prohibits the production, acquisition, and retention of biological weapons. There are significant concerns that a terrorist group may someday use a biological weapon.

Budget Control Act – A 2011 law that raised the federal debt limit. The Budget Control Act mandates $1.2 trillion in cuts across all the branches of the federal government (including the Department of Defense) over the course of 10 years, unless Congress is able to come up with a budget that reduces the deficit by $1.2 trillion by January 2013 (extended to March 2013). These mandatory cuts are also known as sequestration. Sequestration took effect on March 1, 2013.

Butanetriol (BT) – A chemical precursor to Butanetriol trinitrate, a propellant used in HELLFIRE missiles.

Butanetriol trinitrate (BTTN) – A propellant used in HELLFIRE missiles.

Buy American Act (BAA) – A portion of the U.S. Code (41 U.S.C 10a through 10d) that requires the U.S. government to acquire goods made in the United States. A good is considered "American" if at least 50 percent of the cost of its components was incurred in the United States. BAA restrictions are waived in a variety of circumstances.

Commercial-off-the-shelf (COTS) – In U.S. federal acquisitions regulations, COTS refers to items purchased commercially in the same form as they would be available to other customers. The alternative is for the government to buy goods specifically designed for government use. The federal government's effort to save on costs has led to greater reliance on COTS.

Copper-Nickel (Cu-Ni) tubing – A component used on all U.S. Navy ships. Military Cu-Ni tubing has special properties that prevent corrosion and the growth of micro-organisms. Cu-Ni tubing is integrated into hydraulic control, lubrication, and high-pressure air injection systems.

Defense Federal Acquisition Regulations Supplement (DFARS) – Department of Defense-specific regulations that govern the acquisition of defense goods.

Defense industrial base – Refers to the vast array of private firms that produce and maintain U.S. military capabilities, including vehicles, weapons, and devices. In the United States, the defense industrial base is comprised entirely of privately owned firms of all shapes and sizes, including many small businesses, many of which also do business (partially or primarily) in the commercial sector.

Defense Logistics Agency (DLA) – A Department of Defense agency that provides significant logistics support to the U.S. military and procures goods such as fuel, spare parts, and uniforms for distribution to U.S. troops. DLA is an important buyer of goods produced by the defense industrial base.

Defense Production Act (DPA) – Provides authority to the United States to require U.S. defense industrial base firms to prioritize critical defense orders, also called "rated" orders, over commercial orders.

Defense Production Act Committee (DPAC) – An inter-agency body established in 2009 and made up of the heads of federal acquisition departments and agencies. Its purpose is to identify risks to the defense industrial base and make recommendations to address those risks.

Defense Standardization Program – A program to promote standardization of components, decrease costs, and increase operational effectiveness across the Department of Defense.

Department of Defense (DoD) – The government agency responsible preserving U.S. national security. DoD includes the military services, as well as a range of supporting agencies, laboratories, schools, and other institutions. DoD is the primary government agency involved with awarding contracts to the defense industrial base and overseeing its health.

Department of Energy (DoE) – The U.S. government agency responsible for implementing U.S. energy policy. DoE policies support certain sectors of the defense industrial base, such as battery manufacturers. DoE also oversees the maintenance of the U.S. nuclear arsenal.

DoD Manufacturing Technology Program (MANTECH) – A program overseen by the Department of Defense's Office of Manufacturing and Industrial Base Policy to develop technologies and processes to increase the affordability and efficiency of items manufactured by the defense industrial base.

Electric eehicle (EV) – A vehicle powered by electricity stored in batteries. EVs often use Lithium-ion batteries.

Export quota – A policy that limits how much of a particular commodity may be exported. China currently imposes export quotas on rare earth elements produced in China. Products manufactured in China that use rare earth elements do not face similar restrictions.

Fastener – A device that holds larger component parts together. Fasteners are essential lower-tier inputs for a wide range of defense goods.

Fasteners Quality Act (FQA) – A 1990 law intended to ensure that domestic fastener manufacturers do not face unfair competition from substandard foreign fasteners. The FQA establishes standards and includes provisions for the testing and inspection of fasteners. The original FQA was only partially implemented and has been amended several times since its passage, with the most recent changes happening in 1999.

Flame-resistant (FR) rayon – A synthetic fiber with flame-resistant and self-extinguishing properties. It is the main component of Defender M fabric, the fabric currently used in the U.S. Army's Flame-Resistant Army Combat Uniform. Currently, Lenzing (an Austrian company) is the only producer of FR rayon. Flame-Resistant rayon is sometimes referred to as Fire-Resistant rayon.

Flame-Resistant Army Combat Uniform (FR-ACU) – The U.S. Army's fire-resistant uniform, issued to ground troops. It is made from Defender M fabric, which is primarily made of flame-resistant rayon that is produced exclusively in Austria.

Five-Axis machine tool (also known as Five-Axis simultaneous control machine tool) – Sophisticated machines designed to cut, grind, and shape metal for use in other machines. They possess three linear axes and two rotary axes. Five-axis machine tools are important for producing advanced military systems. Japan is currently the largest producer of five-axis machine tools.

Future Combat Systems (FCS) – An ambitious Army modernization program that was cancelled in 2009. The purpose of the program was to build a networked array of systems, including armored combat vehicles and drones, under a single contract. The Army's Ground Combat Vehicle program is a successor to part of the FCS program.

HELLFIRE missile – A single-stage air-to-surface missile fired from a variety of helicopters and fixed-wing aircraft, including unmanned drones. HELLFIRE missile propellant includes Butanetriol Trinitrate (BTTN). BTTN requires Butanetriol, which is produced exclusively in China.

Image intensification (I2) – A technology that magnifies available light from the moon, stars, or nearby cities to enable the wearer to see in low-light environments. The scopes and goggles used by U.S. troops incorporate I2 technology.

Improvised explosive device (IED) – A homemade explosive device widely used in Iraq and, to a lesser extent, in Afghanistan. The emergence of the IED threat in Iraq led to wider deployment of flame-resistant uniforms as well as the rapid fielding of the Mine-Resistant Ambush-Protected Vehicle. Counter-IED activities are led by the Department of Defense's Joint IED Defeat Organization.

International Traffic in Arms Regulations (ITAR) – A section of the U.S. Code of Federal Regulations that governs the export of certain defense goods. ITAR controls items such as the advanced image intensification tubes used in night-vision devices.

Joint Direct Attack Munition (JDAM) – A kit that converts unguided munitions into precision guided, or "smart," munitions.

Lithium ion (Li-ion) battery – A rechargeable battery, made of lithium and carbon, with very high energy density. They are widely used in portable military and civilian applications and are found in electronics, laptops, smartphones, and electric vehicles. Li-ion batteries used in military applications are not currently standardized, leading to a proliferation of different battery sizes and types.

Littoral Combat Ship (LCS) – A multi-mission U.S. Navy ship intended to operate close to shore. There are currently two LCS variants, one of which uses an armored steel hull.

Mine-Resistant Ambush-Protected Vehicle (MRAP) – Armored vehicles that are shaped in such a way that protects them from roadside bombs and improvised explosive devices (IEDs). MRAPs were rapidly acquired and fielded in response to the IED threat in Iraq.

Mountain Pass Mine – The largest U.S. rare earth element mine and the only rare earth mine operating in the United States for the foreseeable future. Located in California and operated by Molycorp Inc., it closed in 2002 but began rare earth production when it reopened in 2012.

National Defense Authorization Act (NDAA) – A bill passed every year by Congress that lays out defense spending priorities for the coming fiscal year. NDAAs also contain certain policies that affect the defense industrial base.

National Defense Stockpile (NDS) – A branch of the U.S. Defense Logistics Agency (DLA). It holds commodities such as zinc, cobalt, and chromium, as well as more precious metals such as platinum, palladium, and iridium. In 2010 it was renamed DLA Strategic Materials.

Neodymium iron boron (NdFeB) magnet – A permanent rare earth magnet made from the rare earth element neodymium. NdFeB magnets are used in a variety of military electronics applications.

Night vision device (NVD) – A device designed to allow the wearer to see in low-light environments. NVDs are also used for civilian commercial and recreational applications.

Office of Manufacturing and Industrial Base Policy (MIBP) – The office within the Department of Defense responsible for monitoring the defense industrial base and intervening when necessary to preserve capabilities. MIBP administers the Sector-by-Sector, Tier-by-Tier survey program.

Original equipment manufacturer (OEM) – The firm responsible for manufacturing an item purchased by the Department of Defense. This entity is also known as a prime contractor.

Personal Armor System for Ground Troops (PASGT) – An armor system that consisted of a vest and helmet previously used by the U.S. military. Interceptor body armor replaced the PASGT vest. The helmet has been replaced as well.

Platinum group metals (PGM) – Refers to six metallic elements in the periodic table, including ruthenium, rhodium, palladium, osmium, iridium, and platinum. They are resistant to high temperatures and are able to catalyze chemical reactions.

Quadrennial Defense Review (QDR) – A review of the Department of Defense's strategy and priorities intended to be released every four years. The most recent QDR, in 2010, warned that the U.S. government had "not adequately addressed the changes" in the U.S. defense industrial base.

Prime contractor – The contractor that the Department of Defense awards a particular contract and that has responsibility for delivering a particular good or capability. Prime contractors typically rely on an array of subcontractors.

Rare earth element (REE) – Seventeen elements with special physical properties that are widely used in defense and civilian electronics. Although REE deposits are found all around the world, including in the United States, approximately 97 percent of REE extraction currently takes place in China.

Rare earth magnet (REM) – Permanent, high-strength magnets made from rare earth elements. REMs are used in a wide range of defense applications.

Rated order - Defense acquisition orders from a U.S. manufacturer that are assigned a priority under the Defense Priorities and Allocation System and that have priority over the manufacturer's non-rated orders.

Research and Development (R&D) – The process of discovering new technologies and processes. Defense R&D is conducted both by private defense industrial base firms and by U.S. government labs and other facilities.

Sector-by-Sector, Tier-by-Tier Industrial Base Review (S2T2) – A Department of Defense (DoD) effort to map out the network of suppliers in key sectors including aircraft; shipbuilding; space; ground vehicles; missiles; missile defense; services; and command, control, communications, computers, intelligence, surveillance, and reconnaissance (C4ISR) systems. DoD is working with the Department of Commerce Bureau of Industry and Security to send out surveys to firms in those industries. Data collected in these surveys will enhance DoD's awareness of defense supply chains.

Semiconductor – The basic building block of modern electronics. Semiconductors operate microprocessors as well as transistors. They are usually made of silicon, which conducts electricity and helps control the flow of electrical current.

Sequestration – Automatic cuts to the entire federal government, including the Department of Defense, under the 2011 Budget Control Act. After the deadline was extended, these cuts took effect on March 1, 2013, when Congress was unable to reach a compromise to reduce the federal budget deficit by $1.2 trillion over the next decade.

Sintered magnets – Magnets, including NdFeB magnets, made from powders compacted under pressure. Sintering is a process of heating powdered oxides to the point where they adhere to each other. The method creates stronger magnets.

Small business set asides – Contracts set aside for small businesses when military services are awarding contracts. This method encourages the participation of small businesses in the defense industrial base.

Specialty Metals Clause (SMC) – A section of the U.S. Code (10 U.S.C. 2553b) requiring the Department of Defense to procure specialty metals, including high-grade steel, titanium, and zirconium from domestic sources. The SMC can be waived under certain circumstances.

Steel armor plate – High-grade plate steel designed to withstand bullets and explosive attacks. Many ground vehicles (such as tanks) and navy ships (such as carriers, submarines, and destroyers) employ steel armor plate.

Strategic Material Security Program (SMSP) – A proposed program to replace the National Defense Stockpile with a program that better coordinates critical material stockpiles across the Department of Defense.

Strategic Materials Act (SMA) – The 1939 act that first established a reserve of critical materials and was a precursor to the National Defense Stockpile.

Strategic Materials Protection Board (SMPB) – An inter-service board, established by the FY2007 National Defense Authorization Act, that determines strategies for assuring a supply of materials critical to U.S. national security.

Synthetic fabric – A fabric made from chemically produced fibers. Synthetic fabrics often have superior performance characteristics compared to fabrics made from natural fibers, and they are used in a variety of defense and civilian applications.

LEGISLATIVE FRAMEWORKS

The U.S. defense industrial base is affected by legislative and administrative frameworks that support U.S. firms and regulate what they can do and with whom they are allowed to trade. Descriptions of several of the most important of these frameworks follow below.

The Berry Amendment – Passed in 1941, the Berry Amendment mandates that the Department of Defense purchase food as well as uniforms and other textiles from domestic sources. From 1973 to 2001, the Berry Amendment covered specialty metals. The FY2002 National Defense Authorization Act made the Berry Amendment part of the U.S. Code (10 U.S.C. 2533a). Several Berry Amendment waivers are currently in effect, including one that allows for foreign-produced flame-resistant rayon fibers. The Secretary of Defense has the authority to waive Berry Amendment restrictions under certain circumstances.

The Buy American Act (BAA) – Enacted in 1933, the BAA is the main statute governing the federal government's procurement of domestic goods. The act's restrictions can be waived in some circumstances, including by other laws. A manufactured good is BAA-compliant if at least 50 percent of the costs of its components are incurred domestically.

Defense Federal Acquisition Regulation Supplement (DFARS) – DFARS is a set of rules in the U.S. Code of Federal Regulations that governs the Department of Defense's acquisition of goods and services.

The Fasteners Quality Act (FQA) – Signed into law in November 1990, the FQA requires that certain fasteners meet quality specifications; provides for the accreditation of test laboratories; and establishes inspection, testing, and certification requirements. The original FQA was never fully implemented and was amended most recently in 1999.

International Traffic in Arms Regulation (ITAR) – ITAR regulates exports of certain weapons and defense items such as advanced night vision devices. ITAR's purpose is to prevent the unauthorized spread of U.S. defense technology.

The Specialty Metals Clause (SMC) – The SMC mandates that the Department of Defense use domestically produced specialty metals in "aircraft, missile and space systems, ships, tank and automotive items, weapon systems, or ammunition." These metals include high-grade steel, titanium, zirconium, and certain other alloys. The SMC was originally passed as part of the Berry Amendment in 1973 but was made a separate part of the U.S. Code (10 U.S.C. 2533b) by the FY2007 National Defense Authorization Act. SMC restrictions can be waived under certain conditions, including domestic non-availability.

The Strategic Materials Protection Board (SMPB) – Established by the FY2007 National Defense Authorization Act, the SMPB is an inter-service board that determines strategies for ensuring adequate supplies of materials critical to national security. Law requires the SMPB to meet at least once every two years.

RECENT AAM PUBLICATIONS

Preparing for 21st Century Risks: Revitalizing American Manufacturing to Protect, Respond and Recover
July 2012
By Former Director of Homeland Security Tom Ridge and retired USAF Colonel Robert B. Stephan

Manufacturing a Better Future for America
July 2009
Edited by Richard McCormack

An Assessment of Environmental Regulation of the Steel Industry in China
March 2009
Alliance for American Manufacturing

How Infrastructure Investments Support the U.S. Economy: Employment, Productivity and Growth
January 2009
By James Heintz and Robert Pollin

Shedding Light on Energy Subsidies in China: An Analysis of China's Steel Industry from 2000-2007
January 2008
By Usha C. V. Haley, Ph.D.

Buyer's Remorse: How America Has Failed to See the Threat Posed by Dangerous Chinese Goods and the Case for "Safe Trade"
December 2007
By Richard Miniter

Enforcing the Rules
May 2007
By Greg Mastel, Andrew Szamosszegi, John Magnus and Lawrence Chimerine

AAM publications are available for download or purchase at
www.americanmanufacturing.org/publications.